Lecture Notes in Physics

Volume 984

The Lecture Notes in Physics

The series Lecture Notes in Physics (LNP), founded in 1969, reports new developments in physics research and teaching-quickly and informally, but with a high quality and the explicit aim to summarize and communicate current knowledge in an accessible way. Books published in this series are conceived as bridging material between advanced graduate textbooks and the forefront of research and to serve three purposes:

- to be a compact and modern up-to-date source of reference on a well-defined topic;
- to serve as an accessible introduction to the field to postgraduate students and nonspecialist researchers from related areas;
- to be a source of advanced teaching material for specialized seminars, courses and schools.

Both monographs and multi-author volumes will be considered for publication. Edited volumes should however consist of a very limited number of contributions only. Proceedings will not be considered for LNP.

Volumes published in LNP are disseminated both in print and in electronic formats, the electronic archive being available at springerlink.com. The series content is indexed, abstracted and referenced by many abstracting and information services, bibliographic networks, subscription agencies, library networks, and consortia.

Proposals should be sent to a member of the Editorial Board, or directly to the responsible editor at Springer:

Dr Lisa Scalone
Springer Nature
Physics
Tiergartenstrasse 17
69121 Heidelberg, Germany
lisa.scalone@springernature.com

More information about this series at http://www.springer.com/series/5304

Peter A. Hogan • Dirk Puetzfeld

Frontiers in General Relativity

 Springer

Peter A. Hogan
School of Physics
University College Dublin
Dublin, Ireland

Dirk Puetzfeld
Center of Applied Space Technology
and Microgravity
University of Bremen
Bremen, Germany

ISSN 0075-8450   ISSN 1616-6361   (electronic)
Lecture Notes in Physics
ISBN 978-3-030-69369-5   ISBN 978-3-030-69370-1   (eBook)
https://doi.org/10.1007/978-3-030-69370-1

This Springer imprint is published by the registered company Springer Nature Switzerland AG.
The registered company address is: Gewerbestrasse 11, 6330 Cham, Switzerland

Preface

In our experience, while students may initially be attracted to general relativity by the fact that it is the area of theoretical physics that predicts the existence of black holes, gravitational waves and the big-bang beginning of the universe, and that all three of these phenomena are currently being confirmed by observations, students are attracted to the *theory* of these phenomena because general relativity provides a framework for modelling them using abstract mathematics. When such students are beginning research and looking around for ways to "get involved", we envisage that this book might possibly provide what they are looking for. Our approach is strongly geometrical. Even the speculative explicit models which we describe in order to stimulate further ideas and research are geometrically motivated. These include a model of the gravitational field of a Kerr black hole incorporating matter escaping at the speed of light and diminishing the mass and angular momentum; a light-like charge in electromagnetic theory whose Maxwell field is a light-like analogue of the Liénard-Wiechert field; a magnetic black hole; the analogue in general relativity of a magnetic monopole, moving in external gravitational and electromagnetic fields; run-away motion of Reissner-Nordström particles in the absence of external fields; and a model of colliding gravitational waves leading to a de Sitter or an anti-de Sitter universe. None of these examples could be claimed to be "fully understood" and thereby leave room for development.

We assume a strong mathematical background in differential geometry (and thus in tensor calculus); although we only occasionally refer to it, knowledge of the powerful Cartan calculus is very useful for carrying out some of the involved explicit calculations referred to in the text. Photons and material particles play important roles in general relativity, and this fact is strongly represented in the topics described in this book. Such objects are involved in analysing and measuring gravitational fields and in constructing mathematical models of gravitational fields of various types. This means that from the space-time geometry point of view, the study of congruences of world lines, both time-like and light-like, are of paramount importance in general relativity. For this reason, the book begins with a standard description of both types of congruence. A key role in measuring a gravitational field is played by the time-like congruences and, in particular, the use of deviation equations in this context. These are therefore discussed at the beginning of the book making use of Synge's world function in the process. With an emphasis on tensor calculus, we make extensive use of the theory of bivectors (or skew-symmetric tensors) in general relativity and thus an early chapter is devoted to them. An application of the time-like congruences is provided by the construction of models of Bateman electromagnetic and gravitational waves in the linear approximation using the gauge invariant and covariant formalism demonstrating that the gravitational waves exist owing to the acquisition of shear or distortion by the t-lines. There follows, interspersed with the provocative examples mentioned above, studies of gravitational radiation in the context of de Sitter (or anti-de Sitter) cosmology and the gravitational compass or clock compass (general relativistic gradiometers) for providing an operational way of measuring a gravitational field.

Finally we should point out that among the relativists with whom one or other of us has collaborated are J.L. Synge, I. Robinson, A. Trautman, G.F.R. Ellis, W. Israel, Y.N. Obukhov and F.W. Hehl, and there are identifiable threads running through the text which reflect their influence upon us.

Dublin, Ireland   Peter A. Hogan
Bremen, Germany   Dirk Puetzfeld[1]
October 2020

[1]D.P. acknowledges the support by the Deutsche Forschungsgemeinschaft (DFG) through the grant PU 461/1-2 project number 369402949.

Contents

About the Authors

Peter A. Hogan is emeritus professor at University College Dublin. He received his B.Sc., M.Sc., Ph.D. and D.Sc. degrees from The National University of Ireland and is a member of the Royal Irish Academy. He was a postdoctoral fellow in the School of Theoretical Physics, Dublin Institute for Advanced Studies; the Centre for Relativity, University of Texas at Austin; and the School of Mathematics, Trinity College Dublin. His publications of more than 90 papers and 3 monographs are in electrodynamics, Yang-Mills gauge theory from the fibre bundle point of view and general relativity (equations of motion, gravitational waves, exact solutions of Einstein's equations and cosmology). He has published research carried out in Japan, South Africa, France, Poland and the USA.

Dirk Puetzfeld is a researcher at Bremen University (Germany). He received his Dipl. Phys. and Dr. rer. nat. degrees from the University of Cologne (Germany), both in theoretical physics. He was an assistant professor at Tohoku University (Japan) and a postdoctoral fellow at the Max-Planck-Institute for Gravitational Physics (Germany), University of Oslo (Norway) and Iowa State University (USA). His main research interests include gravitational physics, cosmology and relativistic geodesy as well as computational methods. His publication record encompasses over 50 scientific papers in international journals as well as two books.

Congruences of World Lines

1

Abstract

Material particles and photons are of paramount importance in constructing and analysing models of gravitational fields in general relativity. Their histories in space-time are congruences of time-like and light-like (or null) world lines respectively. Material particles and photons therefore play a central role in this book and we begin by describing the standard theory of congruences of time-like world lines and of null congruences. In the light-like case we shall only require a knowledge of null *geodesic* congruences. For use later we describe geodesic deviation equations with the help of Synge's world function and parallel propagator.

1.1 Time-Like Conguences

A three parameter family of time-like world lines, with one line passing through each point of space-time, constitutes a *time-like congruence*. Such a congruence is the history of a material continuum. A space-like hypersurface Σ intersects the lines of the congruence and the points of Σ, with intrinsic local coordinates ξ^α, $\alpha = 1, 2, 3$ provide the three parameters necessary to label the lines of the congruence. The parametric equations of the congruence take the form

$$x^a = x^a(s, \xi^\alpha) \,, \tag{1.1}$$

with $\xi^\alpha = $ constants on each line of the congruence (and so, henceforth, we shall refer to "the line ξ^α" for convenience) and s is taken as arc length or proper time along each line, c.f. Fig. 1.1. Thus

$$u^a = \frac{\partial x^a}{\partial s} \quad \text{with} \quad u^a u_a = 1 \,, \tag{1.2}$$

P. A. Hogan, D. Puetzfeld, *Frontiers in General Relativity*, Lecture Notes in Physics 984, https://doi.org/10.1007/978-3-030-69370-1_1

1

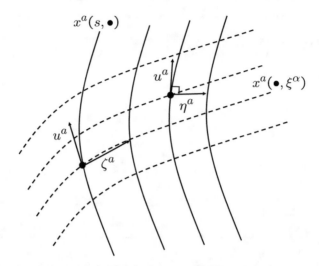

Fig. 1.1 Sketch of the congruence $x^a = x^a(s, \xi^a)$, the infinitesimal connecting vector ζ^a, and the orthogonal connecting vector η^a. Here we denote fixed parameters by the "•" symbol

is the unit tangent vector field to the lines of the congruence and represents the 4-velocity of the particle with world line ξ^α. If ξ^α and $\xi^\alpha + \delta\xi^\alpha$, with $\delta\xi^\alpha$ infinitesimal constants, label neighbouring lines of the congruence then the *infinitesimal connecting vector*

$$\zeta^a = \frac{\partial x^a}{\partial \xi^\alpha} \delta\xi^\alpha \,, \tag{1.3}$$

defined along ξ^α, *joins points on ξ^α and $\xi^\alpha + \delta\xi^\alpha$ of equal parameter value s on each line*. We have

$$\frac{\partial \zeta^a}{\partial s} = \frac{\partial^2 x^a}{\partial s\, \partial \xi^\alpha} \delta\xi^\alpha = \frac{\partial}{\partial \xi^\alpha} \left(\frac{\partial x^a}{\partial s} \right) \delta\xi^\alpha = \frac{\partial u^a}{\partial \xi^\alpha} \delta\xi^\alpha \,. \tag{1.4}$$

But $u^a = u^a(x(s, \xi))$ and so, by the chain rule,

$$\frac{\partial u^a}{\partial \xi^\alpha} = \frac{\partial u^a}{\partial x^b} \frac{\partial x^b}{\partial \xi^\alpha} \,, \tag{1.5}$$

giving us

$$\frac{\partial \zeta^a}{\partial s} = \frac{\partial u^a}{\partial x^b} \frac{\partial x^b}{\partial \xi^\alpha} \delta\xi^\alpha = \frac{\partial u^a}{\partial x^b} \zeta^b \,, \tag{1.6}$$

or, since $\zeta^a = \zeta^a(x(s, \xi))$,

$$\frac{\partial \zeta^a}{\partial x^b} u^b = \frac{\partial u^a}{\partial x^b} \zeta^b \quad \Leftrightarrow \quad \zeta^a{}_{,b} u^b = u^a{}_{,b} \zeta^b , \tag{1.7}$$

with the comma, as always, denoting partial differentiation. Since the Riemannian connection is symmetric, this can be written in the manifestly covariant form

$$\dot{\zeta}^a = u^a{}_{;b} \zeta^b , \tag{1.8}$$

with $\dot{\zeta}^a = \zeta^a{}_{;b} u^b$ with the semicolon denoting, as always, covariant differentiation with respect to the Riemannian connection calculated with the metric g_{ab} of the space-time. In general we will use a dot to indicate the covariant derivative in the direction u^a of any quantity defined along ξ^α. Hence in particular $\dot{u}^a = u^a{}_{;b} u^b$ is the 4-acceleration of the particle with world line ξ^α. This is orthogonal to u^a, that is to say $\dot{u}^a u_a = 0$, on account of the second of (1.2).

The infinitesimal connecting vector ζ^a is not invariantly defined. It clearly depends on the choice of origin of s on each world line ξ^α. To translate the origin of s by differing amounts on each world line we make the transformation $s \to s' = s + f(\xi^\alpha)$, for some function $f(\xi^\alpha)$ which, of course, is constant on each world line ξ^α. The reader can show that under this transformation

$$\zeta^a \to \zeta'^a = \zeta^a + \frac{\partial f}{\partial \xi^\alpha} \delta\xi^\alpha u^a , \tag{1.9}$$

so that the connecting vector acquires a component in the direction of the tangent u^a. Hence to work with a connecting vector which is invariant under (1.9) we define the *orthogonal connecting vector*

$$\eta^a = h^a{}_b \zeta^b , \tag{1.10}$$

with $h^a{}_b = \delta^a_b - u^a u_b$ the projection tensor. The projection tensor projects quantities orthogonal to u^a (and thus in particular $\eta^a u_a = 0$) and satisfies $h^a{}_b u^b = 0, h^a{}_b h^b{}_c = h^a{}_c$ and $h^a{}_a = 3$. The second equation here is characteristic of a projection, namely, the effect of applying the projection twice is the same as applying it once! Using (1.8) and (1.10) we can deduce *the transport law for η^a along ξ^α*:

$$h^a{}_b \dot{\eta}^b = A^a{}_b \eta^b \quad \text{with} \quad A^a{}_b = u^a{}_{;c} h^c{}_b . \tag{1.11}$$

The connecting vector η^a is the position vector of $\xi^\alpha + \delta\xi^\alpha$ relative to ξ^α for any value of s (in other words at all points along ξ^α). Let $\Omega(s)$ be the 3-space orthogonal to u^a at $x^a(s, \xi)$. Then η^a lies in $\Omega(s)$ and at any $s = s_0$ (say) we have $\Omega(s_0)$ as the instantaneous rest frame of the particle with world line ξ^α. The transport law (1.11) means that if we move along ξ^α infinitesimally by replacing s by $s + ds$ then η^a

in $\Omega(s)$ is transformed by infinitesimal, linear transformation to $(\delta_b^a + A^a{}_b\, ds)\eta^b$ in $\Omega(s+ds)$. This linear transformation is generated by $A^a{}_b$. We see from (1.11) that $A_{ab} = u_{a;b} - \dot{u}_a u_b$ and so we immediately have $A_{ab}\, u^a = 0 = A_{ab}\, u^b$. We can decompose A_{ab} into linearly independent parts by first identifying its symmetric and skew-symmetric parts:

$$A_{ab} = \vartheta_{ab} + \omega_{ab}\,, \tag{1.12}$$

with

$$\vartheta_{ab} = A_{(ab)} = \frac{1}{2}(A_{ab} + A_{ba}) = \vartheta_{ba}\,, \tag{1.13}$$

and

$$\omega_{ab} = A_{[ab]} = \frac{1}{2}(A_{ab} - A_{ba}) = -\omega_{ba}\,. \tag{1.14}$$

Then we can subtract the trace from the symmetric part, remembering to preserve the orthogonality with u^a, by defining

$$\sigma_{ab} = \vartheta_{ab} - \frac{1}{3}\vartheta\, h_{ab}\,, \tag{1.15}$$

with $\vartheta = \vartheta^a{}_a$. Then $\sigma_{ab} = \sigma_{ba}$ and $\sigma^a_a = 0$. Hence we have

$$A_{ab} = \sigma_{ab} + \frac{1}{3}\vartheta\, h_{ab} + \omega_{ab} = u_{a;b} - \dot{u}_a u_b\,, \tag{1.16}$$

with

$$\vartheta = u^a{}_{;a}\,, \tag{1.17}$$

$$\sigma_{ab} = u_{(a;b)} - \dot{u}_{(a} u_{b)} - \frac{1}{3}\vartheta\, h_{ab}\,, \tag{1.18}$$

$$\omega_{ab} = u_{[a;b]} - \dot{u}_{[a} u_{b]}\,. \tag{1.19}$$

We emphasise that here, as in (1.13) and (1.14), round brackets denote symmetrisation while square brackets denote skew-symmetrisation. It is also important to point out that (1.16) can be viewed as a decomposition of $u_{a;b}$. We have mentioned above that A_{ab} generates a linear map from $\Omega(s)$ to $\Omega(s+ds)$. It is interesting to interpret geometrically the influence of ϑ, ω_{ab} and σ_{ab} separately on this linear map.

Taking $\omega_{ab} \neq 0$, $\vartheta = 0$, $\sigma_{ab} = 0$ the transport law (1.11) becomes

$$h^a{}_b\, \dot{\eta}^b = \omega^a{}_b\, \eta^b\,. \tag{1.20}$$

To separate from this a propagation law for the *direction* of η^a along ξ^α and a propagation law for the *length* of η^a along ξ^α we simply write $\eta^a = l\,n^a$ with $n^a\,u_a = 0$ and $n^a\,n_a = -1$. Then from (1.20) we obtain the two propagation laws:

$$\dot{l} = 0 \quad \text{and} \quad h^a{}_b\,\dot{n}^b = \omega^a{}_b\,n^b \,. \tag{1.21}$$

The first of these means, of course, that the orthogonal connecting vector is transported along ξ^α without change of length. Since $\omega_{ab} = -\omega_{ba}$ we can define a covariant vector field

$$\omega_a = \frac{1}{2}\eta_{abcd}\,u^b\,\omega^{cd} \,. \tag{1.22}$$

Here $\eta_{abcd} = \sqrt{-g}\,\epsilon_{abcd}$ where $g = \det(g_{ab})$ and ϵ_{abcd} is the four dimensional Levi-Civita permutation symbol. We note that $\omega_a\,u^a = 0$ and so ω_a is space-like and thus $\omega_a\,\omega^a \leq 0$. Using the properties of η_{abcd} (see the Appendix A) we have

$$\omega_{ab} = \eta_{abcd}\,u^c\,\omega^d \,, \tag{1.23}$$

and thus we find that

$$\frac{1}{2}\omega_{ab}\,\omega^{ab} = -\omega^a\,\omega_a = \omega^2 \geq 0 \,, \tag{1.24}$$

for some scalar ω. An indication of what the transport law encapsulated in the second equation in (1.21) means can be found by noting that in the instantaneous rest-frame $\Omega(s_0)$ at $s = s_0$ we can choose coordinates such that $g_{ab} = \eta_{ab} = \mathrm{diag}(1, -1, -1, -1)$, the components of the Riemannian connection $\Gamma^a_{bc} = 0$ and $u^a = \delta^a_0$ and then, at $s = s_0$ the second of (1.21) reduces to

$$\frac{\partial n^\alpha}{\partial s} = \epsilon_{0\alpha\beta\gamma}\,\omega^\beta\,n^\gamma \,, \tag{1.25}$$

with the Greek indices, as always, taking values 1, 2, 3. This can be written in the familiar 3-vector form of a rigid rotation:

$$\frac{\partial \mathbf{n}}{\partial s} = \boldsymbol{\omega} \times \mathbf{n} \,. \tag{1.26}$$

Hence we refer to the tensor ω_{ab} as the *twist tensor* or *vorticity tensor* of the congruence of integral curves of u^a and to ω^a as the *vorticity vector*.

Next assume that $\vartheta \neq 0$, $\omega_{ab} = 0$, $\sigma_{ab} = 0$ and (1.11) becomes

$$h^a{}_b\,\dot{\eta}^b = \frac{1}{3}\vartheta\,\eta^b \,. \tag{1.27}$$

Putting $\eta^a = l\, n^a$ again yields

$$\frac{\dot{l}}{l} = \frac{1}{3}\vartheta \quad \text{and} \quad h^a{}_b\, \dot{n}^b = 0 .\tag{1.28}$$

Hence ϑ measures the rate of change of length per unit length along ξ^α of the connecting vector η^a and the second equation in (1.28) represents propagation along ξ^α without rotation and thus n^a is Fermi transported [1, 2] along ξ^α. If a set of neighbouring particles to ξ^α initially form a 3-sphere of radius l centred on ξ^α then under (1.27) the sphere remains a sphere and its volume V satisfies

$$\frac{\dot{V}}{V} = \vartheta .\tag{1.29}$$

We call the scalar ϑ the *expansion* (if $\vartheta > 0$) or *contraction* (if $\vartheta < 0$) *scalar* of the congruence.

Finally consider the case $\sigma_{ab} \neq 0$, $\vartheta = 0$, $\omega_{ab} = 0$ and (1.11) is now

$$h^a{}_b\, \dot{\eta}^b = \sigma^a{}_b\, \eta^b .\tag{1.30}$$

Since $\sigma_{ab} = \sigma_{ba}$ and $\sigma_{ab}\, u^b = 0$ we see that u^a is the unit time-like eigenvector of σ_{ab} with zero eigenvalue. Let $n^a_{(\alpha)}$ be the three unit space-like eigenvectors with corresponding eigenvalues $\mu_{(\alpha)}$ for $\alpha = 1, 2, 3$. These latter are mutually orthogonal and each are orthogonal to u^a. Thus we have

$$\sigma^a{}_b\, n^b_{(\alpha)} = \mu_{(\alpha)}\, n^a_{(\alpha)} \quad (\alpha = 1, 2, 3) ,\tag{1.31}$$

and, since $\sigma^a{}_a = 0$,

$$\sum_{\alpha=1}^{3} \mu_{(\alpha)} = 0 .\tag{1.32}$$

In addition the orthonormality conditions satisfied by $n^a_{(\alpha)}$ are

$$u_a\, n^a_{(\alpha)} = 0 \quad \text{and} \quad g_{ab}\, n^a_{(\alpha)}\, n^b_{(\beta)} = -\delta_{\alpha\beta} .\tag{1.33}$$

Now choose three orthogonal connecting vectors in the directions $n^a_{(\alpha)}$ given by

$$\eta^a_{(\alpha)} = l_{(\alpha)}\, n^a_{(\alpha)} \quad (\alpha = 1, 2, 3) .\tag{1.34}$$

Each of these must satisfy (1.30) and so we find that

$$\frac{\dot{l}_{(\alpha)}}{l_{(\alpha)}} = \mu_{(\alpha)} \quad \text{and} \quad h^a{}_b\, \dot{n}^b_{(\alpha)} = 0 ,\tag{1.35}$$

for $\alpha = 1, 2, 3$. Thus each $n^a_{(\alpha)}$ is Fermi transported along ξ^α and the rate of change of length per unit length of $\eta^a_{(\alpha)}$ is direction dependent with at least one rate of change negative and at least one rate of change positive. If initially, at s (say), the neighbouring particles to ξ^α form a sphere of radius l centred on ξ^α, then if one moves to $s + ds$ the sphere is deformed into an ellipsoid with principal axes of lengths $(1 + \mu_{(\alpha)} ds)l$ for $\alpha = 1, 2, 3$. The volume of this ellipsoid is approximately $4\pi l^3 (1 + \sum_{\alpha=1}^3 \mu_{(\alpha)} ds)/3 = 4\pi l^3/3$ on account of (1.32). Hence the effect of σ_{ab} is to shear or distort the sphere without changing its volume. We consequently refer to σ_{ab} as the *shear tensor* of the congruence. The reader can check that $\sigma^{ab} \sigma_{ab} \geq 0$ with equality if and only if $\sigma_{ab} = 0$. Hence we can write

$$\frac{1}{2}\sigma^{ab}\sigma_{ab} = \sigma^2 \,, \tag{1.36}$$

for some scalar σ. Early work on the kinematics of a continuum is described in [2] and this has been developed further in [3,4].

Finally we note that the transport law (1.11) for the orthogonal connecting vector η^a simplifies to

$$\dot{\eta}^a = u^a_{;b}\,\eta^b \,, \tag{1.37}$$

if the time-like congruence is geodesic (i.e. $\dot{u}^a = u^a_{;b}u^b = 0$). This is an example of a first order *geodesic deviation equation*. The standard second order *geodesic deviation equation* is satisfied by this η^a too. It is obtained using $\dot{u}^a = 0$ and the Ricci identities as follows:

$$
\begin{aligned}
\ddot{\eta}^a &= \dot{\eta}^a_{;b}\,u^b \\
&= (u^a_{;c}\,\eta^c)_{;b}\,u^b \quad \text{by (1.37)} \\
&= u^a_{;cb}\,\eta^c\,u^b + u^a_{;c}\,\dot{\eta}^c \\
&= (u^a_{;bc} + u_d\,R^{da}_{cb})\,\eta^c\,u^b + u^a_{;c}\,\dot{\eta}^c \quad \text{by the Ricci identities} \\
&= u^a_{;bc}\,u^b\,\eta^c - R^a_{dcb}\,u^d\,\eta^c\,u^b + u^a_{;c}\,\dot{\eta}^c \\
&= -u^a_{;b}\,u^b_{;c}\,\eta^c - R^a_{dcb}\,u^d\,\eta^c\,u^b + u^a_{;c}\,\dot{\eta}^c \quad \text{since } \dot{u}^a = 0 \\
&= -u^a_{;b}\,\dot{\eta}^b - R^a_{dcb}\,u^d\,\eta^c\,u^b + u^a_{;c}\,\dot{\eta}^c \quad \text{by (1.37)} \\
&= -R^a_{dcb}\,u^d\,\eta^c\,u^b \,, \tag{1.38}
\end{aligned}
$$

and thus η^a satisfies the second order geodesic deviation equation

$$\ddot{\eta}^a + R^a_{dcb}\,u^d\,\eta^c\,u^b = 0 \,. \tag{1.39}$$

However this equation holds in a more fundamental geometrical setting than that of a congruence.

1.2   Null Geodesic Congruences

Let $x^i = x^i(r, y^\alpha)$ be the parametric equations of a congruence of null geodesics with $y^\alpha = (y^1, y^2, y^3)$ labeling the curves of the congruence and r an affine parameter along them. The geodesic curves of the congruence are the integral curves of a vector field k^i satisfying

$$k^i = \frac{\partial x^i}{\partial r} \quad \text{with} \quad k^i k_i = 0 \quad \text{and} \quad k^i_{;j} k^j = 0 . \tag{1.40}$$

As in the time-like case above, an infinitesimal connecting vector defined along any curve C of the congruence (specified by a choice of y^α and referred to simply as the curve y^α) joining points of equal parameter value r on neighbouring curves of the congruence, y^α and $y^\alpha + \delta y^\alpha$, is given by

$$\zeta^i = \frac{\partial x^i}{\partial y^\alpha} \delta y^\alpha . \tag{1.41}$$

With the same argument as in the time-like case, ζ^i is transported along C according to the transport law

$$\zeta^i_{;j} k^j = k^i_{;j} \zeta^j \quad \Rightarrow \quad \frac{\partial}{\partial r}(k_i \zeta^i) = 0 , \tag{1.42}$$

with the final equation here a consequence of (1.40). Hence $k_i \zeta^i$ is constant along the null geodesic C. At any point P on C let e^i be a unit time-like vector. The physical significance of introducing e^i will appear later. Thus $e_i e^i = 1$ and we extend e^i to a vector field along C by parallel transport. Thus $e^i_{;j} k^j = 0$. Next we normalise k^i by requiring $k^i e_i = 1$. Now the vector field

$$l^i = e^i - \frac{1}{2} k^i , \tag{1.43}$$

defined along C, satisfies

$$l^i l_i = 0 , \; l^i k_i = 1 , \tag{1.44}$$

and l^i is parallel transported along C. At P on C we have the null vectors k^i and l^i and we can add to them a complex null vector m^i, and its complex conjugate $\bar{m}^i$, both orthogonal to k^i and l^i, with

$$m_i \bar{m}^i = -1 . \tag{1.45}$$

Now $k^i, l^i, m^i, \bar{m}^i$ constitute a null tetrad at P which will continue to be a null tetrad along C if we take m^i to be parallel transported along C so that

$$m^i{}_{;j}\, k^j = 0 = \bar{m}^i{}_{;j}\, k^j \ . \tag{1.46}$$

In terms of this null tetrad we can write the components g_{ij} of the metric tensor as

$$g_{ij} = -m_i\, \bar{m}_j - \bar{m}_i\, m_j + k_i\, l_j + l_i\, k_j \ . \tag{1.47}$$

We now specialise the infinitesimal connecting vector ζ^i by requiring it to satisfy the following equations:

$$\zeta_i\, e^i = 0 \quad \text{and} \quad (\delta^i_j - e^i\, e_j)\, k^j\, \zeta_i = 0 \ . \tag{1.48}$$

Thus ζ^i is orthogonal to e^i and is also orthogonal to the projection of k^i orthogonal to e^i. The physical interpretation of these conditions will appear below. As a consequence of them ζ^i is orthogonal to k^i and to l^i in (1.43) and so can be expressed on the null tetrad as

$$\zeta^i = \bar{\zeta}\, m^i + \zeta\, \bar{m}^i \ , \tag{1.49}$$

with ζ a complex valued function of the coordinates x^i. Substituting this into the transport law (1.42) results in

$$\frac{\partial \bar{\zeta}}{\partial r}\, m^i + \frac{\partial \zeta}{\partial r}\, \bar{m}^i = \bar{\zeta}\, k^i{}_{;j}\, m^j + \zeta\, k^i{}_{;j}\, \bar{m}^j \ . \tag{1.50}$$

Taking the scalar product of this equation with m^i leads to the propagation law for $\zeta(x^i)$ along C:

$$\frac{\partial \zeta}{\partial r} = -\sigma\, \bar{\zeta} - \rho\, \zeta \ , \tag{1.51}$$

with

$$\sigma = k_{i;j}\, m^i\, m^j \quad \text{and} \quad \rho = k_{i;j}\, m^i\, \bar{m}^j \ . \tag{1.52}$$

Although it appears that σ and ρ depend upon the choice of tetrad, in fact $|\sigma|$ and ρ are independent of the choice of tetrad and can be constructed solely from a knowledge of g_{ij} and k^i. It is easy to see that the *argument* of the complex variable σ is dependent upon the choice of tetrad since it is obviously *not* invariant under the simple tetrad transformation $m^i \rightarrow e^{i\psi}\, m^i$, for some real constant ψ. To see explicitly that $|\sigma|$ and ρ can be derived from g_{ij} and k^i we first expand $k_{i;j}$ on the null tetrad taking into consideration that $k_{i;j}$ is real, $k^i\, k_{i;j} = 0$ and $k^j\, k_{i;j} = 0$ and also the definitions (1.52) of σ and ρ. The result is

$$k_{i;j} = a\, k_i\, k_j + b\, k_i\, m_j + \bar{b}\, k_i\, \bar{m}_j + c\, m_i\, k_j + \bar{c}\, \bar{m}_i\, k_j$$
$$+ \sigma\, \bar{m}_i\, \bar{m}_j + \bar{\sigma}\, m_i\, m_j + \rho\, \bar{m}_i\, m_j + \bar{\rho}\, m_i\, \bar{m}_j \,, \tag{1.53}$$

with a a real-valued function of the coordinates and b, c complex-valued functions of the coordinates with complex conjugation denoted by a bar as usual. Making use of the scalar products among the tetrad vectors we derive from (1.53) the following equations:

$$k^i{}_{;i} = -\rho - \bar{\rho} \,, \tag{1.54}$$

$$k_{i;j}\, k^{i;j} = 2\sigma\,\bar{\sigma} + 2\rho\,\bar{\rho} \,, \tag{1.55}$$

$$k_{j;i}\, k^{i;j} = 2\sigma\,\bar{\sigma} + \rho^2 + \bar{\rho}^2 \,. \tag{1.56}$$

From these we conclude that $\rho = -\theta - i\,\omega$ with

$$\theta = \frac{1}{2}\, k^i{}_{;i} \quad \text{and} \quad \omega^2 = \frac{1}{2}\, k_{[i;j]}\, k^{i;j} \geq 0 \,, \tag{1.57}$$

with $k_{[i;j]} = (k_{i;j} - k_{j;i})/2$, and

$$|\sigma| = \sqrt{\frac{1}{2}\, k_{(i;j)}\, k^{i;j} - \theta^2} \,, \tag{1.58}$$

with $k_{(i;j)} = (k_{i;j} + k_{j;i})/2$. We see explicitly from (1.57) and (1.58) that ρ and $|\sigma|$ are determined from a knowledge of g_{ij} and k^i and are therefore independent of the choice of null tetrad. It therefore makes sense to use the local geometrical construction leading to (1.51) to provide a geometrical interpretation of ρ and $|\sigma|$.

At a point P on the null geodesic C consider e^i to be the 4-velocity of a small plane circular opaque disk located in the path of a small bundle of photons having world lines members of the null geodesic congruence in the neighbourhood of C. Since $\zeta_i\, k^i = 0$ we have ζ^i in the rest-frame of the disk and we can take ζ^i to be the position vector of points on the boundary of the disk relative to the centre of the disk. The second condition on ζ^i in (1.48) requires the disk to be oriented relative to the paths of the photons so that the photons strike the disk at right angles when observed in the rest-frame of the disk. If Q is a second point on C to the future of the point P and a small affine distance dr from P then, at Q we consider e^i to be the 4-velocity of a small plane screen on which a shadow of the disk is projected. The position vector of points on the boundary of the shadow, relative to the centre of the shadow, is given by ζ^i which lies in the rest-frame of the screen on account of the first equation in (1.48) and the second equation in (1.42). Again following from the second equation in (1.48) the photons strike the screen at right angles to the screen as observed in the rest-frame of the screen. The passage from the disk to the neighbouring screen corresponds to $r \rightarrow r + dr$ and this corresponds to $\zeta \rightarrow \zeta + d\zeta = \zeta'$ where

$$\zeta' = (1 - \rho\, dr)\, \zeta - \sigma \bar{\zeta}\, dr\,, \tag{1.59}$$

on account of (1.51). We assume that the disk is circular so that $\zeta = e^{i\phi}$, with $0 \le \phi < 2\pi$. With this substituted into (1.59), ζ' specifies points on the boundary of the shadow relative to points on the boundary of the disk. With $\rho = -\theta - i\,\omega$ we consider first the case $\omega = 0 = \sigma$. Now $\zeta' = (1 + \theta\, dr)\zeta$ and thus the shadow is a circle with a radius larger than the disk radius if $\theta > 0$, or smaller than the disk radius if $\theta < 0$. Hence we interpret the scalar θ to represent the *expansion* of the congruence if $\theta > 0$ or the *contraction* of the congruence if $\theta < 0$. Considering next the case $\theta = 0 = \sigma$ we have from (1.59), $\zeta' = (1 + i\,\omega\, dr)\,\zeta = e^{i\,\omega\, dr}\, \zeta$, neglecting $O(dr^2)$-terms. In this case the shadow is a circle but the points on the boundary of the shadow are rotated relative to points on the boundary of the disk through a small angle $\omega\, dr$. Hence the scalar ω is interpreted as the *twist* of the congruence. Finally the case $\theta = 0 = \omega$ results in (1.59) becoming $\zeta' = e^{i\phi} - \sigma e^{-i\phi}\, dr$. Now the shadow is approximately elliptical with semi-major and semi-minor axes corresponding to ϕ given by $e^{2i\phi} = \pm(\sigma/\bar{\sigma})^{1/2}$. The lengths of these axes are then $l_\pm = 1 \pm |\sigma|\, dr$ and their ratio is thus $l_+/l_- = 1 + 2|\sigma|\, dr$ approximately. The area of the shadow is the same as the area of the disk, neglecting $O(dr^2)$-terms, and thus we interpret $|\sigma|$ as the *shear* of the congruence. We note that σ is usually referred to as the *complex shear* of the congruence.

The theory of null geodesic congruences originated in [5, 6] while the presentation given here is strongly influenced by the elegant treatment by Pirani [7] (see also [8]).

1.3   The World Function and Deviation Equations

Our study of congruences above has involved in particular the deviation of neighbouring lines of a congruence. The concept of deviation can be generalised to time-like curves $Y(t)$ and $X(\tilde{t})$ (say) where the parameters t and $\tilde{t}$ along the curves are no longer necessarily proper time or arc length.

As sketched in Fig. 1.2, we assume that there exists a unique geodesic Z joining two points $x \in X$ and $y \in Y$ on the two curves. Along this geodesic we introduce the so-called world function as an integral

$$\sigma(x, y) := \frac{\epsilon}{2}\left(\int_x^y d\tau \right)^2 \tag{1.60}$$

connecting the space-time points x and y. Here $d\tau$ is the differential of the proper time along the geodesic Z, and $\epsilon = \pm 1$ for time-like/space-like curves. The world function thus gives half the squared geodesic interval along Z.

Fig. 1.2 Sketch of the two arbitrarily parametrized world lines $Y(t)$ and $X(\tilde{t})$, and the geodesic Z connecting two points on these world lines. The squared length of the interval between the two points x and y is proportional to the world function $\sigma(x, y)$. The (generalized) deviation vector along the reference world line Y is denoted by η^y

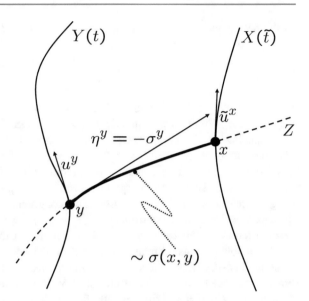

As becomes clear by the definition, the world function is a two-point function, or biscalar. One can also introduce generalizations of ordinary space-time tensor, called bitensors, which depend on not one but two space-times points. Pioneering references in this respect are [2, 9], and a comprehensive review can also be found in [10]. Here we only review some of the essential bitensor concepts.

We start by the introduction of a condensed notation, in which the space-time point to which an index of a bitensor belongs can be directly read from the index itself, for example $B^{x_0 x_1}{}_{y_0 y_1 y_2}(x, y)$ would denote a bitensor of second rank at x and of third rank at y. Indices attached to the world function always denote covariant derivatives, at the given point, i.e. $\sigma_y := \nabla_y \sigma$, hence we do not make explicit use of the semicolon in case of the world function. For any bitensor, covariant derivatives at different points commute with each other, for instance in case of the world function $\sigma_{y_0 y_1 x_0 y_2 x_1} = \sigma_{y_0 y_1 y_2 x_0 x_1} = \sigma_{x_0 x_1 y_0 y_1 y_2}$.

In many calculations the limiting behaviour of a bitensor $B_{...}(x, y)$ as x approaches the reference point y is required. This so-called coincidence limit of a bitensor $B_{...}(x, y)$ is a tensor

$$[B_{...}] = \lim_{x \to y} B_{...}(x, y), \qquad (1.61)$$

at y and will be denoted by square brackets. In particular, for a bitensor B with arbitrary indices at different points (here just denoted by dots), we have the rule [2]

$$[B_{...}]_{;y} = [B_{...;y}] + [B_{...;x}]. \qquad (1.62)$$

We collect the following useful identities for the world function σ:

$$[\sigma] = [\sigma_x] = [\sigma_y] = 0, \tag{1.63}$$

$$[\sigma_{x_0 x_1}] = [\sigma_{y_0 y_1}] = g_{y_0 y_1}, \tag{1.64}$$

$$[\sigma_{x_0 y_1}] = [\sigma_{y_0 x_1}] = -g_{y_0 y_1}, \tag{1.65}$$

$$[\sigma_{x_0 x_1 x_2}] = [\sigma_{x_0 x_1 y_2}]$$

$$= [\sigma_{x_0 y_1 y_2}] = [\sigma_{y_0 y_1 y_2}] = 0, \tag{1.66}$$

$$[\sigma_{(n \text{ indices})}] \sim \nabla^{n-4} R_{abcd} \text{ for } n \geq 4. \tag{1.67}$$

Apart from the world function $\sigma(x, y)$, another important bitensor is the parallel propagator $g^y{}_x(x, y)$—not to be confused with the metric—that allows for the parallel transportation of objects along the unique geodesic that links the points x and y. For example, given a vector V^x at x, the corresponding vector at y is obtained by means of the parallel transport along the geodesic curve as $V^y = g^y{}_x(x, y) V^x$. The coincidence limits of the parallel propagator and its first derivative are given by:

$$\left[g^{x_0}{}_{y_1} \right] = \delta^{y_0}{}_{y_1}, \tag{1.68}$$

$$\left[g^{x_0}{}_{y_1; x_2} \right] = \left[g^{x_0}{}_{y_1; y_2} \right] = 0, \tag{1.69}$$

$$\left[g^{x_0}{}_{y_1; x_2 x_3} \right] = -\left[g^{x_0}{}_{y_1; x_2 y_3} \right] = \left[g^{x_0}{}_{y_1; x_2 x_3} \right]$$

$$= -\left[g^{x_0}{}_{y_1; y_2 y_3} \right] = \frac{1}{2} R^{y_0}{}_{y_1 y_2 y_3}, \tag{1.70}$$

$$\left[g^{x_0}{}_{x_1; (n \text{ indices})} \right] \sim \nabla^{n-2} R_{abcd} \text{ for } n \geq 2. \tag{1.71}$$

With these bitensor related concepts in mind we now return to the deviation of the two curves X and Y depicted in Fig. 1.2. Conceptually the closest object to the connecting vector between the two points x and y is the covariant derivative of the world function σ^y. Note that σ^y is tangent to the geodesic Z at y (its length being the geodesic length between y and x) and only in flat space-time it coincides with the connecting vector. Keeping in mind such an interpretation, let us now work out a propagation equation for this "generalized" connecting vector along the reference curve, cf. Fig. 1.2. Following our conventions the reference curve will be $Y(t)$ and we define the generalized connecting vector to be:

$$\eta^y := -\sigma^y. \tag{1.72}$$

Taking its covariant total derivative, we have

$$\frac{D}{dt}\eta^{y_1} = -\frac{D}{dt}\sigma^{y_1}\left(Y(t), X(\tilde{t})\right)$$

$$= -\sigma^{y_1}{}_{y_2}\frac{\partial Y^{y_2}}{\partial t} - \sigma^{y_1}{}_{x_2}\frac{\partial X^{x_2}}{\partial \tilde{t}}\frac{d\tilde{t}}{dt}$$

$$= -\sigma^{y_1}{}_{y_2}u^{y_2} - \sigma^{y_1}{}_{x_2}\tilde{u}^{x_2}\frac{d\tilde{t}}{dt}, \tag{1.73}$$

where in the last line we defined the velocities along the two curves Y and X. As usual, $\sigma^y{}_{x_1...y_2...} := \nabla_{x_1}\ldots\nabla_{y_2}\ldots(\sigma^y)$ denote the higher order covariant derivatives of the world function. This is the *exact* equivalent of Eq. (1.37) in bitensor language for general curves. As becomes apparent by the indices it still contains quantities defined along the world line X.

Taking the second derivative of (1.73) yields

$$\frac{D^2}{dt^2}\eta^{y_1} = -\sigma^{y_1}{}_{y_2y_3}u^{y_2}u^{y_3} - 2\sigma^{y_1}{}_{y_2x_3}u^{y_2}\tilde{u}^{x_3}\frac{d\tilde{t}}{dt}$$

$$-\sigma^{y_1}{}_{y_2}a^{y_2} - \sigma^{y_1}{}_{x_2x_3}\tilde{u}^{x_2}\tilde{u}^{x_3}\left(\frac{d\tilde{t}}{dt}\right)^2$$

$$-\sigma^{y_1}{}_{x_2}\tilde{a}^{x_2}\left(\frac{d\tilde{t}}{dt}\right)^2 - \sigma^{y_1}{}_{x_2}\tilde{u}^{x_2}\frac{d^2\tilde{t}}{dt^2}, \tag{1.74}$$

where we introduced the accelerations $a^y := Du^y/dt$, and $\tilde{a}^x := D\tilde{u}^x/d\tilde{t}$. Equation (1.74) is already the deviation equation, i.e. the generalization of (1.38), but the goal is to have all the quantities therein defined along the reference word line Y.

We derive some auxiliary formulas, by introduction of the inverse of the second derivative of the world function via the following equations:

$$\overset{-1}{\sigma}{}^{y_1}{}_x\sigma^x{}_{y_2} = \delta^{y_1}{}_{y_2}, \qquad \overset{-1}{\sigma}{}^{x_1}{}_y\sigma^y{}_{x_2} = \delta^{x_1}{}_{x_2}. \tag{1.75}$$

Multiplication of (1.73) by $\overset{-1}{\sigma}{}^{x_3}{}_{y_1}$ results in

$$\tilde{u}^{x_3}\frac{d\tilde{t}}{dt} = -\overset{-1}{\sigma}{}^{x_3}{}_{y_1}\sigma^{y_1}{}_{y_2}u^{y_2} + \overset{-1}{\sigma}{}^{x_3}{}_{y_1}\frac{D\sigma^{y_1}}{dt}$$

$$= K^{x_3}{}_{y_2}u^{y_2} - H^{x_3}{}_{y_1}\frac{D\sigma^{y_1}}{dt}. \tag{1.76}$$

In the last line we defined two auxiliary quantities $K^x{}_y$ and $H^x{}_y$—these are the so-called Jacobi propagators, and the notation follows the terminology of Dixon [11]. Equation (1.76) allows us to formally express the velocity along the curve X

in terms of the quantities which are defined at Y and then "propagated" by $K^x{}_y$ and $H^x{}_y$.

Multiplication of (1.76) with $dt/d\tilde{t}$ yields

$$\tilde{u}^{x_3} = K^{x_3}{}_{y_2}u^{y_2}\frac{dt}{d\tilde{t}} - H^{x_3}{}_{y_1}\frac{D\sigma^{y_1}}{dt}\frac{dt}{d\tilde{t}}, \tag{1.77}$$

and insertion into (1.74) results in:

$$\begin{aligned}
\frac{D^2}{dt^2}\eta^{y_1} = &-\sigma^{y_1}{}_{y_2y_3}u^{y_2}u^{y_3} - \sigma^{y_1}{}_{y_2}a^{y_2} - \sigma^{y_1}{}_{x_2}\tilde{a}^{x_2}\left(\frac{d\tilde{t}}{dt}\right)^2 \\
&-2\sigma^{y_1}{}_{y_2x_3}u^{y_2}\left(K^{x_3}{}_{y_4}u^{y_4} - H^{x_3}{}_{y_4}\frac{D\sigma^{y_4}}{dt}\right) \\
&-\sigma^{y_1}{}_{x_2x_3}\left(K^{x_2}{}_{y_4}u^{y_4} - H^{x_2}{}_{y_4}\frac{D\sigma^{y_4}}{dt}\right) \\
&\times\left(K^{x_3}{}_{y_5}u^{y_5} - H^{x_3}{}_{y_5}\frac{D\sigma^{y_5}}{dt}\right) \\
&-\sigma^{y_1}{}_{x_2}\frac{dt}{d\tilde{t}}\frac{d^2\tilde{t}}{dt^2}\left(K^{x_2}{}_{y_3}u^{y_3} - H^{x_2}{}_{y_3}\frac{D\sigma^{y_3}}{dt}\right). \tag{1.78}
\end{aligned}$$

Note that we may determine the factor $d\tilde{t}/dt$ by requiring that the velocity along the curve X is normalized, i.e. $\tilde{u}^x\tilde{u}_x = 1$, in which case (1.76) yields

$$\frac{d\tilde{t}}{dt} = \tilde{u}_{x_1}K^{x_1}{}_{y_2}u^{y_2} - \tilde{u}_{x_1}H^{x_1}{}_{y_2}\frac{D\sigma^{y_2}}{dt}. \tag{1.79}$$

Up to this point the derivation has been completely general, i.e. (1.78) is the *exact* form of the deviation equation for two general arbitrarily parametrized curves $Y(t)$ and $X(\tilde{t})$. In particular, it allows for a comparison of two general, i.e. not necessarily geodetic, world lines in space-time. Before we consider any approximate version of (1.78), we switch to a synchronous parametrization by rewriting the velocity as

$$u^y = \frac{dY^y}{dt} = \frac{d\tilde{t}}{dt}\frac{dY^y}{d\tilde{t}}. \tag{1.80}$$

Thus we parametrize the first curve with the same parameter $\tilde{t}$ that is used on the second curve. Accordingly, we denote

$$\underset{\sim}{u}^y = \frac{dY^y}{d\tilde{t}}. \tag{1.81}$$

By differentiation, we then derive

$$a^y = \frac{Du^y}{dt} = \frac{D}{dt}\left(\frac{d\tilde{t}}{dt}\underset{\sim}{u}^y\right)$$

$$= \frac{d^2\tilde{t}}{dt^2}\underset{\sim}{u}^y + \left(\frac{d\tilde{t}}{dt}\right)^2 \underset{\sim}{a}^y, \qquad (1.82)$$

where

$$\underset{\sim}{a}^y = \frac{D}{d\tilde{t}}\underset{\sim}{u}^y = \frac{D^2 Y^y}{d\tilde{t}^2}. \qquad (1.83)$$

Analogously we obtain for the derivative of the deviation vector

$$\frac{D^2\eta^y}{dt^2} = \frac{d^2\tilde{t}}{dt^2}\frac{D\eta^y}{d\tilde{t}} + \left(\frac{d\tilde{t}}{dt}\right)^2 \frac{D^2\eta^y}{d\tilde{t}^2}. \qquad (1.84)$$

With this re-parametrisation the exact deviation equation (1.78) is recast into the simpler form

$$\frac{D^2}{d\tilde{t}^2}\eta^{y_1} = -\sigma^{y_1}{}_{y_2}\underset{\sim}{a}^{y_2} - \sigma^{y_1}{}_{x_2}\tilde{a}^{x_2} - \sigma^{y_1}{}_{y_2 y_3}\underset{\sim}{u}^{y_2}\underset{\sim}{u}^{y_3}$$

$$-2\sigma^{y_1}{}_{y_2 x_3}\underset{\sim}{u}^{y_2}\left(K^{x_3}{}_{y_4}\underset{\sim}{u}^{y_4} - H^{x_3}{}_{y_4}\frac{D\sigma^{y_4}}{d\tilde{t}}\right)$$

$$-\sigma^{y_1}{}_{x_2 x_3}\left(K^{x_2}{}_{y_4}\underset{\sim}{u}^{y_4} - H^{x_2}{}_{y_4}\frac{D\sigma^{y_4}}{d\tilde{t}}\right)$$

$$\times\left(K^{x_3}{}_{y_5}\underset{\sim}{u}^{y_5} - H^{x_3}{}_{y_5}\frac{D\sigma^{y_5}}{d\tilde{t}}\right). \qquad (1.85)$$

Now everything is synchronous in the sense that both curves are parametrized by $\tilde{t}$. It should be stressed that the generalized deviation equation qualitatively reproduces several other results for generalized deviation equations in the literature [12–18].

The generalized exact deviation equation (1.85) contains quantities which are not defined along the reference curve Y, in particular the covariant derivatives of the world function. In the next step we expand all quantities in powers of the deviation around Y.

For a general bitensor $B_{...}$ with a given index structure, we have the following general expansion, up to the third order (in powers of σ^y):

$$B_{y_1...y_n} = A_{y_1...y_n} + A_{y_1...y_{n+1}}\sigma^{y_{n+1}}$$

$$+ \frac{1}{2}A_{y_1...y_{n+1}y_{n+2}}\sigma^{y_{n+1}}\sigma^{y_{n+2}} + O\left(\sigma^3\right), \qquad (1.86)$$

$$A_{y_1 \ldots y_n} := \left[B_{y_1 \ldots y_n} \right], \tag{1.87}$$

$$A_{y_1 \ldots y_{n+1}} := \left[B_{y_1 \ldots y_n; y_{n+1}} \right] - A_{y_1 \ldots y_n; y_{n+1}}, \tag{1.88}$$

$$A_{y_1 \ldots y_{n+2}} := \left[B_{y_1 \ldots y_n; y_{n+1} y_{n+2}} \right] - A_{y_1 \ldots y_n y_0} \left[\sigma^{y_0}{}_{y_{n+1} y_{n+2}} \right]$$
$$- A_{y_1 \ldots y_n; y_{n+1} y_{n+2}} - 2 A_{y_1 \ldots y_n (y_{n+1}; y_{n+2})}. \tag{1.89}$$

With the help of (1.86) we are able to iteratively expand any bitensor to any order, provided the coincidence limits entering the expansion coefficients can be calculated. The expansion for bitensors with mixed index structure can be obtained from transporting the indices in (1.86) by means of the parallel propagator. We do not give the explicit form of the expansions here (they can be found in [19]) but only state the final result in the form of the expanded version of (1.85) up to the second order:

$$\frac{D^2}{d\tilde{t}^2} \eta^{y_1} = \tilde{a}^{y_1} - \underaccent{\tilde}{a}^{y_1} - \eta^{y_4} R^{y_1}{}_{y_2 y_3 y_4} \left(\underaccent{\tilde}{u}^{y_2} \underaccent{\tilde}{u}^{y_3} + 2 \underaccent{\tilde}{u}^{y_3} \frac{D \eta^{y_2}}{d\tilde{t}} \right)$$
$$+ \eta^{y_4} \eta^{y_5} \left\{ \underaccent{\tilde}{u}^{y_2} \underaccent{\tilde}{u}^{y_3} \left(\frac{1}{2} \nabla_{y_2} R^{y_1}{}_{y_4 y_5 y_3} - \frac{1}{3} \nabla_{y_4} R^{y_1}{}_{y_2 y_3 y_5} \right) \right.$$
$$\left. + \frac{1}{3} R^{y_1}{}_{y_4 y_5 y_2} \left(\underaccent{\tilde}{a}^{y_2} + \frac{1}{2} \tilde{a}^{y_2} \right) \right\} + O(\sigma^3). \tag{1.90}$$

It becomes clear by specialization to geodesic world lines that the use of systematic expansions of the exact deviation equation in powers of the derivatives of the world function allows for a recovery of the well-known geodesic deviation equation, as given in (1.39), which is represented by the third term in first line on the rhs of (1.90).

Finally we note that the explicit form of the expanded generalized deviation equation up to the third order can be found in [19]. There is even a generalization of the deviation equation available for Riemann-Cartan geometries [20].

References

1. E. Fermi, Atti. Accad. Naz. Lincei Cl. Sci. Fis. Mat. Nat. Rend. **31**, 2151101 (1922)
2. J.L. Synge, *Relativity: The General Theory* (North-Holland, Amsterdam, 1960)
3. J. Ehlers, Akad. Wiss. Mainz Abh. Math.-Naturw. Kl. **11**, 763 (1961)
4. J. Ehlers, Gen. Rel. Grav. **25**, 1225 (1993)
5. P. Jordan, J. Ehlers, R.K. Sachs, Akad. Wiss. Mainz, Abh. Math.-Naturw. Kl. **1**, 1 (1961)
6. P. Jordan, J. Ehlers, R.K. Sachs, Gen. Rel. Grav. **45**, 2691 (2013)
7. F.A.E. Pirani, *Brandeis Lectures* (Prentice Hall, New Jersey, 1965), p. 249
8. S. Chandrasekhar, *The Mathematical Theory of Black Holes* (Oxford University Press, Oxford, 1983)
9. B.S. DeWitt, R.W. Brehme, Ann. Phys. **9**, 220 (1960)
10. E. Poisson, A. Pound, I. Vega, Living Rev. Relativ. **14**(7) (2011)
11. W.G. Dixon, Nuovo Cimento **34**, 317 (1964)

12. D.E. Hodgkinson, Gen. Rel. Grav. **3**, 351 (1972)
13. S.L. Bazański, Ann. H. Poin. A **27**, 115 (1977)
14. A.N. Aleksandrov, K.A. Piragas, Theor. Math. Phys. **38**, 48 (1978)
15. B. Schutz, *Galaxies, Axisymmetric Systems, and Relativity*, ed. by M.A.H. MacCallum, vol. 17 (Cambridge University Press, Cambridge, 1985), p. 237
16. C. Chicone, B. Mashhoon, Class. Quant. Grav. **19**, 4231 (2002)
17. T. Mullari, R. Tammelo, Class. Quant. Grav **23**, 4047 (2006)
18. J. Vines, Gen. Rel. Grav **47**, 59 (2015)
19. D. Puetzfeld, Y.N. Obukhov, Phys. Rev. D **93**, 044073 (2016)
20. Y.N. Obukhov, D. Puetzfeld, *Relativistic Geodesy: Foundations and Application*, ed. by D. Puetzfeld et al. Fundamental Theories of Physics, vol. 196 (Springer, Cham, 2019), p. 87

Bivector Formalism in General Relativity

2

Abstract

The use of bivectors, or skew symmetric tensors, is a powerful research tool in general relativity. Such objects arise naturally in the context of electromagnetism. We therefore introduce them via electromagnetic test fields on arbitrary space-times. All details are provided including an important application to electromagnetic radiation due to Ivor Robinson. This is followed by the extension of bivector theory to gravitational fields. An explicit illustration of their use in relation to Kerr space-time is given.

2.1   Bivectors and Electromagnetic Fields

A source-free electromagnetic test field on a flat or curved background space-time is described by a skew-symmetric tensor with real components $F_{ab} = -F_{ba}$ satisfying Maxwell's source-free field equations

$$F^{ab}{}_{;b} = 0 \quad \text{and} \quad {}^{*}F^{ab}{}_{;b} = 0 \,, \tag{2.1}$$

where

$$ {}^{*}F_{ab} = \frac{1}{2}\eta_{abcd} \, F^{cd} \,, \tag{2.2}$$

are the components of the *dual* of F_{ab}. The covariant derivative with respect to the Riemannian connection calculated with the metric tensor g_{ab} is indicated by a semi-colon, $\eta_{abcd} = \sqrt{-g}\,\epsilon_{abcd}$, with $g = \det(g_{ab})$ and ϵ_{abcd} the Levi-Civita permutation symbol in four dimensions, is the covariant permutation tensor and indices are raised with g^{ab}, where $g^{ab}g_{bc} = \delta^a_c$, as usual. From the properties of η_{abcd} (see Appendix A) it is useful to note that the dual of the dual leads back to the

P. A. Hogan, D. Puetzfeld, *Frontiers in General Relativity*, Lecture Notes
in Physics 984, https://doi.org/10.1007/978-3-030-69370-1_2

original skew-symmetric tensor in the sense that

$$^{**}F_{ab} = \frac{1}{2}\eta_{abcd} {}^{*}F^{cd} = \frac{1}{4}\eta_{abcd} \, \eta^{cdpq} \, F_{pq} = -F_{ab} \, . \tag{2.3}$$

If a time-like vector field with components u^a is chosen such that $u^a \, u_a = 1$ then u^a is the 4-velocity of an observer. Relative to this observer the electromagnetic field described by F_{ab} consists of an *electric vector* $E_a = F_{ab} \, u^b$ and a *magnetic vector* $H_a = {}^{*}F_{ab} \, u^b$. Both of these vectors are orthogonal to u^a and thus each has three independent components. There is a one to one correspondence between the six independent components of the electric and magnetic vectors and the six independent components of F_{ab}. A knowledge of F_{ab}, given u^a, clearly determines E_a and H_a. The converse is also true in the sense that a knowledge of E_a and H_a, given u^a, determines F_{ab}. This follows from the formula

$$F_{ab} = E_a \, u_b - E_b \, u_a + \eta_{abcd} \, u^c \, H^d \, , \tag{2.4}$$

which the reader may like to deduce. In this sense a pair of vectors E_a, H_a is encoded in the skew-symmetric tensor F_{ab}. For this reason such an F_{ab} is referred to as a (real) *bivector*. We can write Maxwell's equations (2.1) in the complex form

$$\mathfrak{F}^{ab}{}_{;b} = 0 \, , \tag{2.5}$$

with $\mathfrak{F}^{ab} = F^{ab} + i {}^{*}F^{ab}$. Thus $\mathfrak{F}_{ab} = -\mathfrak{F}_{ba}$ is a complex bivector which satisfies

$$^{*}\mathfrak{F}_{ab} = -i \, \mathfrak{F}_{ab} \, . \tag{2.6}$$

There are two quadratic invariants, $F_{ab} \, F^{ab}$ and $F_{ab} \, {}^{*}F^{ab}$, which can be constructed from the Maxwell bivector F_{ab}. The reader can verify that the invariant $^{*}F_{ab} \, {}^{*}F^{ab}$ is proportional to $F_{ab} \, F^{ab}$. In terms of the electric and magnetic vectors introduced above these invariants are given by

$$\frac{1}{2} F_{ab} \, F^{ab} = E_a \, E^a - H_a \, H^a \quad \text{and} \quad \frac{1}{2} F_{ab} \, {}^{*}F^{ab} = 2 \, E_a \, H^a \, . \tag{2.7}$$

Hence if the electric and magnetic vectors are equal in magnitude and orthogonal to each other then these invariants both vanish. In this case the electromagnetic field is a pure electromagnetic radiation field. The simplest such case one can imagine is that of plane electromagnetic waves. Such waves are light waves and thus travel with the speed of light. The histories of the plane wave fronts are null hypersurfaces in space-time. Such hypersurfaces have equations of the form

$$\varphi(x^a) = \text{constant} \, , \tag{2.8}$$

with

$$g^{ab} \, \varphi_{,a} \, \varphi_{,b} = 0 \,, \tag{2.9}$$

and with the comma, as always, denoting partial differentiation with respect to the coordinates x^a. Hence the normal to each hypersurface $\varphi_{,a}$ is orthogonal to itself and is therefore tangent to the null hypersurface and is a null vector field whose direction is the propagation direction in space-time of the histories of the waves. This null vector field is determined by the Maxwell bivector F_{ab} at each point of space-time when the two invariants (2.7) vanish and further properties of this null vector field will follow from Maxwell's equations (2.5). More generally, without assuming the vanishing of the two invariants, a Maxwell bivector determines *two* null vectors at each point of space-time. To demonstrate this we will use a basis of four null vector fields, a *null tetrad*, at each point of space-time.

Let $\lambda^a_{(0)}, \lambda^a_{(1)}, \lambda^a_{(2)}, \lambda^a_{(3)}$ be an orthonormal tetrad at a point of space-time, with $\lambda^a_{(0)}$ a unit time-like vector, and thus $g_{ab} \lambda^a_{(0)} \lambda^b_{(0)} = 1$, and $\lambda^a_{(1)}, \lambda^a_{(2)}, \lambda^a_{(3)}$ mutually orthogonal unit space-like vectors, each orthogonal to $\lambda^a_{(0)}$, and thus

$$g_{ab} \lambda^a_{(\alpha)} \lambda^b_{(\beta)} = -\delta_{\alpha\beta} \quad \text{and} \quad g_{ab} \lambda^a_{(0)} \lambda^b_{(\alpha)} = 0 \,, \tag{2.10}$$

with Greek indices taking values 1, 2, 3. These orthonormality conditions can be inverted to express the metric tensor components in terms of the components of the vectors in the orthonormal tetrad as

$$g^{ab} = \lambda^a_{(0)} \lambda^b_{(0)} - \lambda^a_{(1)} \lambda^b_{(1)} - \lambda^a_{(2)} \lambda^b_{(2)} - \lambda^a_{(3)} \lambda^b_{(3)} \,. \tag{2.11}$$

This piece of algebra is an interesting challenge to the reader. In place of the orthonormal tetrad we will work with an equivalent null tetrad consisting of two real null vectors and a complex null vector and its complex conjugate defined by

$$k^a = \frac{1}{\sqrt{2}} (\lambda^a_{(0)} + \lambda^a_{(1)}) \,, \quad l^a = \frac{1}{\sqrt{2}} (\lambda^a_{(0)} - \lambda^a_{(1)}) \,,$$

$$m^a = \frac{1}{\sqrt{2}} (\lambda^a_{(2)} + i\lambda^a_{(3)}) \,, \quad \bar{m}^a = \frac{1}{\sqrt{2}} (\lambda^a_{(2)} - i\lambda^a_{(3)}) \,, \tag{2.12}$$

with complex conjugation denoted by a bar. All of the scalar products involving these vectors vanish except

$$k^a l_a = 1 \quad \text{and} \quad m^a \bar{m}_a = -1 \,. \tag{2.13}$$

Writing the orthonormal tetrad vectors in terms of the null tetrad vectors $k^a, l^a, m^a, \bar{m}^a$ and substituting into (2.11) results in

$$g^{ab} = k^a l^b + k^b l^a - m^a \bar{m}^b - \bar{m}^a m^b \,. \tag{2.14}$$

Hence the line element of the space-time, written in terms of the null tetrad vectors, takes the form

$$ds^2 = 2 k_a \, dx^a \, l_b \, dx^b - 2 \, |m_a \, dx^a|^2 \, , \tag{2.15}$$

where the final term involves the squared modulus of the complex 1-form $m_a \, dx^a$.

From the null tetrad $k^a, l^a, m^a, \bar{m}^a$ we can construct a basis for the complex bivectors $\mathfrak{F}_{ab}$ satisfying (2.6). This basis will consist of three elements and expansion of any such $\mathfrak{F}_{ab}$ on the basis will involve three complex coefficients (the components of $\mathfrak{F}_{ab}$ on the basis) from which one can deduce the six real components of the real bivector F_{ab}. The basis we are looking for will consist of skew-symmetric products of the tetrad vectors. To find the appropriate skew-symmetric products we can start by seeing if the complex bivector $2 \, m^{[a} k^{b]} = (m^a \, k^b - m^b \, k^a)$ satisfies (2.6). We have

$$ {}^*(m^a \, k^b - m^b \, k^a) = \frac{1}{2} \eta^{abcd} \, (m_c \, k_d - m_d \, k_c) = \eta^{abcd} \, m_c \, k_d \, . \tag{2.16}$$

This quantity is skew-symmetric in a, b and it can be written in terms of the tetrad as

$$\eta^{abcd} \, m_c \, k_d = a_1(m^a \, \bar{m}^b - \bar{m}^a \, m^b) + a_2(m^a \, k^b - m^b \, k^a) + a_3(m^a \, l^b - m^b \, l^a)$$
$$+ a_4(\bar{m}^a \, k^b - \bar{m}^b \, k^a) + a_5(\bar{m}^a \, l^b - \bar{m}^b \, l^a) + a_6(k^a \, l^b - k^b \, l^a) \, . \tag{2.17}$$

Multiplying this by m_b yields

$$0 = -a_1 \, m^a + a_4 \, k^a + a_5 \, l^a \, , \tag{2.18}$$

from which we conclude that $a_1 = a_4 = a_5 = 0$. Multiplying (2.17) by k_b similarly results in $a_1 = a_4 = a_5 = 0$. Hence (2.17) reduces to

$$\eta^{abcd} \, m_c \, k_d = a_2 \, (m^a \, k^b - m^b \, k^a) \, . \tag{2.19}$$

From this we have immediately that

$$a_2 = \eta^{abcd} \, \bar{m}_a \, m_b \, l_c \, k_d \, . \tag{2.20}$$

On account of the skew-symmetry of the permutation tensor we see that a_2 is pure imaginary and so we can write (2.19) as

$$ {}^*(m^a \, k^b - m^b \, k^a) = i \, V (m^a \, k^b - m^b \, k^a) \, , \tag{2.21}$$

where the scalar V is real-valued and is given by

$$V = -i\, \eta^{abcd}\, \bar{m}_a\, m_b\, l_c\, k_d = -i\, \eta_{abcd}\, \bar{m}^a\, m^b\, l^c\, k^d \,. \tag{2.22}$$

Hence it follows from (2.21) that $(m^a k^b - m^b k^a)$ is one of the basis complex bivectors we seek provided we can demonstrate that we can choose $V = -1$. In similar fashion the reader can verify that

$$*(\bar{m}^a\, l^b - \bar{m}^b\, l^a) = i\, V(\bar{m}^a\, l^b - \bar{m}^b\, l^a)\,. \tag{2.23}$$

Also

$$*(\bar{m}^a\, m^b - \bar{m}^b\, m^a) = i\, V(k^a\, l^b - k^b\, l^a)\ \text{and}\ *(k^a\, l^b - k^b\, l^a) = i\, V(\bar{m}^a\, m^b - \bar{m}^b\, m^a)\,, \tag{2.24}$$

from which we have

$$*(\bar{m}^a\, m^b - \bar{m}^b\, m^a + k^a\, l^b - k^b\, l^a) = i\, V(\bar{m}^a\, m^b - \bar{m}^b\, m^a + k^a\, l^b - k^b\, l^a)\,. \tag{2.25}$$

Hence the three candidates for basis complex bivectors are

$$L_{ab} = m_a\, k_b - m_b\, k_a = -L_{ba}\,,\quad M_{ab} = \bar{m}_a\, l_b - \bar{m}_b\, l_a = -M_{ba}\,,$$
$$N_{ab} = \bar{m}_a\, m_b - \bar{m}_b\, m_a + k_a\, l_b - k_b\, l_a = -N_{ba}\,. \tag{2.26}$$

The choice $V = -1$ for (2.22) arises when we establish that in general $V^2 = 1$ and then a choice of sign for V corresponds to a choice of orientation of the null tetrad. To show that $V^2 = 1$ we first note that $\eta_{abcd} = \sqrt{-g}\,\epsilon_{abcd}$, and we take $\epsilon_{0123} = +1$, and ϵ_{abcd} is skew-symmetric under interchange of any neighbouring pair of indices. By writing out explicitly the sums implied by the repeated indices (the Einstein summation convention) we have

$$\begin{aligned}
\epsilon_{abcd}\, \bar{m}^a\, m^b\, l^c\, k^d &= (\bar{m}^0 m^1 - m^0\bar{m}^1)(l^2 k^3 - l^3 k^2) + (\bar{m}^0 m^2 - m^0\bar{m}^2)(k^1 l^3 \\
&\quad - k^3 l^1) + (\bar{m}^0 m^3 - m^0\bar{m}^3)(l^1 k^2 - l^2 k^1) + (\bar{m}^1 m^2 \\
&\quad - m^1\bar{m}^2)(l^0 k^3 - k^0 l^3) + (\bar{m}^1 m^3 - m^1\bar{m}^3)(k^0 l^2 - l^0 k^2) \\
&\quad + (\bar{m}^2 m^3 - m^2\bar{m}^3)(l^0 k^1 - l^1 k^0)\,. \tag{2.27}
\end{aligned}$$

Next we also have

$$\epsilon_{abcd}\, g^{a0} g^{b1} g^{c2} g^{d3} = \det(g^{ab}) = g^{-1}\,, \tag{2.28}$$

remembering that $g = \det(g_{ab})$. This is easily verified by writing out the left hand side explicitly, unpacking the summations. Substituting into the left hand side of (2.28) for g^{ab} from (2.14) gives

$$
\begin{aligned}
\epsilon_{abcd}\, g^{a0} g^{b1} g^{c2} g^{d3} = \epsilon_{abcd}\, \bar{m}^a\, m^b\, l^c\, k^d \{ & (\bar{m}^0 m^1 - m^0 \bar{m}^1)(l^2 k^3 - l^3 k^2) + (\bar{m}^0 m^2 \\
& - m^0 \bar{m}^2)(k^1 l^3 - k^3 l^1) + (\bar{m}^0 m^3 - m^0 \bar{m}^3)(l^1 k^2 - l^2 k^1) \\
& + (\bar{m}^1 m^2 - m^1 \bar{m}^2)(l^0 k^3 - k^0 l^3) + (\bar{m}^1 m^3 - m^1 \bar{m}^3)(k^0 l^2 \\
& - k^2 l^0) + (\bar{m}^2 m^3 - m^2 \bar{m}^3)(l^0 k^1 - l^1 k^0) \}\,.
\end{aligned}
\tag{2.29}
$$

Now (2.28) with (2.27) simplifies to

$$
g^{-1} = (\epsilon_{abcd}\, \bar{m}^a\, m^b\, l^c\, k^d)^2 = -g^{-1}(\eta_{abcd}\, \bar{m}^a\, m^b\, l^c\, k^d)^2 = g^{-1} V^2\,,
\tag{2.30}
$$

with the final equality coming from (2.22). Hence we have $V^2 = 1$ and so $V = \pm 1$. From the definition (2.22) of V we see that a choice of sign for V corresponds to a choice of orientation of the tetrad $k^a, l^a, m^a, \bar{m}^a$. We shall choose $V = -1$. Hence the three complex bivectors (2.26) satisfy the conditions

$$
{}^*L_{ab} = -i\, L_{ab}\,,\quad {}^*M_{ab} = -i\, M_{ab}\ \text{ and }\ {}^*N_{ab} = -i\, N_{ab}\,,
\tag{2.31}
$$

and any complex bivector $\mathfrak{F}_{ab}$ satisfying (2.6) can be written

$$
\mathfrak{F}_{ab} = f_1\, L_{ab} + f_2\, M_{ab} + f_3\, N_{ab}\,,
\tag{2.32}
$$

for some complex scalars f_1, f_2, f_3.

The following list of scalar products involving the bivector basis is useful:

$$
\begin{aligned}
& L_{ab}\, L^{bc} = 0\,,\ \ L_{ab}\, M^{bc} = -m_a\, \bar{m}^c + k_a\, l^c\,,\ \ L_{ab}\, N^{bc} = -L_a{}^c\,, \\
& M_{ab}\, M^{bc} = 0\,,\ \ M_{ab}\, N^{bc} = M_a{}^c\,,\ \ N_{ab}\, N^{bc} = \delta_a^c\,,
\end{aligned}
\tag{2.33}
$$

Hence in particular we have

$$
\begin{aligned}
& L_{ab}\, L^{ab} = 0\,,\ \ L_{ab}\, M^{ab} = -2\,,\ \ L_{ab}\, N^{ab} = 0\,, \\
& M_{ab}\, M^{ab} = 0\,,\ \ M_{ab}\, N^{ab} = 0\,,\ \ N_{ab}\, N^{ab} = -4\,.
\end{aligned}
\tag{2.34}
$$

It thus follows from (2.32) that

$$
\mathfrak{F}_{ab}\, \mathfrak{F}^{ab} = -4\, f_1\, f_2 - 4\, f_3^2\,.
\tag{2.35}
$$

But $\mathfrak{F}_{ab} \mathfrak{F}^{ab} = 2 F_{ab} F^{ab} + 2i F_{ab}{}^* F^{ab}$ and so the quadratic invariants (2.7) are given in terms of f_1, f_2, f_3 by

$$\frac{1}{2} F_{ab} F^{ab} + \frac{i}{2} F_{ab}{}^* F^{ab} = -f_1 f_2 - f_3^2 . \tag{2.36}$$

It will also prove useful to note, again from (2.32), that

$$\mathfrak{F}_{ab} k^b = f_2 \bar{m}_a + f_3 k_a , \quad \mathfrak{F}_{ab} m^b = f_2 l_a + f_3 m_a ,$$
$$\mathfrak{F}_{ab} l^b = f_1 m_a - f_3 l_a , \quad \mathfrak{F}_{ab} \bar{m}^b = f_1 k_a - f_3 \bar{m}_a . \tag{2.37}$$

From these we have

$$f_1 = \mathfrak{F}_{ab} l^a \bar{m}^b , \quad f_2 = \mathfrak{F}_{ab} k^a m^b , \quad f_3 = \mathfrak{F}_{ab} l^a k^b = \mathfrak{F}_{ab} m^a \bar{m}^b . \tag{2.38}$$

Establishing the second equality directly in the expression for f_3 is an interesting exercise for the reader.

In (2.14) we have the metric tensor expressed in terms of the null tetrad basis. There is an analogue of this for the complex bivector basis (2.26) which will be useful later. It takes the form

$$g_{abcd} + i \eta_{abcd} = -2 \left(M_{ab} L_{cd} + L_{ab} M_{cd} \right) - N_{ab} N_{cd} , \tag{2.39}$$

where

$$g_{abcd} = g_{ac} g_{bd} - g_{ad} g_{bc} . \tag{2.40}$$

One way to establish this is to first note that each of Eqs. (2.31) is equivalent to

$$i \eta^{abcd} L_{af} = \delta_f^b L^{cd} + \delta_f^d L^{bc} + \delta_f^c L^{db} , \tag{2.41}$$

$$i \eta^{abcd} M_{af} = \delta_f^b M^{cd} + \delta_f^d M^{bc} + \delta_f^c M^{db} , \tag{2.42}$$

$$i \eta^{abcd} N_{af} = \delta_f^b N^{cd} + \delta_f^d N^{bc} + \delta_f^c N^{db} , \tag{2.43}$$

respectively. Multiplying (2.41) by M^{fg} and (2.42) by L^{fg}, adding and using (2.14) and (2.33), results in (after relabelling the indices for convenience)

$$i \eta^{abcd} = -M^{ab} L^{cd} - M^{ad} L^{bc} - M^{ac} L^{db} - L^{ab} M^{cd} - L^{ad} M^{bc} - L^{ac} M^{db} . \tag{2.44}$$

Multiplying (2.43) by N^{fg} and using (2.33) results in (after relabelling the indices)

$$i \eta^{abcd} = -N^{ab} N^{cd} - N^{ad} N^{bc} - N^{ac} N^{db} . \tag{2.45}$$

Equating (2.44) and (2.45) and multiplying the result by $N_b{}^t N_d{}^q$ using (2.33) gives

$$g^{at} g^{cq} - g^{aq} g^{ct} + N^{ac} N^{tq} = -M^{at} L^{cq} - M^{aq} L^{tc} - L^{at} M^{cq}$$
$$-L^{aq} M^{tc} - (L^{tq} M^{ac} + L^{ac} M^{tq}) . \quad (2.46)$$

Now substituting for the first four terms on the right hand side with (2.44) we have

$$g^{at} g^{cq} - g^{aq} g^{ct} + N^{ac} N^{tq} = i \, \eta^{atcq} + 2 \, (M^{ac} L^{qt} + L^{ac} M^{qt}) , \quad (2.47)$$

and this can be rearranged to yield (2.39). The reader may wish to derive (2.39) by first writing the left hand side as a sum of the six products of the basis bivectors, which are symmetric under interchange of the pair of indices (a, b) with the pair of indices (c, d), and then determining the coefficients of the six terms using (2.31) and (2.34).

The null tetrad k^a, l^a, m^a, and the complex conjugate of m^a, is obviously not unique. It can be replaced by $\hat{k}^a, \hat{l}^a, \hat{m}^a$, and the complex conjugate of $\hat{m}^a$, where

$$\hat{k}^a = k^a , \; \hat{l}^a = l^a , \; \hat{m}^a = e^{i\theta} m^a , \quad (2.48)$$

and $\theta(x^a)$ is a real-valued function of the coordinates, or by

$$\hat{k}^a = A \, k^a , \; \hat{l}^a = A^{-1} l^a , \; \hat{m}^a = m^a , \quad (2.49)$$

and $A(x^a)$ is a real-valued function of the coordinates, or by

$$\hat{k}^a = k^a + \beta \, \bar{\beta} \, l^a + \beta \, \bar{m}^a + \bar{\beta} \, m^a ,$$
$$\hat{l}^a = l^a , \quad (2.50)$$
$$\hat{m}^a = m^a + \beta \, l^a ,$$

and $\beta(x^a)$ is a complex-valued function of the coordinates, or by

$$\hat{k}^a = k^a ,$$
$$\hat{l}^a = l^a + \alpha \, \bar{\alpha} \, k^a + \bar{\alpha} \, m^a + \alpha \, \bar{m}^a , \quad (2.51)$$
$$\hat{m}^a = m^a + \alpha \, k^a ,$$

and $\alpha(x^a)$ is a complex-valued function of the coordinates. The transformations (2.50) and (2.51) are examples of *null rotations* which leave one real null direction invariant. Under (2.50) in particular the components of the complex bivector $\mathfrak{F}_{ab}$ on the null tetrad listed in (2.38) transform as

$$\hat{f}_1 = f_1 , \; \hat{f}_2 = f_2 - 2 \beta \, f_3 - \beta^2 f_1 , \; \hat{f}_3 = f_3 + \beta \, f_1 . \quad (2.52)$$

We can assume that $f_1 \neq 0$ because if f_1 vanishes then a transformation (2.51) can make $\hat{f}_1 \neq 0$. In view of the second equation in (2.52) we can choose β to make $\hat{f}_2 = 0$. Such a β satisfies the quadratic equation over the field of complex numbers:

$$f_1 \beta^2 + 2 f_3 \beta - f_2 = 0 . \tag{2.53}$$

In general this equation has two (complex) roots and the corresponding null vectors $\hat{k}^a$, given by (2.50), are the *principal null directions* of the Maxwell bivector F_{ab}. If the roots of (2.53) are equal then β also satisfies

$$f_1 \beta + f_3 = 0 . \tag{2.54}$$

In this case the two principal null directions coincide and $\hat{f}_3 = 0$. Hence using (2.37) we can conclude that with respect to a tetrad basis in which $f_2 = 0$ there are two null vectors k^a satisfying

$$\mathfrak{F}_{ab} k^b = f_3 k_a \Leftrightarrow k_{[c}\mathfrak{F}_{a]b} k^b = 0 \Leftrightarrow k_{[c} F_{a]b} k^b = 0 = k_{[c}{}^* F_{a]b} k^b . \tag{2.55}$$

where, as always, the square brackets denote skew-symmetrisation (for example $w_{[ab]} = (w_{ab} - w_{ba})/2$). If k^a is the *only* null vector satisfying $f_2 = 0$ then $f_3 = 0$ and such a *degenerate* principal null direction satisfies

$$\mathfrak{F}_{ab} k^b = 0 \Leftrightarrow F_{ab} k^b = 0 = {}^* F_{ab} k^b . \tag{2.56}$$

In this case $F_{ab} F^{ab} = 0 = F_{ab}{}^* F^{ab}$ on account of (2.36). This corresponds to pure electromagnetic radiation, mentioned following (2.7) above, if F_{ab} satisfies Maxwell's equations (2.1). The degenerate principal null vector k^a is the propagation direction of the radiation in space-time.

2.2 Electromagnetic Radiation

From ${}^* F_{ab} k^b = 0$ the reader can show that

$$F_{ab} k_c + F_{ca} k_b + F_{bc} k_a = 0 , \tag{2.57}$$

from which it follows that

$$F_{ab} = \xi_a k_b - \xi_b k_a \quad (\Leftrightarrow \mathfrak{F}_{ab} = f_1 L_{ab}) , \tag{2.58}$$

where $\xi_a = F_{ab} l^b$ and so $\xi_a k^a = 0$ since $F_{ab} k^b = 0$. The electromagnetic energy-momentum tensor E_{ab} is given by

$$E_{ab} = F_{ac} F_b{}^c - \frac{1}{4} g_{ab} F_{dc} F^{dc} , \tag{2.59}$$

and, as a consequence of Maxwell's equations (2.1), satisfies

$$E^{ab}{}_{;b} = 0 . \tag{2.60}$$

For the field F_{ab} in (2.58) the electromagnetic energy-momentum tensor takes the simple radiative form

$$E_{ab} = (\xi_c \, \xi^c) \, k_a \, k_b , \tag{2.61}$$

assuming of course that $\xi_c \, \xi^c \neq 0$. In this case (2.60) specialises to

$$k^a{}_{;b} \, k^b = \lambda \, k^a , \tag{2.62}$$

where λ is a real scalar function of the coordinates x^a given by $\lambda(x^a) = -(\xi_c \, \xi^c)^{-1} \{ (\xi_d \, \xi^d)_{,b} \, k^b + \xi_d \, \xi^d \, k^b{}_{;b} \}$. Hence the integral curves of the vector field k^a are *null geodesics* as a consequence of Maxwell's equations [1].

   In 1956 Ivor Robinson made the important discovery that the integral curves of k^a, in addition to being geodesic satisfying (2.62), are *shear-free* (published later in [2]) and in this natural way introduced into general relativity the concept of *shear* of a congruence of null geodesics. With the present formalism, and with the benefit of hindsight, we can readily obtain this additional information from Maxwell's equations

$$(f_1 \, L^{ab})_{;b} = 0 \quad \Leftrightarrow \quad f_{1,b} \, L^{ab} + f_1 \, L^{ab}{}_{;b} = 0 . \tag{2.63}$$

Using the first of (2.33) when multiplying this by L_{ac} we find, since $f_1 \neq 0$, that

$$L_{ac} \, L^{ab}{}_{;b} = 0 . \tag{2.64}$$

Writing this out explicitly using L_{ab} given in (2.26), and the scalar products among the tetrad vectors (see (2.13)), we have

$$(k_{a;b} \, m^a \, k^b) m_c - (k_{a;b} \, m^a \, m^b) k_c = 0 , \tag{2.65}$$

from which it follows that

$$k_{a;b} \, m^a \, k^b = 0 \quad \text{and} \quad k_{a;b} \, m^a \, m^b = 0 . \tag{2.66}$$

Expressing $k_{a;b} \, k^b$ in terms of the tetrad, and using the fact that this real covariant vector is orthogonal to k^a (identically since k^a is null) and to m^a (on account of the first of (2.66)), we see that the first of (2.66) is equivalent to the geodesic equations (2.62). To exhibit the significance of the second equation in (2.66) as a condition on the congruence of null geodesic integral curves of k^a, independently of the choice of l^a, m^a, $\bar{m}^a$, we first write, for convenience,

$$\sigma = k_{a;b}\, m^a\, m^b \,, \tag{2.67}$$

and express $k_{a;b}$ in terms of the null tetrad to arrive at (remembering that $k_{a;b}$ is real-valued)

$$k_{a;b} = c_1\, k_a\, k_b + c_2\, k_a\, l_b + c_3\, k_a\, m_b + \bar{c}_3\, k_a\, \bar{m}_b + c_4\, m_a\, k_b$$
$$+ \bar{c}_4\, \bar{m}_a\, k_b + \bar{\sigma}\, m_a\, m_b + \sigma\, \bar{m}_a\, \bar{m}_b + c_5\, m_a\, \bar{m}_b + \bar{c}_5\, \bar{m}_a\, m_b \,, \tag{2.68}$$

where c_1, c_2 are real-valued while c_3, c_4, σ and c_5 are in general complex-valued. Using this we can conclude, with the help of (2.14), that there exists a real-valued function $\lambda(x^a)$ and a covariant vector ξ_a such that

$$k_{a;b} + k_{b;a} = \lambda\, g_{ab} + k_a\, \xi_b + k_b\, \xi_a + 2(\bar{\sigma}\, m_a\, m_b + \sigma\, \bar{m}_a\, \bar{m}_b) \,. \tag{2.69}$$

Multiplying this by g^{ab} yields $\xi_a\, k^a = k^a{}_{;a} - 2\,\lambda$. Also if an affine parameter is used along the null geodesics tangent to k^a then $\xi_a\, k^a = -\lambda$ or equivalently $\lambda = k^a{}_{;a}$. It is straightforward to obtain from (2.69) the equation

$$k_{(a;b)}\, k^{a;b} - \frac{1}{2}(k^a{}_{;a})^2 = \lambda\,(\lambda - k^a{}_{;a}) + 2\,|\sigma|^2 \,, \tag{2.70}$$

where round brackets enclosing indices denote symmetrisation (for example, $w_{(ab)} = (w_{ab} + w_{ba})/2$). If we use an affine parameter along the null geodesics then this becomes

$$k_{(a;b)}\, k^{a;b} - \frac{1}{2}(k^a{}_{;a})^2 = 2\,|\sigma|^2 \,. \tag{2.71}$$

Robinson's shear-free condition was originally given by (2.70) with $\sigma = 0$, reducing to (2.71) with $\sigma = 0$ if an affine parameter is used. We also note that if k^a is geodesic and shear-free then it follows from (2.69) that there exists λ and ξ_a such that [3]

$$k_{a;b} + k_{b;a} = \lambda\, g_{ab} + k_a\, \xi_b + k_b\, \xi_a \,. \tag{2.72}$$

This Robinson-Trautman test is a practical way of checking if a null congruence is geodesic and shear-free. It also led Robinson and Trautman to give an interesting geometrical interpretation of *geodesic and shear-free* which does not involve the propagation of shadows, as described in Chap. 1, but exploits the fact that the right hand side of (2.72) is a special algebraic form for the Lie derivative of the metric with respect to the vector field k^a.

The geodesic and shear-free conditions (2.66) ensure that the integrability conditions for Maxwell's equations (2.63), as differential equations for the complex-valued function f_1, are satisfied. To demonstrate this we can write Maxwell's equations (2.63), with (2.66) satisfied, in the form

$$X_1 \, f_1 = 0 \quad \text{and} \quad X_2 \, f_1 = 0 \,, \tag{2.73}$$

where the operators X_1, X_2 are given by

$$X_1 = k^a \, \frac{\partial}{\partial x^a} + A \,, \quad X_2 = m^a \, \frac{\partial}{\partial x^a} + B \,, \tag{2.74}$$

with

$$A = -\bar{m}_a \, L^{ab}{}_{;b} \,, \quad B = -l_a \, L^{ab}{}_{;b} \,. \tag{2.75}$$

We note that we can write

$$A \, m^a - B \, k^a = L^{ab}{}_{;b} = \frac{1}{\sqrt{-g}} \frac{\partial}{\partial x^b} (\sqrt{-g} \, L^{ab}) \,, \tag{2.76}$$

with $g = \det(g_{ab})$ as always. It then follows that

$$(A \, m^a - B \, k^a)_{;a} = \frac{1}{\sqrt{-g}} \frac{\partial^2}{\partial x^a \partial x^b} (\sqrt{-g} \, L^{ab}) = 0 \,. \tag{2.77}$$

The integrability conditions for (2.73) require that the operator $[X_1, X_2]$, defined by $[X_1, X_2] \, h = X_1(X_2 h) - X_2(X_1 h)$ for any scalar function h, is a linear combination of the operators X_1 and X_2 (see Eisenhart [4], p.70). The reader can now verify, using the geodesic and shear-free conditions (2.66), that

$$[X_1, X_2] = (A - k^a{}_{;a}) X_2 - (B - m^a{}_{;a}) X_1 - (A \, m^a - B \, k^a)_{;a} \,, \tag{2.78}$$

and the final term here vanishes on account of (2.77). This integrability property is the basis of a seminal result known as the Robinson Theorem [2]. Many significant studies have been carried out on this topic which the interested reader can find in [5, 6].

2.3   Bivectors and Gravitational Fields

In the previous section the key field variable for describing electromagnetic fields was the Maxwell bivector with components $F_{ab} = -F_{ba}$. The analogous field variable for describing gravitational fields is the Weyl conformal curvature tensor with components C_{abcd}. This is defined in terms of the components R_{abcd} of the

Riemann curvature tensor, the components R_{ab} of the Ricci tensor and the Ricci scalar R by the formula:

$$C_{abcd} = R_{abcd} + \frac{1}{2}(g_{ad}\, R_{bc} + g_{bc}\, R_{ad} - g_{ac}\, R_{bd} - g_{bd}\, R_{ac}) + \frac{1}{6}R(g_{ac}\, g_{bd} - g_{ad}\, g_{bc})\,. \tag{2.79}$$

The mathematical model of a vacuum gravitational field is a space-time for which Einstein's vacuum field equations, $R_{ab} = 0$ ($\Rightarrow R = 0$), hold and in this case $C_{abcd} = R_{abcd}$. Thus the Riemann tensor describes a vacuum gravitational field while a non-vacuum gravitational field is described by the Weyl tensor. Following from the algebraic symmetries of the Riemann tensor, the Weyl tensor has the algebraic symmetries

$$C_{abcd} = C_{cdab}\,, \quad C_{abcd} = -C_{bacd}\,, \quad C_{abcd} = -C_{abdc}\,, \tag{2.80}$$

and

$$C_{abcd} + C_{adbc} + C_{acdb} = 0\,. \tag{2.81}$$

In addition the Weyl tensor has the symmetry

$$g^{bd}\, C_{abcd} = 0\,, \tag{2.82}$$

which the reader can easily verify from (2.79). Since the Riemann tensor components R_{abcd} are skew-symmetric in a, b and in c, d we can define a *left dual*

$$^{*}R_{abcd} = \frac{1}{2}\eta_{abpq}\, R^{pq}{}_{cd}\,, \tag{2.83}$$

and a *right dual*

$$R^{*}_{abcd} = \frac{1}{2}\eta_{cdpq}\, R_{ab}{}^{pq}\,, \tag{2.84}$$

and similarly for the Weyl tensor. Using the properties of the permutation tensor η_{abcd} the reader can demonstrate that

$$\begin{aligned}
^{*}R^{*}_{abcd} &= \frac{1}{4}\eta_{abpq}\, \eta_{cdrs}\, R^{pqrs} \\
&= -R_{abcd} - g_{bc}\, R_{ad} - g_{ad}\, R_{bc} + g_{bd}\, R_{ac} + g_{ac}\, R_{bd} \\
&\quad + \frac{1}{2}R(g_{ad}\, g_{bc} - g_{ac}\, g_{bd})\,.
\end{aligned} \tag{2.85}$$

It is now clear that if this left *and* right dual is calculated for the Weyl tensor the result, making use of the symmetry (2.82), is

$$^*C^*_{abcd} = -C_{abcd} \,.$$

(2.86)

Taking the left dual of both sides of this equation and using the fact, mentioned following (2.2) of the previous section, that the dual of the dual (the "double dual") of a skew-symmetric tensor recovers the tensor with a minus sign, so that in particular $^{**}C^*_{abcd} = -C^*_{abcd}$, we arrive at

$$C^*_{abcd} = {}^*C_{abcd} \,.$$

(2.87)

Hence we can say that for a tensor having the algebraic symmetries of the Weyl tensor the left and right duals are equal. On account of (2.85) this statement is true of the Riemann tensor if the vacuum field equations are satisfied.

By analogy with the complex bivector $\mathfrak{F}_{ab}$ satisfying (2.6) we define

$$\mathfrak{C}_{abcd} = C_{abcd} + i\,{}^*C_{abcd} \,,$$

(2.88)

and this satisfies

$$^*\mathfrak{C}_{abcd} = -i\,\mathfrak{C}_{abcd} \quad \text{and} \quad \mathfrak{C}^*_{abcd} = -i\,\mathfrak{C}_{abcd} \,,$$

(2.89)

and thus can be expressed in terms of products of the bivectors (2.26). It is useful to note that the cyclic algebraic symmetry (2.81) and the algebraic symmetry (2.82) can be written as the complex equation

$$g^{bd}\,\mathfrak{C}_{abcd} = 0 \,.$$

(2.90)

The real part of this equation is clearly (2.82) while the imaginary part is

$$\eta_a{}^{dpq}\,C_{cdpq} = 0 \,,$$

(2.91)

and it is a useful exercise to show, using the properties of η_{abcd}, that this is equivalent to the cyclic symmetry (2.81).

To express $\mathfrak{C}_{abcd}$ in terms of the basis bivectors L_{ab}, M_{ab}, N_{ab} we start by multiplying (2.39) of the previous chapter by $C^{cd}{}_{pq}$ to obtain

$$2\,\mathfrak{C}_{abpq} = -2\,M_{ab}\,L_{cd}\,C^{cd}{}_{pq} - 2\,L_{ab}\,M_{cd}\,C^{cd}{}_{pq} - N_{ab}\,N_{cd}\,C^{cd}{}_{pq} \,.$$

(2.92)

Now each $L_{cd} C^{cd}{}_{pq}$, $M_{cd} C^{cd}{}_{pq}$ and $N_{cd} C^{cd}{}_{pq}$ is a complex bivector satisfying an equation of the form of (2.6). Hence each can be expressed as a linear combination of the basis bivectors L_{ab}, M_{ab}, N_{ab} thus:

$$L_{cd} C^{cd}{}_{pq} = \alpha_1 L_{pq} + \alpha_2 M_{pq} + \alpha_3 N_{pq} , \tag{2.93}$$

$$M_{cd} C^{cd}{}_{pq} = \beta_1 L_{pq} + \beta_2 M_{pq} + \beta_3 N_{pq} , \tag{2.94}$$

$$N_{cd} C^{cd}{}_{pq} = \gamma_1 L_{pq} + \gamma_2 M_{pq} + \gamma_3 N_{pq} , \tag{2.95}$$

with

$$\alpha_1 = -\frac{1}{2} L_{cd} C^{cd}{}_{pq} M^{pq} = -2 C_{abcd} m^a k^b \bar{m}^c l^d , \tag{2.96}$$

$$\alpha_2 = -\frac{1}{2} L_{cd} C^{cd}{}_{pq} L^{pq} = -2 C_{abcd} m^a k^b m^c k^d , \tag{2.97}$$

$$\alpha_3 = -\frac{1}{4} L_{cd} C^{cd}{}_{pq} N^{pq} = -C_{abcd} m^a k^b \bar{m}^c m^d - C_{abcd} m^a k^b k^c l^d , \tag{2.98}$$

$$\beta_1 = -\frac{1}{2} M_{cd} C^{cd}{}_{pq} M^{pq} = -2 C_{abcd} \bar{m}^a l^b \bar{m}^c l^d , \tag{2.99}$$

$$\beta_2 = -\frac{1}{2} M_{cd} C^{cd}{}_{pq} L^{pq} = -2 C_{abcd} \bar{m}^a l^b m^c k^d , \tag{2.100}$$

$$\beta_3 = -\frac{1}{4} M_{cd} C^{cd}{}_{pq} N^{pq} = -C_{abcd} \bar{m}^a l^b \bar{m}^c m^d - C_{abcd} \bar{m}^a l^b k^c l^d , \tag{2.101}$$

and

$$\gamma_1 = -\frac{1}{2} N_{cd} C^{cd}{}_{pq} M^{pq} = -2 C_{abcd} \bar{m}^a m^b \bar{m}^c l^d - 2 C_{abcd} k^a l^b \bar{m}^c l^d ,$$

$$\gamma_2 = -\frac{1}{2} N_{cd} C^{cd}{}_{pq} L^{pq} = -2 C_{abcd} \bar{m}^a m^b m^c k^d - 2 C_{abcd} k^a l^b m^c l^d ,$$

$$\gamma_3 = -\frac{1}{4} N_{cd} C^{cd}{}_{pq} N^{pq} = -C_{abcd} \bar{m}^a m^b \bar{m}^c m^d - C_{abcd} k^a l^b k^c l^d$$
$$- 2 C_{abcd} \bar{m}^a m^b k^c l^d . \tag{2.102}$$

The nine functions (2.96)–(2.102) are not all independent on account of the algebraic symmetries (2.80)–(2.82). As a guide to finding the relations between some of these functions we can make use of (2.90) and also the symmetry

$$\mathfrak{C}_{abcd} = \mathfrak{C}_{cdab} , \tag{2.103}$$

which follows from the first of (2.80) and also from

$$*C_{abcd} = \frac{1}{2} \eta_{ab}{}^{rs} C_{rscd} = \frac{1}{2} \eta_{ab}{}^{rs} C_{cdrs} = C^*_{cdab} = *C_{cdab} ,$$

(2.104)

with the final equality coming from (2.87). If we now substitute (2.93)–(2.95) into (2.92) we see immediately that (2.103) is satisfied provided

$$\alpha_1 = \beta_2 , \quad 2\alpha_3 = \gamma_2 , \quad 2\beta_3 = \gamma_1 .$$

(2.105)

With these satisfied we readily see that (2.90) reduces to $(\gamma_3 + 2\beta_2)g_{ab} = 0$, with g_{ab} given in terms of the null tetrad by (2.14), and so we must also have

$$\gamma_3 + 2\beta_2 = 0 .$$

(2.106)

We can verify (2.105) and (2.106) directly from (2.96)–(2.102) using the algebraic symmetries of C_{abcd} to arrive at

$$\alpha_1 = -2 C_{abcd} m^a k^b \bar{m}^c l^d = \beta_2 = -\frac{1}{2}\gamma_3 ,$$

(2.107)

$$\alpha_2 = -2 C_{abcd} m^a k^b m^c k^d ,$$

(2.108)

$$\alpha_3 = -2 C_{abcd} m^a k^b k^c l^d = \frac{1}{2}\gamma_2 ,$$

(2.109)

$$\beta_1 = -2 C_{abcd} \bar{m}^a l^b \bar{m}^c l^d ,$$

(2.110)

$$\beta_3 = -2 C_{abcd} k^a l^b \bar{m}^c l^d = \frac{1}{2}\gamma_1 .$$

(2.111)

Finally (2.92) can be written as

$$\frac{1}{2}\mathfrak{C}_{abpq} = \Psi_0 M_{ab} M_{pq} + \Psi_1 (M_{ab} N_{pq} + N_{ab} M_{pq}) + \Psi_2 (-N_{ab} N_{pq}$$
$$+ M_{ab} L_{pq} + L_{ab} M_{pq}) + \Psi_3 (L_{ab} N_{pq} + N_{ab} L_{pq}) + \Psi_4 L_{ab} L_{pq} ,$$

(2.112)

where the five complex-valued coefficients here are given by

$$\Psi_0 = -\frac{1}{2}\alpha_2 = C_{abcd} m^a k^b m^c k^d ,$$

(2.113)

$$\Psi_1 = -\frac{1}{2}\alpha_3 = C_{abcd} m^a k^b k^c l^d ,$$

(2.114)

$$\Psi_2 = -\frac{1}{2}\alpha_1 = C_{abcd} m^a k^b \bar{m}^c l^d ,$$

(2.115)

$$\Psi_3 = -\frac{1}{2}\beta_3 = C_{abcd}\, k^a\, l^b\, \bar{m}^c\, l^d\,, \tag{2.116}$$

$$\Psi_4 = -\frac{1}{2}\beta_1 = C_{abcd}\, \bar{m}^a\, l^b\, \bar{m}^c\, l^d\,. \tag{2.117}$$

The (in general) five complex components of the Weyl tensor on the null tetrad $k^a, l^a, m^a, \bar{m}^a$ are listed in (2.113)–(2.117). These are generally referred to as the Newman-Penrose [7] components of the Weyl tensor. If we transform the tetrad by the null rotation given in (1.50) of the previous chapter then these components transform to $\hat{\Psi}_0, \hat{\Psi}_1, \hat{\Psi}_2, \hat{\Psi}_3$ and $\hat{\Psi}_4$ related to (2.113)–(2.117) by

$$\hat{\Psi}_0 = \Psi_0 - 4\,\Psi_1\,\beta - 6\,\Psi_2\,\beta^2 + 4\,\Psi_3\,\beta^3 + \Psi_4\,\beta^4\,, \tag{2.118}$$

$$\hat{\Psi}_1 = \Psi_1 + 3\,\Psi_2\,\beta - 3\,\Psi_3\,\beta^2 - \Psi_4\,\beta^3\,, \tag{2.119}$$

$$\hat{\Psi}_2 = \Psi_2 - 2\,\Psi_3\,\beta - \Psi_4\,\beta^2\,, \tag{2.120}$$

$$\hat{\Psi}_3 = \Psi_3 + \Psi_4\,\beta\,, \tag{2.121}$$

$$\hat{\Psi}_4 = \Psi_4. \tag{2.122}$$

This sequence of equations has the interesting property that the derivative of $\hat{\Psi}_0$ with respect to β is a constant multiple of $\hat{\Psi}_1$, the derivative of $\hat{\Psi}_1$ with respect to β is a constant multiple of $\hat{\Psi}_2$, the derivative of $\hat{\Psi}_2$ with respect to β is a constant multiple of $\hat{\Psi}_3$ and the derivative of $\hat{\Psi}_3$ with respect to β is $\hat{\Psi}_4$. We can make immediate use of this fact. We first assume that $\Psi_4 \neq 0$. If this is not the case then a null rotation (2.51) of the previous chapter will make it so. Now $\hat{\Psi}_0 = 0$ is a quartic equation for β over the field of complex numbers and therefore, by the fundamental theorem of algebra, has at most four (complex) roots. Thus there are in general four null vectors $\hat{k}^a$, given by (2.50) of the previous chapter, for which $\hat{\Psi}_0 = 0$. These are *principal null directions* of the Weyl tensor. On account of the properties just mentioned of the sequence (2.118)–(2.122), it follows that if two roots of $\hat{\Psi}_0 = 0$ coincide then the coincident root is a solution of the equation $\hat{\Psi}_1 = 0$. If three roots of $\hat{\Psi}_0 = 0$ coincide then the coincident root is a solution of the equations $\hat{\Psi}_1 = \hat{\Psi}_2 = 0$ and if the four roots of $\hat{\Psi}_0 = 0$ coincide then this root is a solution of $\hat{\Psi}_1 = \hat{\Psi}_2 = \hat{\Psi}_3 = 0$. Since $L_{ab}\,k^b = 0$, $M_{ab}\,k^b = \bar{m}_a$, $N_{ab}\,k^b = k_a$ we have from (2.112) the equations

$$\frac{1}{2}\mathfrak{C}_{abpq}\, k^b\, k^p = -\Psi_0\,\bar{m}_a\,\bar{m}_q - \Psi_1(\bar{m}_a\, k_q + k_a\,\bar{m}_q) + \Psi_2\, k_a\, k_q\,, \tag{2.123}$$

and

$$\frac{1}{2}\mathfrak{C}_{abpq}\, k^b = \Psi_0\,\bar{m}_a\, M_{pq} + \Psi_1\,(\bar{m}_a\, N_{pq} + k_a\, M_{pq}) + \Psi_2(-k_a\, N_{pq} + \bar{m}_a\, L_{pq})$$

$$+\Psi_3\, k_a\, L_{pq}\,. \tag{2.124}$$

From (2.123) we can readily deduce that if k^a is a simple principal null direction of the Weyl tensor then $\Psi_0 = 0$ and it satisfies

$$k_{[f}\, \mathfrak{C}_{a]bp[q}k_{g]}\, k^b\, k^p = 0 \,, \tag{2.125}$$

with the square brackets, as always, denoting skewsymmetrisation. If k^a is a doubly degenerate principal null direction then $\Psi_0 = \Psi_1 = 0$ and so k^a satisfies

$$k_{[f}\, \mathfrak{C}_{a]bpq}\, k^b\, k^p = 0 \,. \tag{2.126}$$

If k^a is a triply degenerate principal null direction then $\Psi_0 = \Psi_1 = \Psi_2 = 0$ and, from (2.124), we have

$$k_{[f}\, \mathfrak{C}_{a]bpq}\, k^b = 0 \,. \tag{2.127}$$

If k^a is a quadruply degenerate principal null direction then $\Psi_0 = \Psi_1 = \Psi_2 = \Psi_3 = 0$ and by (2.124) the vector k^a satisfies

$$\mathfrak{C}_{abpq}\, k^b = 0 \,. \tag{2.128}$$

If any of (2.126)–(2.128) hold we say that the Weyl tensor is *algebraically special*. This Weyl tensor algebra constitutes the so-called Petrov-Pirani [8, 9] classification while the principal null directions are referred to as Debever-Penrose [10, 11] directions. Finally the complex components (2.113)–(2.117) of the Weyl tensor on the null tetrad are the Newman-Penrose [7] components of the Weyl tensor.

The Bianchi identities satisfied by the Riemann tensor read

$$R_{abcd;e} + R_{abec;d} + R_{abde;c} = 0 \,. \tag{2.129}$$

An equivalent way of writing this is

$$\eta^{fcde}\, R_{abcd;e} = 0 \quad \Leftrightarrow \quad R^*_{ab}{}^{fe}{}_{;e} = 0 \,. \tag{2.130}$$

On multiplication by η^{rsab} this reads

$$*R^{*rsfe}{}_{;e} = 0 \,. \tag{2.131}$$

With $*R^*{}_{abcd}$ given by (2.85) we can write (2.131) as

$$R_{abcd}{}^{;d} = R_{ac;b} - R_{bc;a} \,. \tag{2.132}$$

Writing the Riemann tensor in terms of the Weyl tensor via (2.79) this equation takes the form

$$C_{abcd}{}^{;d} = \frac{1}{2}(R_{ac;b} - R_{bc;a}) + \frac{1}{12}(g_{bc} R_{,a} - g_{ac} R_{,b}) \, . \tag{2.133}$$

In a vacuum space-time $R_{ab} = 0$ and $C_{abcd} = R_{abcd}$ and by (2.130) and (2.133) we have

$$\mathcal{R}_{abcd}{}^{;d} = 0 \quad \text{with} \quad \mathcal{R}_{abcd} = R_{abcd} + i\, R^{*}_{abcd} \, . \tag{2.134}$$

We note that in a vacuum space-time the left and right duals of the Riemann tensor are equal. Now if R_{abcd} is algebraically special with k^{a} as degenerate principal null direction then $\Psi_0 = \Psi_1 = 0$ and (2.112) reduces to

$$\frac{1}{2}\mathcal{R}_{abpq} = -\Psi_2\, N_{ab}\, N_{pq} + A_{ab}\, L_{pq} + L_{ab}\, A_{pq} \, , \tag{2.135}$$

with

$$A_{ab} = \Psi_2\, M_{ab} + \Psi_3\, N_{ab} + \frac{1}{2}\Psi_4\, L_{ab} \, . \tag{2.136}$$

Substituting (2.135) into (2.134) and multiplying the result by L^{ps} yields

$$- \Psi_2{}^{;q} N_{ab}\, L^{s}{}_{q} - \Psi_2\, N_{ab}{}^{;q}\, L^{s}{}_{q} - \Psi_2\, N_{ab}\, L^{ps}\, N_{pq}{}^{;q}$$
$$+ A_{ab}\, L^{ps}\, L_{pq}{}^{;q} + L_{ab}{}^{;q}\, A_{pq}\, L^{ps} + L_{ab}\, A_{pq}{}^{;q}\, L^{ps}$$
$$= 0 \, . \tag{2.137}$$

Multiplying this successively by L^{ab}, N^{ab} and M^{ab} results in the equations:

$$\Psi_2\, L^{ab}\, N_{ab}{}^{;q}\, L^{s}{}_{q} + 2\, \Psi_2\, L^{ps}\, L_{pq}{}^{;q} = 0 \, , \tag{2.138}$$

$$4\, \Psi_2{}^{;q}\, L^{s}{}_{q} + 4\, \Psi_2\, L^{ps}\, N_{pq}{}^{;q} - 4\, \Psi_3\, L^{ps}\, L_{pq}{}^{;q} + N^{ab}\, L_{ab}{}^{;q}\, A_{pq}\, L^{ps} = 0 \, , \tag{2.139}$$

and

$$-\Psi_2\, M^{ab}\, N_{ab}{}^{;q}\, L^{s}{}_{q} - \Psi_4\, L^{ps}\, L_{pq}{}^{;q} + M^{ab}\, L_{ab}{}^{;q}\, A_{pq}\, L^{ps} - 2\, A_{pq}{}^{;q}\, L^{ps} = 0 \, . \tag{2.140}$$

The reader can check that

$$L^{ab}\, N_{ab}{}^{;q}\, L^{s}{}_{q} = 4\, L^{ps}\, L_{pq}{}^{;q} \, , \tag{2.141}$$

and so (2.138) reduces to

$$\Psi_2\, L^{ps}\, L_{pq}{}^{;q} = 0\,.\tag{2.142}$$

If $L^{ps}\, L_{pq}{}^{;q} \neq 0$ then this requires $\Psi_2 = 0$. Putting $\Psi_2 = 0$ in (2.139) and using (2.141) again simplifies (2.139) to read

$$\Psi_3\, L^{ps}\, L_{pq}{}^{;q} = 0\,,\tag{2.143}$$

from which, if $L^{ps}\, L_{pq}{}^{;q} \neq 0$, we must have $\Psi_3 = 0$. Now (2.140) with $\Psi_2 = \Psi_3 = 0$ becomes

$$\Psi_4\, L^{ps}\, L_{pq}{}^{;q} = 0\,.\tag{2.144}$$

Hence we see that if $L^{ps}\, L_{pq}{}^{;q} \neq 0$ then $\Psi_2 = \Psi_3 = \Psi_4 = 0$ and the vacuum space-time is flat. Therefore if $R_{abcd} \neq 0$ we must have

$$L^{ps}\, L_{pq}{}^{;q} = 0\,,\tag{2.145}$$

and so the degenerate principal null direction k^a is geodesic and shear-free as in (2.64). This result, which is analogous to that of Robinson described in the previous chapter, leads to the important Goldberg–Sachs [12] theorem for vacuum algebraically special space-times which has been generalised in a natural way by Robinson and Schild [13].

2.4   The Kerr Space-Time

As an illustration of the use of the theory of bivectors and gravitational fields we choose the axially symmetric vacuum space-time describing the gravitational field of a rotating black hole of mass m and angular momentum per unit mass a about its symmetry axis, discovered by Roy Kerr [14]. We start with Kerr's line element in the form

$$
\begin{aligned}
ds^2 &= g_{ab}\, dx^a\, dx^b\\
&= -(r^2 + a^2 \cos^2\theta)\,(d\theta^2 + \sin^2\theta\, d\phi^2) + 2\,(du + a\,\sin^2\theta\, d\phi)\,\times\\
&\quad \left\{ dr - a\,\sin^2\theta\, d\phi + \left(\frac{1}{2} - \frac{m\,r}{r^2 + a^2\cos^2\theta}\right)(du + a\,\sin^2\theta\, d\phi) \right\},
\end{aligned}
$$

$$\tag{2.146}$$

giving the components g_{ab} of the metric tensor in coordinates $x^a = (\theta, \phi, r, u)$. When $a = 0$ this reduces to the Eddington–Finkelstein form of the Schwarzschild line element:

$$ds^2 = -r^2(d\theta^2 + \sin^2\theta \, d\phi^2) + 2 \, du \, dr + \left(1 - \frac{2m}{r}\right) du^2 \,. \tag{2.147}$$

To apply to (2.146) the transformation (see [14, 15]) $x^a \to X^i = (x, y, z, t)$ given by

$$x + i \, y = (r + i \, a) \, e^{i\phi} \sin\theta \,, \quad z = r \, \cos\theta \,, \quad u = t - r \,, \tag{2.148}$$

we first note that r is a function of x, y, z given by

$$\frac{x^2 + y^2}{r^2 + a^2} + \frac{z^2}{r^2} = 1 \,, \tag{2.149}$$

and hence

$$dr - a \, \sin^2\theta \, d\phi = \left(\frac{r \, x + a \, y}{r^2 + a^2}\right) dx + \left(\frac{r \, y - a \, x}{r^2 + a^2}\right) dy + \frac{z}{r} dz \,, \tag{2.150}$$

from which it follows that

$$du + a \, \sin^2\theta \, d\phi = dt - (dr - a \, \sin^2\theta \, d\phi)$$

$$= dt - \left(\frac{r \, x + a \, y}{r^2 + a^2}\right) dx - \left(\frac{r \, y - a \, x}{r^2 + a^2}\right) dy - \frac{z}{r} dz$$

$$= k_i \, dX^i \quad \text{(say)} \,. \tag{2.151}$$

Also we have

$$dx^2 + dy^2 + dz^2 = (dr - a \, \sin^2\theta \, d\phi)^2 + (r^2 + a^2 \cos^2\theta)(d\theta^2 + \sin^2\theta \, d\phi^2) \,, \tag{2.152}$$

and so, using (2.151),

$$- dx^2 - dy^2 - dz^2 + dt^2 = -(r^2 + a^2 \cos^2\theta)(d\theta^2 + \sin^2\theta \, d\phi^2)$$

$$+ 2 \, (du + a \, \sin^2\theta \, d\phi) \left\{ dr - a \, \sin^2\theta \, d\phi + \frac{1}{2}(du + a \, \sin^2\theta \, d\phi) \right\} \,.$$

$$\tag{2.153}$$

Now with the help of (2.151) and (2.153) we can write the Kerr line element (2.146) in the coordinates X^i as

$$ds^2 = -dx^2 - dy^2 - dz^2 + dt^2 - \frac{2\,m\,r^3}{r^4 + a^2 z^2}\,(k_i\,dX^i)^2 = g_{ij}\,dX^i\,dX^j\,. \quad (2.154)$$

We will return to the form (2.154) of the Kerr line element below but first we write the metric tensor components, given via the line element (2.146), in terms of the null tetrad defined via the 1-forms:

$$m_a\,dx^a = \frac{1}{\sqrt{2}}(r + i\,a\,\cos\theta)\,(d\theta + i\,\sin\theta\,d\phi)\,, \quad (2.155)$$

$$\bar{m}_a\,dx^a = \frac{1}{\sqrt{2}}(r - i\,a\,\cos\theta)\,(d\theta - i\,\sin\theta\,d\phi)\,, \quad (2.156)$$

$$k_a\,dx^a = du + a\,\sin^2\theta\,d\phi\,, \quad (2.157)$$

$$l_a\,dx^a = dr - a\,\sin^2\theta\,d\phi + \left(\frac{1}{2} - \frac{m\,r}{r^2 + a^2\cos^2\theta}\right)(du + a\,\sin^2\theta\,d\phi)\,. \quad (2.158)$$

We note that

$$g_{ab} = -m_a\,\bar{m}_b - m_b\,\bar{m}_a + k_a\,l_b + k_b\,l_a\,, \quad (2.159)$$

and all scalar products among $m_a, \bar{m}_a, k_a, l_a$ vanish except $m_a\,\bar{m}^a = -1$ and $k_a\,l^a = +1$. The Kerr metric tensor is a solution of Einstein's vacuum field equations $R_{ab} = 0$ and so the Weyl conformal curvature tensor C_{abcd} reduces to the Riemann curvature tensor R_{abcd} which, using (2.112), can be written in terms of the bivector basis L_{ab}, M_{ab}, N_{ab} as

$$\frac{1}{2}(R_{abcd} + i^*R_{abcd}) = \Psi_0\,M_{ab}\,M_{cd} + \Psi_1\,(M_{ab}\,N_{cd} + N_{ab}\,M_{cd})$$

$$+ \Psi_2\,(-N_{ab}\,N_{cd} + M_{ab}\,L_{cd} + L_{ab}\,M_{cd})$$

$$+ \Psi_3\,(L_{ab}\,N_{cd} + N_{ab}\,L_{cd}) + \Psi_4\,L_{ab}\,L_{cd}\,. \quad (2.160)$$

Calculation of the coefficients Ψ_A, $A = 0, 1, 2, 3, 4$ here for the Kerr space-time yields $\Psi_0 = \Psi_1 = 0$ and

$$\Psi_2 = -\frac{m}{(r + i\,a\,\cos\theta)^3}\,,\quad \Psi_3 = \frac{3\,i\,m\,a\,\sin\theta}{\sqrt{2}\,(r + i\,a\,\cos\theta)^4}\,,\quad \Psi_4 = -\frac{3\,m\,a^2\,\sin^2\theta}{(r + i\,a\,\cos\theta)^5}\,. \quad (2.161)$$

We note from these that

$$3\,\Psi_2\,\Psi_4 + 2\,\Psi_3^2 = 0\,. \tag{2.162}$$

The significance of this condition is that it allows us to transform the null tetrad consisting of k^i, l^i, m^i, and the complex conjugate of m^i, to a null tetrad $\hat{k}^i$, $\hat{l}^i$, $\hat{m}^i$, and the complex conjugate of $\hat{m}^i$, having the property that the transformed Ψ_A, denoted $\hat{\Psi}_A$, for $A = 0, 1, 2, 3, 4$, all vanish except for $\hat{\Psi}_2 \neq 0$. The transformation of the tetrad takes the form

$$\hat{k}^i = k^i\,,\ \hat{l}^i = l^i + \alpha\,\bar{\alpha}\,k^i + \bar{\alpha}\,m^i + \alpha\,\bar{m}^i\,,\ \hat{m}^i = m^i + \alpha\,k^i\,, \tag{2.163}$$

where α is a complex-valued function of the coordinates $x^a = (\theta, \phi, r, u)$. Under this change of tetrad the transformations of Ψ_A are given by

$$\hat{\Psi}_0 = \Psi_0\,, \tag{2.164}$$

$$\hat{\Psi}_1 = \Psi_1 - \bar{\alpha}\,\Psi_0\,, \tag{2.165}$$

$$\hat{\Psi}_2 = \Psi_2 + 2\,\bar{\alpha}\,\Psi_1 - \bar{\alpha}^2\Psi_0\,, \tag{2.166}$$

$$\hat{\Psi}_3 = \Psi_3 - 3\,\bar{\alpha}\,\Psi_2 - 3\,\bar{\alpha}^2\Psi_1 + \bar{\alpha}^3\Psi_0\,, \tag{2.167}$$

$$\hat{\Psi}_4 = \Psi_4 + 4\,\bar{\alpha}\,\Psi_3 - 6\,\bar{\alpha}^2\,\Psi_2 - 4\,\bar{\alpha}^3\Psi_1 + \bar{\alpha}^4\Psi_0\,. \tag{2.168}$$

Applying this to the Kerr case above we see that

$$\hat{\Psi}_0 = 0\,,\ \hat{\Psi}_1 = 0\,,\ \hat{\Psi}_2 = \Psi_2 = -\frac{m}{(r + i\,a\,\cos\theta)^3}\,, \tag{2.169}$$

and $\hat{\Psi}_3 = 0$ provided

$$\bar{\alpha} = \frac{\Psi_3}{3\,\Psi_2} = -\frac{i\,a\,\sin\theta}{\sqrt{2}(r + i\,a\,\cos\theta)}\,. \tag{2.170}$$

With α given by (2.170) we see from (2.168) that $\hat{\Psi}_4 = 0$ on account of the condition (2.162). With this new tetrad (2.160) simplifies to

$$\frac{1}{2}(R_{abcd} + i^*R_{abcd}) = -\frac{m}{(r + i\,a\,\cos\theta)^3}\,(-\hat{N}_{ab}\,\hat{N}_{cd} + \hat{M}_{ab}\,\hat{L}_{cd} + \hat{L}_{ab}\,\hat{M}_{cd})\,, \tag{2.171}$$

where we have put hats on the bivectors to emphasise that they are now calculated with the new tetrad. We now use (1.39) from the previous chapter,

$$g_{abcd} + i\,\eta_{abcd} = -2\,(\hat{M}_{ab}\,\hat{L}_{cd} + \hat{L}_{ab}\,\hat{M}_{cd}) - \hat{N}_{ab}\,\hat{N}_{cd}\,, \tag{2.172}$$

to rewrite (2.171) in the form

$$R_{abcd} + i {}^* R_{abcd} = \frac{m}{(r + i a \, \cos\theta)^3} \, (g_{abcd} + i \, \eta_{abcd} + 3 \, \hat{N}_{ab} \, \hat{N}_{cd}) \,. \tag{2.173}$$

Finally we want to express this in terms of the coordinates $X^i = (x, y, z, t)$ in which the metric tensor components, given via the line element (2.154), have the Kerr–Schild form

$$g_{ij} = \eta_{ij} + 2 H \, k_i \, k_j \quad \text{with} \quad H = -\frac{m \, r^3}{r^4 + a^2 \, z^2} \,, \tag{2.174}$$

with k_i given by (2.151) and $\eta_{ij} = \text{diag}(1, -1, -1, -1)$. We note that for this form of metric $\det(g_{ij}) = \det(\eta_{ij}) = -1$ (see Appendix A) and so $\eta_{ijkl} = \epsilon_{ijkl}$. Under the tetrad transformation (2.163) the bivector N_{ab} transforms as

$$\hat{N}_{ab} = N_{ab} - 2 \, \bar{\alpha} \, L_{ab} \,, \tag{2.175}$$

following from the definition of the bivector basis. Hence we can write the 2-form

$$\hat{N} = \frac{1}{2} \hat{N}_{ab} \, dx^a \wedge dx^b \,,$$
$$= i \, (r^2 + a^2 \cos^2\theta) \, \sin\theta \, d\theta \wedge d\phi + (du + a \, \sin^2\theta \, d\phi) \wedge (dr - a \, \sin^2\theta \, d\phi)$$
$$+ \sqrt{2} \, \bar{\alpha} \, (r + i a \, \cos\theta) \, (du + a \, \sin^2\theta \, d\phi) \wedge (d\theta + i \, \sin\theta \, d\phi) \,, \tag{2.176}$$

and, with α given by (2.170),

$$\sqrt{2} \, \bar{\alpha} \, (r + i a \, \cos\theta) \, (d\theta + i \, \sin\theta \, d\phi) = -i a \, \sin\theta \, (d\theta + i \, \sin\theta \, d\phi) \,. \tag{2.177}$$

Making the transformation to coordinates $X^i = (x, y, z, t)$ given by (2.148), and using (2.150) and (2.151), we find the intermediate results:

$$(r^2 + a^2 \cos^2\theta) \sin\theta \, d\theta \wedge d\phi = \left(\frac{a \, x - r \, y}{r^2 + a^2}\right) dx \wedge dz + \left(\frac{a \, y + r \, x}{r^2 + a^2}\right) dy \wedge dz$$
$$+ \frac{z}{r} \, dx \wedge dy \,, \tag{2.178}$$

and

$$\sin\theta \, (d\theta + i \, \sin\theta \, d\phi) = \frac{r}{r^4 + a^2 z^2} \, (A \, dx + B \, dy + C \, dz) \,, \tag{2.179}$$

with

$$A = zx + i \frac{\{r\,a\,x(r^2 - z^2) - y\,(r^4 + a^2 z^2)\}}{r\,(r^+a^2)} , \tag{2.180}$$

$$B = zy + i \frac{\{r\,a\,y\,(r^2 - z^2) + x\,(r^4 + a^2 z^2)\}}{r\,(r^2 + a^2)} , \tag{2.181}$$

$$C = \frac{(r^2 - z^2)(a\,z - r^2)}{r^2} . \tag{2.182}$$

Using (2.177)–(2.182) in (2.176) results in the following simplified expression for $\hat{N}$:

$$\hat{N} = -\frac{r}{r^2 + i\,a\,z} \left\{ x\,(dx \wedge dt - i\,dy \wedge dz) + y\,(dy \wedge dt - i\,dz \wedge dx) \right.$$

$$\left. + (z + i\,a)\,(dz \wedge dt - i\,dx \wedge dy) \right\} = \frac{1}{2} \hat{N}_{ij}\, dX^i \wedge dX^j . \tag{2.183}$$

Hence we can write (2.173) in coordinates $X^i = (x, y, z, t)$ as

$$R_{ijkl} + i\,{}^*R_{ijkl} = \frac{m\,r^3}{(r^2 + i\,a\,z)^3}\,(g_{ijkl} + i\,\eta_{ijkl} + 3\,\hat{N}_{ij}\,\hat{N}_{kl}) , \tag{2.184}$$

with $\hat{N}_{ij}$ given via the 2-form (2.183). We will refer to $\hat{N}_{ij}$ as the components of the *Kerr complex bivector field*. We note that with k_i given by (2.151), and since $g^{ij} = \eta^{ij} - 2\,H\,k^i\,k^j$ with $k^i = \eta^{ij}\,k_j$ we have $k^i = g^{ij}\,k_j$ and

$$\hat{N}_{ij}\,k^j = k_i . \tag{2.185}$$

It thus follows that

$$\hat{N}^{ij} = g^{ik}\,g^{jl}\,\hat{N}_{kl} = \eta^{ik}\,\eta^{jl}\,\hat{N}_{kl} , \tag{2.186}$$

and therefore

$$\hat{N}_{ij}\,\hat{N}^{ij} = -4 . \tag{2.187}$$

Hence from (2.184) we see that

$$(R_{ijkl} + i\,{}^*R_{ijkl})\,\hat{N}^{kl} = -\frac{4\,m\,r^3}{(r^2 + i\,a\,z)^3}\,g_{ijkl}\,\hat{N}^{kl} . \tag{2.188}$$

In this sense $\hat{N}_{ij}$ is an eigenbivector of the Kerr Riemann tensor. We note in passing that, in coordinates X^i,

$$^*\hat{N}_{ij} = \frac{1}{2}\epsilon_{ijkl}\,\hat{N}^{kl} = -i\,\hat{N}_{ij}\,.\tag{2.189}$$

2.5   Passage to Charged Kerr Space-Time

We can make use of the Kerr eigenbivector $\hat{N}_{ij}$ to obtain the charged Kerr solution of the vacuum Einstein–Maxwell field equations [15, 16] assuming that the metric tensor in the charged case has the Kerr–Schild form (2.174) with H to be determined. This can be achieved without using all of the field equations but with the addition of a boundary condition which states that asymptotically (for $r \rightarrow +\infty$) the electromagnetic field and the gravitational field should both be spherically symmetric. Starting with the electromagnetic case, in coordinates $X^i = (x, y, z, t)$, we look for a Maxwell field in the form

$$\mathfrak{F}^{ij} = F^{ij} + i\,{}^*F^{ij} = f(r, z)\,\hat{N}^{ij}\,,\tag{2.190}$$

from which, using (2.185), we have

$$\mathfrak{F}^{ij}\,k_j = f\,k^i\,.\tag{2.191}$$

Using

$$\mathfrak{F}^{ij}{}_{;i}\,k_j = 0\,,\tag{2.192}$$

which follows from Maxwell's equations ($\mathfrak{F}^{ij}{}_{;i} = 0$) but is weaker than them, we arrive at

$$f_{,i}\,k^i + f\,k^i{}_{;i} = f\,\hat{N}^{ij}\,k_{j;i} = \frac{1}{2}\hat{N}^{ij}\,(k_{j,i} - k_{i,j})\,.\tag{2.193}$$

With k^i given via (2.151) and $\hat{N}^{ij}$ given via (2.183) we find that

$$k^i{}_{;i} = k^i{}_{,i} = \frac{2\,r^3}{r^4 + a^2 z^2}\,,\tag{2.194}$$

and

$$\hat{N}^{ij}\,k_{j;i} = \frac{2\,i\,a\,r\,z}{r^4 + a^2 z^2}\,.\tag{2.195}$$

Substituting into (2.193), and simplifying, results in

$$\frac{\partial f}{\partial r} + \frac{z}{r}\frac{\partial f}{\partial z} = -\frac{2\,r\,f}{r^2 + i\,a\,z}\,. \tag{2.196}$$

Here $\partial/\partial z$ stands for partial differentiation with respect to z keeping r fixed. The general solution of this equation is

$$f(r, z) = \frac{e(w)\,r^2}{(r^2 + i\,a\,z)^2} \quad \text{with} \quad w = \frac{z}{r}\,, \tag{2.197}$$

and $e(w)$ is an arbitrary function of its argument. If we require a spherically symmetric field asymptotically (as $r \to +\infty$) then we must have $e = $ constant which results, asymptotically, in the Coulomb field. Alternatively if we require the remainder of Maxwell's vacuum field equations to be satisfied then we also find that $e = $ constant. The corresponding argument in the Einstein–Maxwell field equations case is to use only the field equations $R = 0$ and $R_{ij}\,k^j = -2\,E_{ij}\,k^j$ which, written out explicitly read as follows:

The vanishing of the Ricci scalar, $g^{ij}\,R_{ij} \equiv R = 0$, which is a consequence of the field equations $R_{ij} = -2\,E_{ij}$ and $g^{ij}\,E_{ij} \equiv 0$, yields

$$H_{,ij}\,k^i\,k^j = -\frac{4\,r^3}{r^4 + a^2 z^2}\,H_{,i}\,k^i - \frac{2\,r^2}{r^4 + a^2 z^2}\,H\,, \tag{2.198}$$

while the field equations

$$R_{ij}\,k^j = -2\,E_{ij}\,k^j = \frac{e^2 r^4}{(r^4 + a^2 z^2)^2}\,k_i\,, \tag{2.199}$$

provide us with

$$H_{,ij}\,k^i\,k^j = -\frac{2\,r^3}{r^4 + a^2 z^2}\,H_{,i}\,k^i - \frac{4\,r^2 a^2 z^2}{(r^4 + a^2 z^2)^2}\,H + \frac{e^2 r^4}{(r^4 + a^2 z^2)^2}\,. \tag{2.200}$$

As a consequence of (2.198) and (2.200) we see that $H(r, z)$ must satisfy

$$H_{,i}\,k^i + \frac{(r^4 - a^2 z^2)}{r(r^4 + a^2 z^2)}\,H = -\frac{e^2 r}{2\,(r^4 + a^2 z^2)}\,. \tag{2.201}$$

We note that with $H = H(r, z)$, $r = r(x, y, z)$ given by (2.149) and $k^i = \eta^{ij}\,k_j$ with k_j given by (2.151) we can write

$$H_{,i}\,k^i = \frac{\partial H}{\partial r} + \frac{z}{r}\frac{\partial H}{\partial z}\,, \tag{2.202}$$

with $\partial/\partial z$ here referring to partial differentiation with respect to z keeping r fixed. The general solution of (2.201) is

$$H(r, z) = \frac{\frac{1}{2}e^2r^2 - r^3m(w)}{r^4 + a^2z^2} \ , \quad \text{with} \ \ w = \frac{z}{r} \ , \tag{2.203}$$

and m is an arbitrary function of its argument. For a spherically symmetric field asymptotically we must have $m = $ constant and thus we have arrived at the Kerr–Newman solution of the Einstein–Maxwell field equations:

$$ds^2 = -dx^2 - dy^2 - dz^2 + dt^2 - \left(\frac{2\,m\,r^3 - e^2r^2}{r^4 + a^2z^2}\right)(k_i\,dX^i)^2 \ . \tag{2.204}$$

2.6 Using the Bianchi Identities

When the electromagnetic field (2.190) with (2.197) (and $e = $ constant) is present the Weyl conformal curvature tensor has the form

$$C_{abcd} + i\,^*C_{abcd} = G(r, z)\,(g_{abcd} + i\,\eta_{abcd} + 3\,N_{ab}\,N_{cd}) \ , \tag{2.205}$$

with N_{ab} given by the 2-form (2.183), dropping the hat for convenience from now on. We shall now use the Bianchi identities, and the Einstein–Maxwell field equations, to determine the complex valued function G in a way analogous to our use of Maxwell's equations to determine the complex valued function $f(r, z)$ in (2.190). In addition to (2.194) and (2.195) we shall require the formula

$$\eta_{abrs}\,k^a\,k^{r;s} = -\frac{2\,a\,r\,z}{r^4 + a^2z^2}\,k_b \ . \tag{2.206}$$

We now find from (2.205) that

$$(C_{abcd} + i\,^*C_{abcd})^{;d}\,k^a\,k^c = \left\{2\,G_{,d}\,k^d + \frac{6\,r}{r^2 + i\,a\,z}\,G\right\}k_b \ . \tag{2.207}$$

With the Einstein–Maxwell field equations

$$R_{ab} = -2\,E_{ab} \ , \tag{2.208}$$

where E_{ab} is the electromagnetic energy-momentum tensor

$$
\begin{aligned}
E_{ab} &= \frac{1}{2} g^{cd} \, \mathfrak{F}_{ac} \, \bar{\mathfrak{F}}_{bd} \\
&= \frac{1}{2} \frac{e^2 r^4}{(r^4 + a^2 z^2)^2} g^{cd} \, N_{ac} \, \bar{N}_{bd} \\
&= \frac{1}{2} \frac{e^2 r^4}{(r^4 + a^2 z^2)^2} \left(\eta^{cd} \, N_{ac} \, \bar{N}_{bd} - 2 \, H \, k_a \, k_b \right) ,
\end{aligned} \tag{2.209}
$$

with the bar denoting complex conjugation, we easily recover the second of (2.199). The Bianchi identities read

$$
C_{abcd}{}^{;d} = E_{cb;a} - E_{ca;b} , \tag{2.210}
$$

from which we have, using the second of (2.199),

$$
C_{abcd}{}^{;d} k^a k^c = \frac{2 e^2 r^7}{(r^4 + a^2 z^2)^3} k_b . \tag{2.211}
$$

Next using (2.210) again we have

$$
{}^* C_{abcd}{}^{;d} = -\eta_{ab}{}^{rs} E_{cr;s} , \tag{2.212}
$$

from which we deduce, using (2.199) and (2.206), that

$$
{}^* C_{abcd}{}^{;d} k^a k^c = -\frac{e^2 a r^5 z}{(r^4 + a^2 z^2)^3} k_b + \eta_{ab}{}^{rs} k^a k^c{}_{;s} E_{cr} . \tag{2.213}
$$

Using (2.209) and the fact that

$$
\bar{N}_{ab} = -i \, {}^* \bar{N}_{ab} , \tag{2.214}
$$

We can write

$$
\begin{aligned}
\eta_{ab}{}^{rs} k^a k^c{}_{;s} E_{cr} &= \frac{e^2 r^4}{2 (r^4 + a^2 z^2)^2} \eta_{ab}{}^{rs} k^a k^c{}_{;s} \left(\eta^{pq} N_{pc} \bar{N}_{qr} - 2 \, H \, k_c \, k_r \right) \\
&= -\frac{i \, e^2 r^4}{2 (r^4 + a^2 z^2)^2} \eta_{ab}{}^{rs} k^a k^c{}_{;s} \eta^{pq} N_{pc} {}^* \bar{N}_{qr} \\
&= -\frac{i \, e^2 r^4}{4 (r^4 + a^2 z^2)^2} \eta_{ab}{}^{rs} \eta_{qrlm} k^a k^c{}_{;s} \eta^{pq} N_{pc} \bar{N}^{lm}
\end{aligned}
$$

$$= -\frac{i\,e^2\,r^4}{2\,(r^4 + a^2 z^2)^2}\,k^{c;p}\,N_{cp}\,k_b$$

$$= -\frac{e^2\,a\,z\,r^5}{(r^4 + a^2 z^2)^3}\,k_b \ . \tag{2.215}$$

Putting (2.211), and (2.213) with (2.215), into (2.207) we finally arrive at

$$G_{,d}\,k^d + \frac{3\,r}{r^2 + i\,a\,z}\,G = \frac{e^2\,r^5\,(r^2 - i\,a\,z)}{(r^4 + a^2 z^2)^3} \ . \tag{2.216}$$

Substituting

$$G = \mathcal{G}(r, z) - \frac{e^2\,r^4}{(r^2 + i\,a\,z)^3 (r^2 - i\,a\,z)} \ , \tag{2.217}$$

we have the following equation for $\mathcal{G}$:

$$\mathcal{G}_{,d}\,k^d + \frac{3\,r}{r^2 + i\,a\,z}\,\mathcal{G} = 0 \ . \tag{2.218}$$

The general solution of this equation is

$$\mathcal{G} = \frac{m(w)\,r^3}{(r^2 + i\,a\,z)^3} \quad \text{with} \quad w = \frac{z}{r} \ , \tag{2.219}$$

where m is an arbitrary function of its argument. For asymptotic spherical symmetry we must have

$$m = \text{constant} \ . \tag{2.220}$$

Combining (2.217) and (2.219) we obtain the function G in (2.205), namely,

$$G = \frac{m\,r^3}{(r^2 + i\,a\,z)^3} - \frac{e^2\,r^4}{(r^2 + i\,a\,z)^3 (r^2 - i\,a\,z)} \ , \tag{2.221}$$

with m, e constants.

References

1. L. Mariot, C. R. Acad. Sci. **238**, 2055 (1954)
2. I. Robinson, J. Math. Phys. **2**, 290 (1961)
3. I. Robinson, A. Trautman, J. Math. Phys. **24**, 1425 (1983)
4. L.P. Eisenhart, *Riemannian Geometry* (Princeton University Press, Princeton, 1966)
5. I. Robinson, A. Schild, J. Math. Phys. **2**, 484 (1963)

6. J. Tafel, Lett. Math. Phys. **10**, 33 (1985)
7. E.T. Newman, R. Penrose, J. Math. Phys. **3**, 566 (1962)
8. A.Z. Petrov, Sci. Not. **114**, 55 (1954)
9. F.A.E. Pirani, Phys. Rev. **105**, 1089 (1957)
10. R. Debever, C. R. Acad. Sci. **249**, 1324 (1959)
11. R. Penrose, Ann. Phys. **10**, 171 (1960)
12. J.N. Goldberg, R.K. Sachs, Acta Phys. Polon. **22**, 13 (1962)
13. I. Robinson, A. Schild, J. Math. Phys. **4**, 484 (1963)
14. R.P. Kerr, J. Math. Phys. **11**, 237 (1963)
15. G.C. Debney, R.P. Kerr, A. Schild, J. Math. Phys. **10**, 1842 (1969)
16. E.T. Newman, E.Couch, K. Chinnapered, A. Exton, A. Prakas, R. Torrence, J. Math. Phys. **6**, 918 (1965)

Hypothetical Objects in Electromagnetism and Gravity

3

Abstract

I. The Maxwell field of a charged light-like particle with a non-geodesic world line (a light-like analogue of the Liénard–Wiechert field) can be constructed utilising the Minkowskian geometry in the neighbourhood of such a world line. In the process a fundamental question regarding the existence of a special parameter along the world line has to be addressed.

II. In the original generalisation of the Schwarzschild black hole with the introduction of a variable mass by Vaidya, the variable mass depends upon a parameter which has a simple geometrical origin. This parameter can also be identified from the geometry in the case of a Kerr black hole. With the mass and angular momentum in this case depending upon this parameter a rotating generalisation of the Vaidya space-time emerges.

3.1 Part I: A Light-Like Charge

From the point of view of Minkowskian geometry the Coulomb field is a solution of Maxwell's equations on Minkowskian space-time which is singular on a time-like geodesic (the history of a point charge). The Liénard–Wiechert field is the generalisation of the Coulomb field which is singular on a non-geodesic world line (the history of an accelerated point charge). When considering a hypothetical charged particle moving with the speed of light there is a Maxwell field available to describe it which is singular on a null geodesic and is a spin-off from the Robinson–Trautman solutions of the Einstein–Maxwell field equations. To describe it we begin with the Minkowskian line element in coordinates $X^i = (T, X, Y, Z)$:

$$ds^2 = \eta_{ij}\, dX^i\, dX^j = dT^2 - dX^2 - dY^2 - dZ^2 \,. \tag{3.1}$$

P. A. Hogan, D. Puetzfeld, *Frontiers in General Relativity*, Lecture Notes
in Physics 984, https://doi.org/10.1007/978-3-030-69370-1_3

We take the world line of the charge to be the null geodesic

$$X^i = u\, v^i \quad \text{with} \quad v^i = (1, 0, 0, 1) \, . \tag{3.2}$$

Clearly v^i is the null tangent to this world line ($\eta_{ij}\, v^i\, v^j = v_i\, v^i = 0$) and u is an affine parameter along it with $-\infty < u < +\infty$. The position 4-vector of a point of Minkowskian space-time relative to this world line can be written

$$X^i = u\, v^i + r\, k^i \, , \tag{3.3}$$

with k^i chosen so that

$$k_i\, k^i = 0 \quad \text{and} \quad k_i\, v^i = 1 \, . \tag{3.4}$$

We see that $r = 0$ corresponds to the world line (3.2) and we shall take $0 \le r < +\infty$. The null vector field k^i defined along the world line $r = 0$ and normalised according to (3.4) can be written in terms of two parameters ξ, η with $-\infty < \xi, \eta < +\infty$ as

$$k^i = \left(\frac{1}{2}(\xi^2 + \eta^2 + 1)\, , \xi\, , \eta\, , \frac{1}{2}(\xi^2 + \eta^2 - 1) \right) \, . \tag{3.5}$$

We notice that if $\xi^2 + \eta^2$ is large then k^i points in the direction of the tangent v^i to the world line $r = 0$. We can view (3.3) with (3.5) as a coordinate transformation from the coordinates $X^i = (T, X, Y, Z)$ to the coordinates $x^i = (\xi, \eta, r, u)$ which results in the Minkowskian line element (3.1) taking the form

$$ds^2 = -r^2(d\xi^2 + d\eta^2) + 2\, du\, dr \, . \tag{3.6}$$

Introducing a basis of 1-forms

$$\vartheta^{(1)} = r\, d\xi \, , \ \vartheta^{(2)} = r\, d\eta \, , \ \vartheta^{(3)} = dr \, , \ \vartheta^{(4)} = du \, , \tag{3.7}$$

we can write

$$ds^2 = -(\vartheta^{(1)})^2 - (\vartheta^{(2)})^2 + 2\, \vartheta^{(3)}\, \vartheta^{(4)} = g_{(a)(b)}\, \vartheta^{(a)}\, \vartheta^{(b)} \, , \tag{3.8}$$

where $g_{(a)(b)}$ are the components of the metric tensor on the half null tetrad defined via the basis 1-forms. Tetrad indices are enclosed in round brackets to distinguish them from coordinate indices. As potential 1-form due to a particle of constant charge e with world line $r = 0$ we take

$$A = \frac{e}{r}\, du = \frac{e}{r}\, \vartheta^{(4)} \, . \tag{3.9}$$

The exterior derivative of this 1-form is the 2-form

$$F = dA = -\frac{e}{r^2} dr \wedge du = -\frac{e}{r^2} \vartheta^{(3)} \wedge \vartheta^{(4)} , \tag{3.10}$$

and its Hodge dual is the 2-form

$$*F = \frac{e}{r^2} \vartheta^1 \wedge \vartheta^2 = e\, d\xi \wedge d\eta . \tag{3.11}$$

It is clear that the exterior derivative of this 2-form vanishes,

$$d*F = 0 , \tag{3.12}$$

and thus F is a Maxwell field. In other words the potential 1-form (3.9) gives rise to the Maxwell field of a charged particle travelling with the speed of light.

3.2   Geometry Based on a Non-geodesic Null World Line

Following the description above of a light-like analogue of the Coulomb field, we now turn our attention to the construction of a light-like analogue of the Liénard–Wiechert field in which the charge has a non-geodesic null world line and its electromagnetic field specialises to the Maxwell field given above when the world line of the charge is a null geodesic [1]. The parametric equations of the non-geodesic null world line are now

$$X^i = w^i(u) \quad \text{with } v^i = \frac{dw^i}{du} \text{ and } v^i v_i = 0 . \tag{3.13}$$

We define

$$a^i = \frac{dv^i}{du} \quad \Rightarrow \quad a^i v_i = 0 . \tag{3.14}$$

If $a^i = 0$ then the world line (3.13) is a null geodesic with u an affine parameter along it while if $a^i = \lambda(u)\, v^i$, for some function $\lambda(u)$, then the world line is a null geodesic but u is not an affine parameter along it. We will exclude these cases from now on. We now generalise the position 4-vector (3.3) to the position 4-vector of a point of Minkowskian space-time relative to the world line (3.13) by writing

$$X^i = w^i(u) + r\, k^i , \tag{3.15}$$

with $k_i k^i = 0$ and $k_i v^i = 1$. If we parametrize the *direction* of k^i with the parameters x, y for which $-\infty < x, y < +\infty$ we can write

$$P_0 k^i = \left(1 + \frac{1}{4}(x^2 + y^2), \; -x, \; -y, \; -1 + \frac{1}{4}(x^2 + y^2)\right), \qquad (3.16)$$

for some function $P_0(x, y, u)$. This function is determined by the normalisation $k_i v^i = 1$ of k^i to read

$$P_0 = \left\{1 + \frac{1}{4}(x^2 + y^2)\right\} v^0(u) + x \, v^1(u) + y \, v^2(u) + \left\{1 - \frac{1}{4}(x^2 + y^2)\right\} v^3(u).$$
$$(3.17)$$

It is useful to note that this function satisfies

$$P_0^2 \left(\frac{\partial^2}{\partial x^2} + \frac{\partial^2}{\partial y^2}\right) \log P_0 = v^i \, v_i = 0. \qquad (3.18)$$

From (3.16) we find that

$$P_0 \, a_i \, k^i = \left\{1 + \frac{1}{4}(x^2 + y^2)\right\} a^0(u) + x \, a^1(u) + y \, a^2(u) + \left\{1 - \frac{1}{4}(x^2 + y^2)\right\} a^3(u),$$
$$(3.19)$$

and thus we have

$$h_0 := a_i \, k^i = \frac{\partial}{\partial u} \log P_0. \qquad (3.20)$$

From now on we shall assume that, for example, $v^0 - v^3 \neq 0$. If $v^0 = v^3$ then since v^i is a null vector we must have $v^1 = 0 = v^2$ and also a^i must be in the same direction (in the T, Z-plane) as v^i. In this case $r = 0$ is a null geodesic and we are back to the case discussed above. Hence assuming $v^0 - v^3 \neq 0$ we can write (3.17) in the form

$$P_0 = \frac{(v^0 - v^3)}{4} \left\{\left(x + \frac{2 v^1}{v_0 - v^3}\right)^2 + \left(y + \frac{2 v^2}{v_0 - v^3}\right)^2\right\}. \qquad (3.21)$$

If we now consider (3.15)–(3.17) as a coordinate transformation from the coordinates X^i to the coordinates x, y, r, u then, under this transformation, the Minkowskian line element (3.1) takes the form

$$ds^2 = -P_0^{-2}(dx^2 + dy^2) + 2 \, du \, dr - 2 \, h_0 \, r \, du^2, \qquad (3.22)$$

with h_0 given by (3.20). The form of P_0 in (3.21) suggests a coordinate transformation from x, y to ξ, η given by

$$\xi = P_0^{-1}\left(x + \frac{2v^1}{v^0 - v^3}\right), \tag{3.23}$$

$$\eta = P_0^{-1}\left(y + \frac{2v^2}{v^0 - v^3}\right). \tag{3.24}$$

These transformations result in the line element (3.22) taking the form

$$ds^2 = -r^2\left\{\left(d\xi + \frac{\partial q}{\partial \eta}\,du\right)^2 + \left(d\eta + \frac{\partial q}{\partial \xi}\,du\right)^2\right\} + 2\,du\,dr - 2\,h_0\,r\,du^2, \tag{3.25}$$

with $q(\xi, \eta, u)$ given by

$$q = \frac{1}{2}A^1\,\eta\,(\xi^2 - \frac{1}{3}\eta^2) + \frac{1}{2}A^2\,\xi\,(\eta^2 - \frac{1}{3}\xi^2) + \left(\frac{a^0 - a^3}{v^0 - v^3}\right)\xi\eta, \tag{3.26}$$

with

$$A^1 = a^1 - \left(\frac{a^0 - a^3}{v^0 - v^3}\right)v^1 \text{ and } A^2 = a^2 - \left(\frac{a^0 - a^3}{v^0 - v^3}\right)v^2. \tag{3.27}$$

Writing the components of k^i given by (3.16) and (3.17) in terms of ξ, η we arrive at

$$k^i = \zeta^i - \frac{1}{2}\zeta_j\,\zeta^j\,v^i, \tag{3.28}$$

with

$$\zeta^i = \left(\frac{1 - \xi\,v^1 - \eta\,v^2}{v^0 - v^3}, -\xi, -\eta, \frac{1 - \xi\,v^1 - \eta\,v^2}{v^0 - v^3}\right). \tag{3.29}$$

This simplifies the calculation of h_0 (in (3.20)) in terms of ξ, η since $a_i\,v^i = 0$. The result is

$$h_0 = a_i\,\zeta^i = A^1\,\xi + A^2\,\eta + \left(\frac{a^0 - a^3}{v^0 - v^3}\right) = \frac{\partial^2 q}{\partial\xi\,\partial\eta}. \tag{3.30}$$

We also note that q in (3.26) is a harmonic function and thus

$$\Delta q := \left(\frac{\partial^2}{\partial\xi^2} + \frac{\partial^2}{\partial\eta^2}\right)q = 0. \tag{3.31}$$

Up to now the parameter u along the world line $r = 0$ is unspecified. We can resolve this issue by making the coordinate transformation

$$\bar{\xi} = \mu\,\xi\;,\;\; \bar{\eta} = \mu\,\eta\;,\;\; \bar{r} = \mu^{-1} r\;,\;\; \bar{u} = \bar{u}(u)\;, \tag{3.32}$$

with $\mu = \mu(u)$ given by

$$\mu^{-1}\frac{d\mu}{du} = \frac{a^0 - a^3}{v^0 - v^3} \quad\text{and}\quad \frac{d\bar{u}}{du} = \mu(u)\;. \tag{3.33}$$

It follows from (3.33) that taking

$$\bar{u} = \int (v^0 - v^3)du\;, \tag{3.34}$$

determines $\bar{u}$ up to a linear transformation $\bar{u} \rightarrow c_1\,\bar{u} + c_2$ with c_1, c_2 two real constants. If we let

$$A^i = a^i - \left(\frac{a^0 - a^3}{v^0 - v^3}\right) v^i\;, \tag{3.35}$$

then the cases $i = 1$ and $i = 2$ are given in (3.27). If the world line $r = 0$ is a null geodesic then, in general, $a^i = \lambda(u)\,v^i$ for some function $\lambda(u)$ and $A^i = 0$. The change of parameter u along the world line $r = 0$ given via (3.33) results in

$$v^i = \mu\,\bar{v}^i\;,\;\; a^i = \mu^2\,\bar{a}^i + \mu\left(\frac{a^0 - a^3}{v^0 - v^3}\right)\bar{v}^i\;, \tag{3.36}$$

where $\bar{v}^i = dw^i/d\bar{u}$ and $\bar{a}^i = d\bar{v}^i/d\bar{u}$. When this is substituted into (3.35) we obtain

$$A^i = \mu^2\,\bar{a}^i\;, \tag{3.37}$$

and thus we have the result that

$$a^i = \lambda(u)\,v^i \;\;\Rightarrow\;\; \bar{a}^i = 0\;. \tag{3.38}$$

Hence we see that the parameter $\bar{u}$ has the important property that if $r = 0$ is a geodesic then $\bar{u}$ is an affine parameter along it (cf. [2]).

The coordinate transformation (3.32) with (3.33) applied to the line element (3.25) transforms it into

$$ds^2 = -\bar{r}^2\left\{\left(d\bar{\xi} + \frac{\partial\bar{q}}{\partial\bar{\eta}}\,d\bar{u}\right)^2 + \left(d\bar{\eta} + \frac{\partial\bar{q}}{\partial\bar{\xi}}\,d\bar{u}\right)^2\right\} + 2\,d\bar{u}\,d\bar{r} - 2\,\bar{h}_0\,\bar{r}\,d\bar{u}^2\;, \tag{3.39}$$

with

$$
\bar{q}(\bar{\xi}, \bar{\eta}, \bar{u}) = \frac{1}{2}\bar{a}^1(\bar{u})\,\bar{\eta}\left(\bar{\xi}^2 - \frac{1}{3}\bar{\eta}^2\right) + \frac{1}{2}\bar{a}^2(\bar{u})\,\bar{\xi}\left(\bar{\eta}^2 - \frac{1}{3}\bar{\xi}^2\right), \tag{3.40}
$$

and

$$
\bar{h}_0 = \frac{\partial^2 \bar{q}}{\partial \bar{\xi} \partial \bar{\eta}}. \tag{3.41}
$$

We see that $\bar{a}^0$ and $\bar{a}^3$ do not appear in (3.40). However from (3.35) we have $A^0 = A^3$ and then (3.37) gives $\bar{a}^0 = \bar{a}^3$. If we use $\bar{u}$ given by (3.34) then $\bar{v}^0 - \bar{v}^3 = 1$ (in general we have $\bar{v}^0 - \bar{v}^3 = $ constant $\neq 0$). Now the orthogonality of $\bar{v}^i$ and $\bar{a}^i$ yields $\bar{a}^0 = \bar{a}^3 = \bar{v}^1 \bar{a}^1 + \bar{v}^2 \bar{a}^2$.

3.3   Maxwell Field of a Charge with Non-geodesic World Line

Guided by the work of Robinson and Trautman [3, 4] on solutions of the vacuum Einstein–Maxwell field equations, and requiring the solution of Maxwell's equations for the electromagnetic field of a charge having a non-geodesic light-like world line to specialise to the case of a charge having a geodesic light-like world line in Sect. 3.1 above, we look for a potential 1-form to describe the Maxwell field in the non-geodesic case given by

$$
A = e\left(\frac{1}{\bar{r}} + G(\bar{\xi}, \bar{\eta}, \bar{u})\right) d\bar{u}, \tag{3.42}
$$

with $G(\bar{\xi}, \bar{\eta}, \bar{u})$ to be determined in order to satisfy Maxwell's vacuum field equations. The following basis 1-forms are suggested by the form of the line element (3.39):

$$
\bar{\vartheta}^{(1)} = \bar{r}\left(d\bar{\xi} + \frac{\partial \bar{q}}{\partial \bar{\eta}} d\bar{u}\right), \quad \bar{\vartheta}^{(2)} = \bar{r}\left(d\bar{\eta} + \frac{\partial \bar{q}}{\partial \bar{\xi}} d\bar{u}\right),
$$

$$
\bar{\vartheta}^{(3)} = d\bar{r} - \bar{h}_0\,\bar{r}\,d\bar{u} \quad \text{and} \quad \bar{\vartheta}^{(4)} = d\bar{u}. \tag{3.43}
$$

The candidate for Maxwell 2-form is the exterior derivative of (3.42):

$$
F = dA = -\frac{e}{\bar{r}^2}\,d\bar{r} \wedge d\bar{u} + e\frac{\partial G}{\partial \bar{\xi}}\,d\bar{\xi} \wedge d\bar{u} + e\frac{\partial G}{\partial \bar{\eta}}\,d\bar{\eta} \wedge d\bar{u}
$$

$$
= -\frac{e}{\bar{r}^2}\,\bar{\vartheta}^{(3)} \wedge \bar{\vartheta}^{(4)} + \frac{e}{\bar{r}}\frac{\partial G}{\partial \bar{\xi}}\,\bar{\vartheta}^{(1)} \wedge \bar{\vartheta}^{(4)} + \frac{e}{\bar{r}}\frac{\partial G}{\partial \bar{\eta}}\,\bar{\vartheta}^{(2)} \wedge \bar{\vartheta}^{(4)}. \tag{3.44}
$$

The Hodge dual of this 2-form is the 2-form

$$
\begin{aligned}
{}^{*}F &= \frac{e}{r^2}\,\vartheta^{(1)} \wedge \vartheta^{(2)} + \frac{e}{r}\frac{\partial G}{\partial \bar{\xi}}\,\vartheta^{(2)} \wedge \vartheta^{(4)} - \frac{e}{r}\frac{\partial G}{\partial \bar{\eta}}\,\vartheta^{(1)} \wedge \vartheta^{(4)} \\
&= e\,d\bar{\xi} \wedge d\bar{\eta} + e\left(\frac{\partial \bar{q}}{\partial \bar{\xi}} - \frac{\partial G}{\partial \bar{\eta}}\right) d\bar{\xi} \wedge d\bar{u} - e\left(\frac{\partial \bar{q}}{\partial \bar{\eta}} - \frac{\partial G}{\partial \bar{\xi}}\right) d\bar{\eta} \wedge d\bar{u}\,,
\end{aligned}
$$

$$(3.45)$$

from which we obtain the exterior derivative:

$$
d^{*}F = e\left(-2\frac{\partial^2 \bar{q}}{\partial \bar{\xi} \partial \bar{\eta}} + \frac{\partial^2 G}{\partial \bar{\xi}^2} + \frac{\partial^2 G}{\partial \bar{\eta}^2}\right) d\bar{\xi} \wedge d\bar{\eta} \wedge d\bar{u}\,.
\tag{3.46}
$$

Hence Maxwell's vacuum field equations $d^{*}F = 0$ require G to satisfy

$$
\Delta G = \frac{\partial^2 G}{\partial \bar{\xi}^2} + \frac{\partial^2 G}{\partial \bar{\eta}^2} = 2\frac{\partial^2 \bar{q}}{\partial \bar{\xi} \partial \bar{\eta}} = 2\,\bar{h}_0\,.
\tag{3.47}
$$

With $\bar{q}$ given by (3.40) we see that $\Delta \bar{h}_0 = 0$ and so G satisfies the biharmonic equation

$$
\Delta \Delta G = 0\,.
\tag{3.48}
$$

It is well known that the general solution of this equation is (a proof due to A. Schild is given in [5])

$$
G(\bar{\xi}, \bar{\eta}, \bar{u}) = \mathrm{Re}\{f(\bar{\xi} + i\bar{\eta}, \bar{u}) + (\bar{\xi} - i\bar{\eta})\,F(\bar{\xi} + i\bar{\eta}, \bar{u})\}\,,
\tag{3.49}
$$

where f, F are arbitrary analytic functions of $\bar{\xi} + i\bar{\eta}$. Thus $f(\bar{\xi} + i\bar{\eta}, \bar{u}) = U(\bar{\xi}, \bar{\eta}, \bar{u}) + i V(\bar{\xi}, \bar{\eta}, \bar{u})$ with

$$
\frac{\partial U}{\partial \bar{\xi}} = \frac{\partial V}{\partial \bar{\eta}} \quad \text{and} \quad \frac{\partial U}{\partial \bar{\eta}} = -\frac{\partial V}{\partial \bar{\xi}}\,,
\tag{3.50}
$$

while $F(\bar{\xi} + i\bar{\eta}, \bar{u}) = W(\bar{\xi}, \bar{\eta}, \bar{u}) + i S(\bar{\xi}, \bar{\eta}, \bar{u})$ with

$$
\frac{\partial W}{\partial \bar{\xi}} = \frac{\partial S}{\partial \bar{\eta}} \quad \text{and} \quad \frac{\partial W}{\partial \bar{\eta}} = -\frac{\partial S}{\partial \bar{\xi}}
\tag{3.51}
$$

Hence (3.49) reads

$$
G(\bar{\xi}, \bar{\eta}, \bar{u}) = U + \bar{\xi}\,W + \bar{\eta}\,S\,.
\tag{3.52}
$$

We note in passing that we could equally well have used the imaginary part of $f + (\bar{\xi} + i\bar{\eta}) F$ for G in (3.49). Clearly not all solutions of (3.52) are solutions of (3.47) and so substituting (3.52) into (3.47) yields

$$\Delta G = 2 \left(\frac{\partial W}{\partial \bar{\xi}} + \frac{\partial S}{\partial \bar{\eta}} \right) = 4 \frac{\partial W}{\partial \bar{\xi}} = 4 \frac{\partial S}{\partial \bar{\eta}} = 2 \frac{\partial^2 \bar{q}}{\partial \bar{\xi} \partial \bar{\eta}} . \tag{3.53}$$

From this we have

$$W = \frac{1}{2} \frac{\partial \bar{q}}{\partial \bar{\eta}} + \alpha(\bar{\eta}) \quad \text{and} \quad S = \frac{1}{2} \frac{\partial \bar{q}}{\partial \bar{\xi}} + \beta(\bar{\xi}) , \tag{3.54}$$

where α, β are functions of integration. But

$$0 = \frac{\partial W}{\partial \bar{\eta}} + \frac{\partial S}{\partial \bar{\xi}} = \frac{1}{2} \Delta \bar{q} + \frac{d\alpha}{d\bar{\eta}} + \frac{d\beta}{d\bar{\xi}} = \frac{d\alpha}{d\bar{\eta}} + \frac{d\beta}{d\bar{\xi}} , \tag{3.55}$$

since $\bar{q}$ is a harmonic function, and hence we must have

$$\frac{d\alpha}{d\bar{\eta}} = C_1 = -\frac{d\beta}{d\bar{\xi}} \quad \Rightarrow \quad \alpha(\bar{\eta}) = C_1 \bar{\eta} + C_2 , \quad \beta(\bar{\xi}) = -C_1 \bar{\xi} + C_3 , \tag{3.56}$$

where C_1 is a separation constant and C_2, C_3 are constants of integration. Substituting (3.54) with (3.56) into (3.52) gives

$$G = \frac{1}{2} \left(\bar{\xi} \frac{\partial \bar{q}}{\partial \bar{\eta}} + \bar{\eta} \frac{\partial \bar{q}}{\partial \bar{\xi}} \right) + U + C_2 \bar{\xi} + C_3 \bar{\eta} . \tag{3.57}$$

The last three terms here constitute an arbitrary harmonic function. When substituted into the Maxwell field (3.44) this harmonic function describes spherical electromagnetic waves which are independent of the light-like particle and so we eliminate them and the electromagnetic field of the light-like particle is described simply by (3.44) with

$$G = \frac{1}{2} \left(\bar{\xi} \frac{\partial \bar{q}}{\partial \bar{\eta}} + \bar{\eta} \frac{\partial \bar{q}}{\partial \bar{\xi}} \right) = \frac{1}{4} (\bar{\xi}^2 + \bar{\eta}^2)\{\bar{a}^1(u) \bar{\xi} + \bar{a}^2(u) \bar{\eta}\} . \tag{3.58}$$

If the world line of the particle is a null geodesic then $\bar{a}^i(\bar{u}) = 0$ and the Maxwell field of the accelerated light-like particle specialises to the case described in Sect. 3.1.

In coordinates $\bar{x}^i = (\bar{\xi}, \bar{\eta}, \bar{r}, \bar{u})$ the components $\bar{F}_{ij}$ of the Maxwell field and the components $*\bar{F}_{ij}$ of its dual can be read off from the 2-forms (3.44) and (3.45) with G given by (3.58). In these coordinates if $\bar{k}^i = \delta^i_3$ then $\bar{k}_i = \delta^4_i$, confirming that $\bar{k}^i \bar{k}_i = 0$, and we easily find that

$$\bar{F}_{ij} \bar{k}^j = \frac{e}{r^2} \bar{k}_i \quad \text{and} \quad {}^*\bar{F}_{ij} \bar{k}^j = 0 \,. \tag{3.59}$$

It follows from these equations that the Maxwell field is algebraically general with $\bar{k}^i$ a principal null direction. Also for large values of $\bar{r}$ the Maxwell field becomes algebraically special (radiative) with degenerate principal null direction $\bar{k}^i$ which then represents the direction in Minkowskian space-time of the history of the electromagnetic radiation emitted by the charged particle on account of the fact that its world line is not a geodesic.

Many years ago Synge [6] suggested a model of a charged particle having a non-geodesic world line in Minkowskian space-time. Synge chose a 4-potential A^i, in coordinates X^i, which mimics the 4-potential for the Liénard–Wiechert field in the time-like case by taking

$$A^i = \frac{e}{r} v^i \quad \Rightarrow \quad A = A_i \, dX^i = \frac{e}{r} v_i \, dX^i \,, \tag{3.60}$$

with r given via (3.15). One can verify (see [1] or [6]) that the 2-form $F = dA$ satisfies Maxwell's equations $d^*F = 0$ and that the Maxwell field F is algebraically special (radiative) for all values of $r > 0$. However the Maxwell field vanishes if the world line $r = 0$ is a geodesic. To see this we note that (3.15) in effect gives u, r, k^i as functions of X^i. Taking the partial derivative of (3.15) with respect to X^j results in

$$\delta^i_j = v^i u_{,j} + k^i r_{,j} + r k^i_{,j} \,. \tag{3.61}$$

Multiplying this by v_i (using $v_i v^i = 0$ and $v^i k_i = 1$) yields

$$v_j = r_{,j} + r v_i k^i_{,j} = r_{,j} - r a_i k^i u_{,j} = r_{,j} - r h_0 u_{,j} \,, \tag{3.62}$$

with h_0 given by (3.20). Thus we have the 1-form

$$v_j \, dX^j = dr - r h_0 \, du \,. \tag{3.63}$$

Hence Synge's potential 1-form can be written

$$A = -e h_0 \, du + e \, d(\log r) \,. \tag{3.64}$$

If the world line $r = 0$ is a geodesic then $a^i = \lambda(u) v^i$ and so $h_0 = a_i k^i = \lambda(u)$. In this case A is an exact differential and therefore the corresponding Maxwell field vanishes.

3.4 Part II: A Kerr Black Hole and Light-Like Matter

In this part of the current chapter we describe a generalisation of the Kerr space-time in the spirit of the Vaidya generalisation of the Schwarzschild space-time. We begin by describing the latter in a form convenient for our purposes. Using Eqs. (2.151) and (2.154) of Chap. 2 we can write the Schwarzschild line element in the form

$$ds^2 = dT^2 - dX^2 - dY^2 - dZ^2 - \frac{2\,m}{r}(k_i\,dX^i)^2 = g_{ij}\,dX^i\,dX^j\,, \qquad (3.65)$$

in coordinates $X^i = (T, X, Y, Z)$ with $i = 0, 1, 2, 3$. The parameter m is the constant mass of the source,

$$r^2 = X^2 + Y^2 + Z^2\,, \qquad (3.66)$$

and the 1-form $k_i\,dX^i$ is given by

$$k_i\,dX^i = dt - \frac{1}{r}(X\,dX + Y\,dY + Z\,dZ)\,. \qquad (3.67)$$

Clearly from (3.65) we have the Kerr–Schild form of metric tensor components

$$g_{ij} = \eta_{ij} - \frac{2\,m}{r}\,k_i\,k_j \quad \text{with} \quad \eta_{ij} = \text{diag}(1, -1, -1, -1)\,. \qquad (3.68)$$

We see from (3.66) and (3.67) that $\eta^{ij}\,k_i\,k_j = 0$ so that $k^i = \eta^{ij}\,k_j$ is a null vector field with respect to the Minkowskian metric η_{ij}. It is also a null vector field with respect to the metric g_{ij} given via (3.65) since

$$g^{ij} = \eta^{ij} + \frac{2\,m}{r}\,k^i\,k^j \quad \text{with} \quad k^i = \eta^{ij}\,k_j = g^{ij}\,k_j\,. \qquad (3.69)$$

We make the simple observation that

$$u \equiv g_{ij}\,k^i\,X^j = \eta_{ij}\,k^i\,X^j = T - r\,, \qquad (3.70)$$

using (3.66). If with this variable u we make the generalisation of (3.65)

$$ds^2 = dT^2 - dX^2 - dY^2 - dZ^2 - \frac{2\,m(u)}{r}(k_i\,dX^i)^2\,, \qquad (3.71)$$

by replacing the constant m in (3.65) by the function $m(u)$ we have arrived at the Vaidya [7] generalisation of the Schwarzschild space-time. For the space-time with line element (3.71) the Ricci tensor components R_{ij} are given by

$$R_{ij} = \frac{2\,\dot{m}}{r^2}\,k_i\,k_j \quad \text{with} \quad \dot{m} = \frac{dm}{du}\,. \qquad (3.72)$$

For comparison purposes later we note that the Einstein tensor components $G_{ij} = R_{ij} - \frac{1}{2} g_{ij} R$, with $R = g^{ij} R_{ij}$ the Ricci scalar, read

$$G_{ij} = \frac{2 \dot{m}}{r^2} k_i k_j = -8 \pi T_{ij} , \qquad (3.73)$$

where T_{ij} are the components of the energy-momentum-stress tensor of a matter distribution. Eq.(3.73) represents Einstein's field equations in the current context. The matter distribution here consists of particles travelling with the speed of light radially away from the isolated spherical source.

Following the pattern established here of starting with the Kerr–Schild form (3.65) of the Schwarzschild line element, identifying a variable u using (3.70), and then using it to generalise the Schwarzschild line element, we wish to apply this to the axisymmetric Kerr line element [8]. There is a literature on the generalisation of the Vaidya space-time to axial symmetry which has been summarised in [9] as follows: "In the axisymmetric case, the complete solution was first found by Herlt [10], using a formalism developed by Vaidya [11, 12]". The Herlt solution includes the so-called "radiating Kerr metric" constructed by Vaidya and Patel [13]. It is importantly emphasised in [9] that none of the non-vacuum solutions of this type can be interpreted as having a pure radiation Maxwell field as source. The Vaidya solution has continued to stimulate research (see for example [14–18]).

3.5   Axial Symmetry

Our starting point in the axisymmetric case is the Kerr line element in the Kerr–Schild form given by Eq. (2.154) of Chap. 2:

$$\begin{aligned}
ds^2 &= dT^2 - dX^2 - dY^2 - dZ^2 - \frac{2 m r^3}{r^4 + a^2 Z^2} (k_i dX^i)^2 , \\
&= \left(\eta_{ij} - \frac{2 m r^3}{r^4 + a^2 Z^2} k_i k_j \right) dX^i dX^j , \\
&= g_{ij} dX^i dX^j , \qquad (3.74)
\end{aligned}$$

with

$$k_i dX^i = dt - \left(\frac{r X + a Y}{r^2 + a^2} \right) dX - \left(\frac{r Y - a X}{r^2 + a^2} \right) dY - \frac{Z}{r} dZ , \qquad (3.75)$$

and r is given as a function of X, Y, Z by

$$\frac{X^2 + Y^2}{r^2 + a^2} + \frac{Z^2}{r^2} = 1 . \qquad (3.76)$$

The constants m and a are the mass and angular momentum per unit mass of the source. As in the Schwarzschild case we have $k^i = \eta^{ij} k_j = g^{ij} k_j$ and thus in this case we find, using (3.75) and (3.76), that

$$u \equiv g_{ij} k^i X^j = \eta_{ij} k^i X^j = T - r \,. \tag{3.77}$$

While this expression has the identical algebraic form to (3.70) we emphasise that the coordinate r in (3.70) is given by (3.66) while the coordinate r in (3.77) is given by (3.76). We now assume that m and a depend upon u (i.e. $m = m(u)$ and $a = a(u)$) in the line element (3.74), the 1-form (3.75) and Eq. (3.76) for r. Denoting as usual partial derivatives with respect to the coordinates X^i by a comma, and derivatives of m and a with respect to u by a dot, we find that

$$r_{,i} = \left(-\frac{a \dot{a} \,(r^2 - Z^2)}{D}, \frac{r^3 X}{D}, \frac{r^3 Y}{D}, \frac{r Z \,(r^2 + a^2)}{D} \right) , \tag{3.78}$$

with

$$D = r^4 + a^2 Z^2 - a \dot{a} r \,(r^2 - Z^2) \,, \tag{3.79}$$

and thus

$$u_{,i} = \left(\frac{r^4 + a^2 Z^2}{D}, -\frac{r^3 X}{D}, -\frac{r^3 Y}{D}, -\frac{r Z \,(r^2 + a^2)}{D} \right) . \tag{3.80}$$

Hence we have

$$u_{,i} k^i = 0 \,, \tag{3.81}$$

$$g^{ij} u_{,i} u_{,j} = \eta^{ij} u_{,i} u_{,j} = -\frac{a^2 \,(r^2 - Z^2) \,(r^4 + a^2 Z^2)}{D^2} \,, \tag{3.82}$$

and thus $u = $ constant are not null hypersurfaces in general if $a \neq 0$. They are asymptotically null, if $a \neq 0$, for large positive values of r since

$$g^{ij} u_{,i} u_{,j} = -\frac{a^2}{r^2} \left(1 - \frac{Z^2}{r^2} \right) + O\left(\frac{1}{r^3} \right) . \tag{3.83}$$

The future-pointing vector field given by the 1-form (3.75) with $a = a(u)$ is null, so that

$$g^{ij} k_i k_j = \eta^{ij} k_i k_j = 0 \,, \tag{3.84}$$

and geodesic so that

$$k^i{}_{;j} k^j = k^i{}_{,j} k^j = 0 \,, \tag{3.85}$$

with the semicolon denoting covariant differentiation with respect to the Riemannian connection calculated with the metric tensor g_{ij}. This null geodesic vector field has expansion

$$\frac{1}{2} k^i{}_{;i} = \frac{1}{2} k^i{}_{,i} = \frac{2 r^3 - a \dot{a} r (r^2 - Z^2)}{2 D} \,, \tag{3.86}$$

and the squared modulus of its complex shear σ is given by

$$|\sigma|^2 = \frac{1}{2} k_{(i;j)} k^{i;j} - \left(\frac{1}{2} k^i{}_{;i}\right)^2 = \frac{1}{2} k_{(i,j)} k^{i,j} - \left(\frac{1}{2} k^i{}_{,i}\right)^2 = \frac{a^2 \dot{a}^2 (r^2 - Z^2)^2}{4 D^2} \,, \tag{3.87}$$

with the round brackets enclosing indices denoting symmetrization. For large positive values of r we thus have

$$k_i = u_{,i} + O\left(\frac{1}{r}\right) \,, \quad \frac{1}{2} k^i{}_{;i} = \frac{1}{r} + O\left(\frac{1}{r^2}\right) \quad \text{and} \quad |\sigma| = O\left(\frac{1}{r^2}\right) \,. \tag{3.88}$$

Thus asymptotically the hypersurfaces $u = \text{constant}$ are future directed null cones generated by expanding, shear-free null geodesics.

3.6   Energy-Momentum-Stress Tensor

With the metric tensor given via the line element (3.74), with (3.75) and (3.76) holding and with $m = m(u)$ and $a = a(u)$, we calculate the energy-momentum-stress tensor components T_{ij} of the matter distribution using Einstein's field equations

$$- 8 \pi T_{ij} = G_{ij} \,. \tag{3.89}$$

We shall calculate the Einstein tensor components G_{ij} here only asymptotically, for large positive values of r, since this will yield T_{ij} with sufficient accuracy to facilitate the calculation of the asymptotic flux of 4-momentum

$$P^i = \lim_{r \to +\infty} r^2 \int_{u_0}^{u_1} du \int T^{ij} r_{,j} \sin\theta \, d\theta \, d\phi \,, \tag{3.90}$$

and the asymptotic flux of angular momentum

$$S^{ij} = \lim_{r \to +\infty} r^2 \int_{u_0}^{u_1} du \int (T^{ki} X^j - T^{kj} X^i) r_{,k} \sin \theta \, d\theta \, d\phi \qquad (3.91)$$

crossing $r = $ constant $\to +\infty$ outwards in the direction of increasing r between the future null cones $u = u_0$ and $u = u_1 > u_0$ (say). The polar angles θ, ϕ arise from the parametrisation of (3.76):

$$X = \sqrt{r^2 + a^2} \sin \theta \cos \phi \, , \quad Y = \sqrt{r^2 + a^2} \sin \theta \sin \phi \, , \quad Z = r \cos \theta \, , \qquad (3.92)$$

with $0 \le \theta \le \pi$ and $0 \le \phi \le 2\pi$. Detailed derivations of the 3-volume elements in (3.90) and (3.91) can be found in [19]. We deduce from (3.90) and (3.91) that $P^i(u)$ and $S^{ij}(u)$ satisfy

$$\frac{dP^i}{du} = \lim_{r \to +\infty} r^2 \int T^{ij} r_{,j} \sin \theta \, d\theta \, d\phi \, , \qquad (3.93)$$

and

$$\frac{dS^{ij}}{du} = \lim_{r \to +\infty} r^2 \int (T^{ki} X^j - T^{kj} X^i) r_{,k} \sin \theta \, d\theta \, d\phi \, , \qquad (3.94)$$

respectively. Denoting by $\hat{T}^{ij}$ the leading term in the expansion of T^{ij} in inverse powers of r which contributes to (3.93) and (3.94) we find that

$$8\pi \hat{T}^{ij} = -\frac{2\dot{m}}{r^2} - \frac{3}{r^3} \frac{d}{du} (m\,a) (\hat{k}^i \hat{\lambda}^j + \hat{k}^j \hat{\lambda}^i) \, , \qquad (3.95)$$

with

$$\hat{k}^i = \left(1, \frac{X}{r}, \frac{Y}{r}, \frac{Z}{r} \right) \quad \text{and} \quad \hat{\lambda}^i = \left(0, \frac{Y}{r}, -\frac{X}{r}, 0 \right) \, . \qquad (3.96)$$

The region of space-time in which (3.95) holds corresponds to large positive values of r and in this region of space-time, in which we are evaluating the integrals (3.95) and (3.96), we can take $X = r \sin \theta \cos \phi$, $Y = r \sin \theta \sin \phi$, $Z = r \cos \theta$ and thus $r^2 = X^2 + Y^2 + Z^2$. We see from (3.78) that if $\hat{v}^i = \delta_0^i$ we can write, for large values of r,

$$r_{,i} = \hat{v}_i - \hat{k}_i \, . \qquad (3.97)$$

We note the useful formulas:

$$\hat{v}^i \hat{v}_i = 1 \, , \quad \hat{k}^i \hat{v}_i = 1 \, , \quad \hat{\lambda}^i \hat{v}_i = 0 \, , \quad \hat{k}^i \hat{\lambda}_i = 0 \, . \qquad (3.98)$$

Now evaluation of the integrals (3.93) and (3.94) yields

$$\frac{dP^i}{du} = (-\dot{m}, 0, 0, 0) , \qquad (3.99)$$

and $S^{ij} \equiv 0$ except $S^{12} = -S^{21} \neq 0$ (the Z-component of the angular momentum of the matter distribution described by T^{ij}) with

$$\frac{dS^{12}}{du} = \frac{d}{du}(m\,a) . \qquad (3.100)$$

Since P^i is the 4-momentum flowing out of the system the matter distribution described by T^{ij} is losing energy provided $\dot{m} < 0$ while the rate at which angular momentum is flowing away from the system is given by (3.100).

3.7   Two Conservation Laws

The part (3.95) of the energy-momentum-stress tensor which contributes to the outward flow of 4-momentum and angular momentum at future null infinity is constructed from:

$$t_{(1)}^{ij} = -\frac{2\dot{m}}{r^2} \hat{k}^i \hat{k}^j \quad \text{and} \quad t_{(2)}^{ij} = -\frac{3}{r^3} \frac{d}{du}(m\,a)\, (\hat{k}^i \hat{\lambda}^j + \hat{k}^j \hat{\lambda}^i) . \qquad (3.101)$$

These tensors are defined in a region of space-time with metric tensor η_{ij}. They each satisfy a *conservation equation*:

$$t_{(1),j}^{ij} = 0 \quad \text{and} \quad t_{(2),j}^{ij} = 0 . \qquad (3.102)$$

To verify this we need

$$\hat{k}_{i,j} = \frac{1}{r}(\eta_{ij} - \hat{k}_i \hat{v}_j - \hat{k}_j \hat{v}_i + \hat{k}_i \hat{k}_j) \quad \Rightarrow \quad \hat{k}^i{}_{,i} = \frac{2}{r} , \qquad (3.103)$$

together with (3.97), $\hat{k}_i = u_{,i}$ on account of (3.88) and

$$\hat{\lambda}^i{}_{,j} \hat{k}^j = 0 = \hat{\lambda}^i{}_{,i} . \qquad (3.104)$$

Since $t_{(A)}^{ij} \hat{k}_j = 0$ for $A = 1, 2$ there is no flux of 4-momentum or angular momentum across the future null cones $u = $ constant in this region of space-time and, on account of (3.102), this means that the 4-momentum or angular momentum that escapes to infinity is independent of r for sufficiently large positive values of r. The type of generalisation of the Kerr space-time described here could play a role as a background space-time in the approach to high frequency gravitational radiation in Kerr–Schild space-times described in an important paper by Taub [20].

References

1. C.G. Böhmer, P.A. Hogan, Phys. Rev. D **100**, 044021 (2019)
2. P.O. Kazinski, A.A. Sharapov, Class. Quant. Grav. **20**, 2715 (2003)
3. I. Robinson, A. Trautman, Phys. Rev. Lett. **4**, 431 (1960)
4. I. Robinson, A. Trautman, Proc. R. Soc. A. **265**, 463 (1962)
5. J.L. Synge, *The Hypercircle in Mathematical Physics* (Cambridge University Press, Cambridge, 1957)
6. J.L. Synge, Tensor **24**, 69 (1972)
7. P.C. Vaidya, Curr. Sci. **12**, 183 (1943)
8. C.G. Böhmer, P.A. Hogan, Mod. Phys. Lett. A. **32**, 1750189 (2017)
9. H. Stephani, D. Kramer, M.A.H. MacCallum, C. Hoenselaers, E. Herlt, *Exact Solutions of Einstein's Field Equations*, 2nd edn. (Cambridge University Press, Cambridge, 2003)
10. E. Herlt, Gen. Rel. Grav. **12**, 1 (1980)
11. P.C. Vaidya, Tensor **27**, 276 (1973)
12. P.C. Vaidya, Proc. Camb. Philos. Soc. **75**, 383 (1974)
13. P.C. Vaidya, L.K. Patel, Phys. Rev. D **7**, 3590 (1973)
14. D. Bini, A. Geralico, R.T. Jantzen, O. Semerák, Class. Quantum Grav. **28**, 245019 (2011)
15. V.A. Berezin, V.I. Dokuchaev, Y.N. Eroshenko, Class. Quantum Grav. **33**, 145003 (2016)
16. V.A. Berezin, V.I. Dokuchaev, Y.N. Eroshenko, J. Exp. Theor. Phys. **124**, 446 (2017)
17. S.M. Ruan, Phys. Rev. D **93**, 064061 (2016)
18. P. Rudra, M. Faizal, A.F. Ali, Nucl. Phys. **909**, 725 (2016)
19. J.L. Synge, Ann. Math. Pur. Appl. IV **84**, 33 (1970)
20. A.H. Taub, Commun. Math. Phys. **47**, 185 (1976)

Bateman Waves in the Linear Approximation

4

Abstract

In Minkowskian space-time in which points are labelled with rectangular Cartesian coordinates x, y, z and time t the so-called t-lines correspond to constant values of x, y, z. These world lines constitute a time-like geodesic congruence which is twist-free, shear-free and expansion-free. They play an important role in encoding in a bivector field on Minkowskian space-time the information contained in the electric and magnetic 3-vectors on three dimensional Euclidean space. A striking illustration of this is found in the study of electromagnetic radiation and the derivation of Bateman electromagnetic waves. The counterpart of the latter in the case of gravitational waves in the linear approximation using the gauge invariant and covariant approach demonstrates that the existence of the gravitational waves is due to perturbations in the shear of the t-lines.

4.1 Electromagnetic Radiation

As coordinates x^i ($i = 0, 1, 2, 3$) in Minkowskian space-time we take $x^0 = t$ as the time coordinate and $x^1 = x, x^2 = y, x^3 = z$ as rectangular Cartesian coordinates, writing for short $x^\alpha = (x, y, z)$ ($\alpha = 1, 2, 3$). The Minkowskian metric tensor in these coordinates has components $\eta_{ij} = \mathrm{diag}(1, -1, -1, -1)$. Indices on the components of tensors will be raised and lowered using η^{ij} and η_{ij} respectively with η^{ij} defined by $\eta^{ij} \eta_{jk} = \delta^i_k$. A vacuum Maxwell field in three dimensional Euclidean space is described by a pair of 3-vectors $\mathbf{E}, \mathbf{B}$ representing the electric and magnetic fields respectively. When viewed as a field on Minkowskian space-time the information in this pair of 3-vectors is encoded in a real bivector field with components $F_{ij} = -F_{ji}$. The relationship between the pair of vector fields and the tensor field is made using the time-like geodesic congruence on Minkowskian space-time consisting of the integral curves of the vector field $u^i = \delta^i_0$. These parallel

© The Author(s), under exclusive license to Springer Nature Switzerland AG 2021
P. A. Hogan, D. Puetzfeld, *Frontiers in General Relativity*, Lecture Notes
in Physics 984, https://doi.org/10.1007/978-3-030-69370-1_4

curves are sometimes referred to as the t-lines on Minkowskian space-time. We begin by defining the 4-vectors

$$E_i = F_{ij} u^j = F_{i0} \quad \text{and} \quad B_i = {}^*F_{ij} u^j = {}^*F_{i0} , \tag{4.1}$$

where the dual of the tensor F_{ij}, denoted ${}^*F_{ij}$, is defined by

$$^*F_{ij} = \frac{1}{2}\epsilon_{ijkl} F^{kl} , \tag{4.2}$$

and ϵ_{ijkl} is the Levi–Civita permutation symbol in four dimensions with $\epsilon_{0123} = -1$. Hence we have

$$E_i = (0, E_\alpha) = (0, F_{\alpha 0}) \quad \Rightarrow \quad E^\alpha = F_{0\alpha} , \tag{4.3}$$

and

$$B_i = (0, B_\alpha) = (0, {}^*F_{\alpha 0}) \quad \Rightarrow \quad B^\alpha = {}^*F_{0\alpha} . \tag{4.4}$$

We shall henceforth take

$$\mathbf{E} = (E^\alpha) = (E^1, E^2, E^3) \quad \text{and} \quad \mathbf{B} = (B^\alpha) = (B^1, B^2, B^3) . \tag{4.5}$$

We note that with our convention for raising and lowering indices $E_\alpha = -E^\alpha$, $B_\alpha = -B^\alpha$ and we have also used the skew symmetry of F_{ij} and ${}^*F_{ij}$. From (4.3) and (4.4) we can write the skew-symmetric tensor F_{ij} in terms of the 3-vectors (4.5) and display its components as the entries in the skew-symmetric matrix

$$(F_{ij}) = \begin{pmatrix} 0 & E^1 & E^2 & E^3 \\ -E^1 & 0 & -B^3 & B^2 \\ -E^2 & B^3 & 0 & -B^1 \\ -E^3 & -B^2 & B^1 & 0 \end{pmatrix} . \tag{4.6}$$

Similarly we find that

$$(^*F_{ij}) = \begin{pmatrix} 0 & B^1 & B^2 & B^3 \\ -B^1 & 0 & E^3 & -E^2 \\ -B^2 & -E^3 & 0 & E^1 \\ -B^3 & E^2 & -E^1 & 0 \end{pmatrix} . \tag{4.7}$$

We shall denote the scalar product of 3-vectors by a dot and the vector product by a multiplication sign. Thus in particular if $\mathbf{P} = (P^\alpha)$ and $\mathbf{Q} = (Q^\alpha)$ are 3-vectors we have $\mathbf{P} \cdot \mathbf{Q} = P^\alpha Q^\alpha$ and $\mathbf{P} \times \mathbf{Q} = ((\mathbf{P} \times \mathbf{Q})^\alpha) = (\epsilon_{\alpha\beta\gamma} P^\beta Q^\gamma)$ where $\epsilon_{\alpha\beta\gamma}$ is the three dimensional Levi–Civita permutation symbol with $\epsilon_{123} = +1$. We shall also write $\mathbf{P} \cdot \mathbf{P} = |\mathbf{P}|^2$. The gradient of a scalar φ will be denoted $\nabla \varphi = ((\nabla \varphi)^\alpha) = (\varphi_{,\alpha})$ with the comma denoting partial differentiation. We also have $\nabla \cdot \mathbf{P} = P^\alpha_{,\alpha}$ and $(\nabla \times \mathbf{P})^\alpha = \epsilon_{\alpha\beta\gamma} P^\gamma_{,\beta}$. We shall not raise the indices on the permutation symbols. In terms of the bivector field F_{ij} on Minkowskian space-time, or the pair of vector fields $\mathbf{E}, \mathbf{B}$ on three dimensional Euclidean space, Maxwell's vacuum field equations read:

$$F^{ij}_{,j} = 0 \;\Leftrightarrow\; \nabla \cdot \mathbf{E} = 0 \text{ and } \frac{\partial \mathbf{E}}{\partial t} = \nabla \times \mathbf{B}, \tag{4.8}$$

and

$$ ^*F^{ij}_{,j} = 0 \;\Leftrightarrow\; \nabla \cdot \mathbf{B} = 0 \text{ and } \frac{\partial \mathbf{B}}{\partial t} = -\nabla \times \mathbf{E}. \tag{4.9}$$

To illustrate the role in this context of a null geodesic congruence we consider the Maxwell field $\mathbf{E}, \mathbf{B}$ or F_{ij} satisfying (4.8) and (4.9) subject to the algebraic conditions

$$F_{ij} F^{ij} = 0 \text{ and } ^*F_{ij} F^{ij} = 0 \;\Leftrightarrow\; |\mathbf{E}|^2 = |\mathbf{B}|^2 \text{ and } \mathbf{E} \cdot \mathbf{B} = 0. \tag{4.10}$$

Such Maxwell fields describe pure electromagnetic radiation (see Chap. 2). With these conditions holding it is straightforward to verify that (4.6) and (4.7) can be written in the forms

$$F_{ij} = q_i k_j - q_j k_i \text{ and } ^*F_{ij} = w_i k_j - w_j k_i, \tag{4.11}$$

with

$$k^i = \left(1, \frac{\mathbf{E} \times \mathbf{B}}{|\mathbf{E}|^2}\right), \; q^i = (0, \mathbf{E}) \text{ and } w^i = (0, \mathbf{B}). \tag{4.12}$$

We see that

$$k^i k_i = 0, \; q^i k_i = 0, \; w^i k_i = 0, \tag{4.13}$$

and in addition

$$q^i w_i = 0, \; q^i q_i = -|\mathbf{E}|^2, \; w^i w_i = -|\mathbf{B}|^2. \tag{4.14}$$

with the first of (4.13) and the first of (4.14) following from (4.10). Hence k^i is a null vector field on Minkowskian space-time and q^i, w^i are space-like vector fields on Minkowskian space-time which are both orthogonal to k^i and to each other. To aid calculation we define the unit vectors on three dimensional Euclidean space:

$$\mathbf{e} = \frac{\mathbf{E}}{|\mathbf{E}|} \ , \ \mathbf{b} = \frac{\mathbf{B}}{|\mathbf{B}|} \ , \ \mathbf{k} = \mathbf{e} \times \mathbf{b} \ . \tag{4.15}$$

These form a right handed triad. The product of two three dimensional permutation symbols can be written in terms of a determinant as

$$\epsilon_{\alpha\beta\gamma} \, \epsilon_{\lambda\rho\sigma} = \begin{vmatrix} \delta_{\alpha\lambda} & \delta_{\alpha\rho} & \delta_{\alpha\sigma} \\ \delta_{\beta\lambda} & \delta_{\beta\rho} & \delta_{\beta\sigma} \\ \delta_{\gamma\lambda} & \delta_{\gamma\rho} & \delta_{\gamma\sigma} \end{vmatrix} \ . \tag{4.16}$$

We also note the useful spacial case

$$\epsilon_{\alpha\beta\gamma} \, \epsilon_{\alpha\rho\sigma} = \begin{vmatrix} \delta_{\beta\rho} & \delta_{\beta\sigma} \\ \delta_{\gamma\rho} & \delta_{\gamma\sigma} \end{vmatrix} \ . \tag{4.17}$$

An immediate consequence of (4.16) is

$$k^\alpha \, k^\lambda = \epsilon_{\alpha\beta\gamma} \, e^\beta \, b^\gamma \, \epsilon_{\lambda\rho\sigma} \, e^\rho \, b^\sigma = \delta_{\alpha\lambda} - e^\alpha \, e^\lambda - b^\alpha \, b^\lambda \ , \tag{4.18}$$

and so the components of the Euclidean metric tensor in rectangular Cartesian coordinates can be written in terms of the orthonormal triad $\mathbf{k}, \mathbf{e}, \mathbf{b}$ as

$$\delta_{\alpha\lambda} = k^\alpha \, k^\lambda + e^\alpha \, e^\lambda + b^\alpha \, b^\lambda \ . \tag{4.19}$$

As a consequence of Maxwell's equations the null vector field k^i on Minkowskian space-time has, as we have seen in Chap. 2, the geometrical properties of being geodesic and shear-free which we now derive in the present context.

With k^i given by (4.12) and (4.15) in the form $k^i = (1, \mathbf{k})$, the first equation satisfied by $\mathbf{k}$ that we require is

$$\frac{\partial \mathbf{k}}{\partial t} = -(\mathbf{k} \cdot \nabla) \, \mathbf{k} \ . \tag{4.20}$$

To establish this we begin by noting from the first of Maxwell's equations in vector form in (4.8) and (4.9) that

$$\nabla \cdot \mathbf{e} = -\frac{1}{2 \, |\mathbf{E}|^2} \, \mathbf{e} \cdot \nabla |\mathbf{E}|^2 \ \text{ and } \ \nabla \cdot \mathbf{b} = -\frac{1}{2 \, |\mathbf{E}|^2} \, \mathbf{b} \cdot \nabla |\mathbf{E}|^2 \ . \tag{4.21}$$

For future reference it will be useful to write these as

$$\frac{1}{2 \, |\mathbf{E}|^2} \, \nabla |\mathbf{E}|^2 = -(\nabla \cdot \mathbf{e}) \, \mathbf{e} - (\nabla \cdot \mathbf{b}) \, \mathbf{b} + \frac{1}{2 \, |\mathbf{E}|^2} (\mathbf{k} \cdot \nabla |\mathbf{E}|^2) \, \mathbf{k} \ . \tag{4.22}$$

The consistency of the second of Maxwell's equations in vector form in (4.8) and (4.9) with the vector form of the algebraic conditions (4.10) requires

$$\frac{\partial}{\partial t}|\mathbf{E}|^2 = -|\mathbf{E}|^2(\nabla \cdot \mathbf{k}) - \mathbf{k} \cdot \nabla |\mathbf{E}|^2 \,, \tag{4.23}$$

$$\mathbf{b} \cdot (\nabla \times \mathbf{b}) - \mathbf{e} \cdot (\nabla \times \mathbf{e}) = 0 \,, \tag{4.24}$$

$$\mathbf{e} \cdot (\nabla \times \mathbf{b}) + \mathbf{b} \cdot (\nabla \times \mathbf{e}) = 0 \,. \tag{4.25}$$

For the derivation of (4.20) we will require (4.22) and (4.23). Using the second of Maxwell's equations in vector form in (4.8) and (4.9) and $\mathbf{k}$ given by (4.12) we have

$$\begin{aligned}\frac{\partial \mathbf{k}}{\partial t} &= -\frac{1}{|\mathbf{E}|^2}\frac{\partial}{\partial t}(|\mathbf{E}|^2)\,\mathbf{k} + \frac{1}{|\mathbf{E}|^2}\left((\nabla \times \mathbf{B}) \times \mathbf{B} + (\nabla \times \mathbf{E}) \times \mathbf{E}\right) \\ &= -\frac{1}{|\mathbf{E}|^2}\frac{\partial}{\partial t}(|\mathbf{E}|^2)\,\mathbf{k} + \frac{1}{|\mathbf{E}|^2}\left(-\nabla|\mathbf{E}|^2 + (\mathbf{B} \cdot \nabla)\mathbf{B} + (\mathbf{E} \cdot \nabla)\mathbf{E}\right) \,.\end{aligned}$$
$$\tag{4.26}$$

But using (4.15)

$$(\mathbf{B} \cdot \nabla)\mathbf{B} + (\mathbf{E} \cdot \nabla)\mathbf{E} = \frac{1}{2}\left(\nabla|\mathbf{E}|^2 - (\mathbf{k} \cdot \nabla|\mathbf{E}|^2)\,\mathbf{k}\right) + |\mathbf{E}|^2\left((\mathbf{b} \cdot \nabla)\mathbf{b} + (\mathbf{e} \cdot \nabla)\mathbf{e}\right) \,, \tag{4.27}$$

and now this and (4.22) with (4.23) in (4.26) yields

$$\frac{\partial \mathbf{k}}{\partial t} = (\nabla \cdot \mathbf{k})\,\mathbf{k} + (\nabla \cdot \mathbf{e})\,\mathbf{e} + (\mathbf{e} \cdot \nabla)\mathbf{e} + (\nabla \cdot \mathbf{b})\,\mathbf{b} + (\mathbf{b} \cdot \nabla)\mathbf{b} \,. \tag{4.28}$$

However using (4.19) we have

$$(\mathbf{e} \cdot \nabla)\mathbf{e} + (\mathbf{b} \cdot \nabla)\mathbf{b} = -(\nabla \cdot \mathbf{e})\,\mathbf{e} - (\nabla \cdot \mathbf{b})\,\mathbf{b} - (\mathbf{k} \cdot \nabla)\mathbf{k} - (\nabla \cdot \mathbf{k})\,\mathbf{k} \,, \tag{4.29}$$

and when this is substituted into (4.28) the result is (4.20). In the context of Minkowskian space-time (4.20) means that

$$k^i{}_{,j}\,k^j = \frac{\partial k^i}{\partial t} + k^\alpha\,k^i{}_{,\alpha} = \left(0, \frac{\partial \mathbf{k}}{\partial t} + (\mathbf{k} \cdot \nabla)\mathbf{k}\right) = 0 \,. \tag{4.30}$$

Hence we see that the integral curves of the vector field k^i constitute a null geodesic congruence on Minkowskian space-time confirming the result of Mariot [1]. We note that $\mathbf{k}$ is a unit 3-vector in the direction of the Poynting vector (energy flux density) and k^i is the light-like propagation direction in Minkowskian space-time of pure electromagnetic radiation.

A further important equation, in addition to (4.30), satisfied by the null geodesic vector field k^i is available by expanding $k_{\alpha,\beta} + k_{\beta,\alpha}$, where $k_\alpha = -k^\alpha$, on the orthonormal triad $\mathbf{k}, \mathbf{e}, \mathbf{b}$. Such an expansion of any $A_{\alpha\beta}$ follows from the identity $A_{\alpha\beta} = \delta_{\mu\alpha}\,\delta_{\nu\beta}\,A_{\mu\nu}$ after substitution for the Kronecker deltas from (4.19). Proceeding in this way we obtain

$$k_{\alpha,\beta} + k_{\beta,\alpha} = (k_\alpha\, e_\beta + k_\beta\, e_\alpha)\, k_{\mu,\lambda}\, e^\mu\, k^\lambda + (k_\alpha\, b_\beta + k_\beta\, b_\alpha)\, k_{\mu,\lambda}\, b^\mu\, k^\lambda$$

$$+ (e_\alpha\, b_\beta + e_\beta\, b_\alpha)(k_{\mu,\lambda}\, b^\mu\, e^\lambda + k_{\mu,\lambda}\, e^\mu\, b^\lambda) + 2\, e_\alpha\, e_\beta\,(k_{\mu,\lambda}\, e^\mu\, e^\lambda)$$

$$+ 2\, b_\alpha\, b_\beta\,(k_{\mu,\lambda}\, b^\mu\, b^\lambda)\,. \tag{4.31}$$

We use (4.24) and (4.25) to simplify this expression. First we note that

$$\mathbf{e}\cdot(\nabla\times\mathbf{b}) = e^\alpha\,\epsilon_{\alpha\beta\gamma}\,b^\gamma{}_{,\beta} = e^\alpha\,\epsilon_{\alpha\beta\mu}\,\delta_{\mu\gamma}\,b^\gamma{}_{,\beta}$$

$$= e^\alpha\,\epsilon_{\alpha\beta\mu}\,(k^\mu\,k^\gamma + e^\mu\,e^\gamma + b^\mu\,b^\gamma)\,b^\gamma{}_{,\beta}\quad\text{by (4.19)}$$

$$= -e^\alpha\,\epsilon_{\alpha\beta\mu}\,k^\mu\,k^\gamma{}_{,\beta}\,b^\gamma\quad\text{(since } b^\gamma\,b^\gamma = 1 \text{ and } k^\gamma\,b^\gamma = 0)$$

$$= -e^\alpha\,\epsilon_{\alpha\beta\mu}\,\epsilon_{\mu\rho\sigma}\,e^\rho\,b^\sigma\,k^\gamma{}_{,\beta}\,b^\gamma\quad\text{(since } k^\mu = \epsilon_{\mu\rho\sigma}\,e^\rho\,b^\sigma)$$

$$= -b^\beta\,b^\gamma\,k^\gamma{}_{,\beta}\quad\text{by (4.17)}$$

$$= k_{\gamma,\beta}\,b^\gamma\,b^\beta\,. \tag{4.32}$$

In similar fashion we find that

$$\mathbf{b}\cdot(\nabla\times\mathbf{e}) = -k_{\gamma,\beta}\,e^\gamma\,e^\beta\,,\quad \mathbf{b}\cdot(\nabla\times\mathbf{b}) = -k_{\gamma,\beta}\,b^\gamma\,e^\beta + k^\gamma\,e^\beta\,b^\beta{}_{,\gamma}\,, \tag{4.33}$$

and

$$\mathbf{e}\cdot(\nabla\times\mathbf{e}) = k_{\gamma,\beta}\,e^\gamma\,b^\beta - k^\gamma\,b^\beta\,e^\beta{}_{,\gamma}\,. \tag{4.34}$$

Substituting these into (4.24) and (4.25) (and using $b^\beta\,e^\beta = 0$ in (4.25)) we have

$$k_{\gamma,\beta}\,e^\gamma\,e^\beta = k_{\gamma,\beta}\,b^\gamma\,b^\beta\quad\text{and}\quad k_{\gamma,\beta}\,b^\gamma\,e^\beta + k_{\gamma,\beta}\,e^\gamma\,b^\beta = 0\,. \tag{4.35}$$

When these are introduced into (4.31) the result is

$$k_{\alpha,\beta} + k_{\beta,\alpha} = k_\alpha\left\{(k_{\mu,\lambda}\, e^\mu\, k^\lambda)\, e_\beta + (k_{\mu,\lambda}\, b^\mu\, k^\lambda)\, b_\beta\right\}$$

$$+ k_\beta\left\{(k_{\mu,\lambda}\, e^\mu\, k^\lambda)\, e_\alpha + (k_{\mu,\lambda}\, b^\mu\, k^\lambda)\, b_\alpha\right\}$$

$$+ 2\, k_{\mu,\lambda}\, e^\mu\, e^\lambda\,(e_\alpha\, e_\beta + b_\alpha\, b_\beta)\,. \tag{4.36}$$

But by the first of (4.35)

$$2\, k_{\mu,\lambda}\, e^\mu\, e^\lambda = k_{\mu,\lambda}\,(e^\mu\, e^\lambda + b^\mu\, b^\lambda) = k_{\mu,\lambda}\,(\delta_{\mu\lambda} - k^\mu\, k^\lambda) = k_{\mu,\mu} = -k^\mu{}_{,\mu}\,, \tag{4.37}$$

and writing $e_\alpha e_\beta + b_\alpha b_\beta = \delta_{\alpha\beta} - k_\alpha k_\beta$ in the final term in (4.36) we finally arrive at the expression

$$k_{\alpha,\beta} + k_{\beta,\alpha} = -k^\mu{}_{,\mu}\, \delta_{\alpha\beta} + \xi_\alpha\, k_\beta + \xi_\beta\, k_\alpha \,, \tag{4.38}$$

with

$$\xi_\alpha = \frac{1}{2}k^\mu{}_{,\mu}\, k_\alpha + (k_{\mu,\lambda}\, e^\mu\, k^\lambda)\, e_\alpha + (k_{\mu,\lambda}\, b^\mu\, k^\lambda)\, b_\alpha \,. \tag{4.39}$$

We note that

$$\boldsymbol{\xi} \cdot \mathbf{k} = \frac{1}{2}k^\mu{}_{,\mu} \,, \tag{4.40}$$

and

$$|\boldsymbol{\xi}|^2 = \frac{1}{4}(k^\mu{}_{,\mu})^2 + (k_{\mu,\lambda}\, e^\mu\, k^\lambda)^2 + (k_{\mu,\lambda}\, b^\mu\, k^\lambda)^2$$

$$= \frac{1}{4}(k^\mu{}_{,\mu})^2 + k_{\mu,\lambda}\, k_{\rho,\sigma}\, k^\lambda\, k^\sigma\, (\delta_{\mu\rho} - k^\mu\, k^\rho)$$

$$= \frac{1}{4}(k^\mu{}_{,\mu})^2 + |(\mathbf{k} \cdot \boldsymbol{\nabla})\mathbf{k}|^2$$

$$= \frac{1}{4}(k^\mu{}_{,\mu})^2 + \left|\frac{\partial \mathbf{k}}{\partial t}\right|^2 \,, \tag{4.41}$$

with the final equality coming from (4.20). To obtain from (4.38) an equation involving only $\mathbf{k}$ (and not $\mathbf{e}$ and $\mathbf{b}$) we use (4.40) and (4.41) to arrive at

$$(k_{\alpha,\beta}+k_{\beta,\alpha})(k^{\alpha,\beta}+k^{\beta,\alpha}) = 2\,(k_{\alpha,\beta}+k_{\beta,\alpha})\, k^{\alpha,\beta} = 2\,(k^\lambda{}_{,\lambda})^2 + 2\left|\frac{\partial \mathbf{k}}{\partial t}\right|^2 . \tag{4.42}$$

To see what equation on Minkowskian space-time this is equivalent to, in terms of $k^i = (1, k^\alpha)$, we put $k_{(i,j)} = (k_{i,j} + k_{j,i})/2$ and calculate

$$k_{(i,j)}\, k^{i,j} - \frac{1}{2}(k^i{}_{,i})^2 = \frac{1}{2}\left((k_{\alpha,\beta} + k_{\beta,\alpha})\, k^{\alpha,\beta} - (k^\alpha{}_{,\alpha})^2 - \left|\frac{\partial \mathbf{k}}{\partial t}\right|^2\right) . \tag{4.43}$$

Hence (4.42) and (4.43) show that in Minkowskian space-time the propagation direction k^i of electromagnetic radiation is not only geodesic, as indicated above by (4.30), but also satisfies Robinson's [2] equation

$$k_{(i,j)}\, k^{i,j} - \frac{1}{2}(k^i{}_{,i})^2 = 0 \,. \tag{4.44}$$

Thus we have confirmed that the integral curves of the vector field k^i constitute a null geodesic congruence on Minkowskian space-time which is "shear-free". In reviewing the derivation of Robinson's equation above it is clear that, in addition to the geodesic condition (4.30), the key further ingredients of the derivation are Eqs. (4.24) and (4.25) (equivalently (4.35)) which ensure that the time evolution of the electromagnetic field via Maxwell's equations preserves the algebraic conditions (4.10).

We have not made use of the electromagnetic energy tensor

$$E^{ij} = F^i{}_k F^{jk} - \frac{1}{4}\eta^{ij} F_{lk} F^{lk} .\tag{4.45}$$

This has the algebraic symmetries $E^{ij} = E^{ji}$ and $\eta_{ij} E^{ij} = 0$ and also, as a consequence of Maxwell's vacuum field equations (4.8) and (4.9), satisfies

$$E^{ij}{}_{,j} = 0 .\tag{4.46}$$

When F_{ij} in (4.11) is substituted into (4.45) the result is

$$E^{ij} = -|\mathbf{E}|^2 k^i k^j .\tag{4.47}$$

Using this in (4.46) gives

$$k^i{}_{,j} k^j = \lambda k^i \quad \text{with} \quad \lambda = -k^i{}_{,i} - \frac{1}{|\mathbf{E}|^2}(|\mathbf{E}|^2)_{,j} k^j .\tag{4.48}$$

With $k^i = (1, \mathbf{k})$ we have

$$\lambda = -\frac{1}{|\mathbf{E}|^2} \frac{\partial}{\partial t}|\mathbf{E}|^2 - \nabla \cdot \mathbf{k} - \frac{1}{|\mathbf{E}|^2} \mathbf{k} \cdot \nabla |\mathbf{E}|^2 = 0 ,\tag{4.49}$$

with the final equality following from (4.23). By this means we have the first of (4.48) with $\lambda = 0$ and so we have recovered (4.30). When the derivation of (4.46) from Maxwell's vacuum field equations is examined carefully it reveals that not all of the content of Maxwell's equations has been utilised. Among the extra information in Maxwell's equations is the Eq. (4.44).

The discussion so far has been restricted to Minkowskian space-time in rectangular Cartesian coordinates and time because this was required in the next section. However it can be presented in covariant form on a general space-time and for this the reader may consult the Appendix B.

4.2 Bateman Electromagnetic Waves

An illustration of the use of the shear-free, null geodesic congruence tangent to the vector field k^i in (4.12) is to demonstrate how Maxwell's equations can be fully integrated for the case of electromagnetic radiation propagating in a vacuum. We begin with the null geodesic equation (4.30) which we write as

$$\frac{\partial}{\partial t}(k^1 - i\,k^2) + (\mathbf{k} \cdot \nabla)(k^i - i\,k^2) = 0\,, \tag{4.50}$$

and

$$\frac{\partial k^3}{\partial t} + (\mathbf{k} \cdot \nabla)k^3 = 0\,. \tag{4.51}$$

Now define the complex-valued function $\mathcal{Y}(t, x, y, z)$, with complex conjugate $\bar{\mathcal{Y}}$, by

$$\mathcal{Y} = \frac{k^1 - i\,k^2}{1 - k^3} \quad \Rightarrow \quad k^3 = \frac{\mathcal{Y}\bar{\mathcal{Y}} - 1}{\mathcal{Y}\bar{\mathcal{Y}} + 1}\,, \tag{4.52}$$

with the implication here following from $\mathbf{k} \cdot \mathbf{k} = 1$. Introducing new coordinates $\zeta, \bar{\zeta}, u, v$ via

$$\zeta = x - i\,y\,, \bar{\zeta} = x + i\,y\,, u = -t - z\,, v = t - z\,, \tag{4.53}$$

we have

$$\mathbf{k} \cdot \nabla = (1 - k^3)\left(\mathcal{Y}\frac{\partial}{\partial \zeta} + \bar{\mathcal{Y}}\frac{\partial}{\partial \bar{\zeta}}\right) - k^3\left(\frac{\partial}{\partial u} + \frac{\partial}{\partial v}\right)\,. \tag{4.54}$$

Now (4.50) reads

$$\mathcal{Y}_v - \mathcal{Y}_u + (1 - k^3)\left(\mathcal{Y}\,\mathcal{Y}_\zeta + \bar{\mathcal{Y}}\,\mathcal{Y}_{\bar{\zeta}}\right) - k^3\left(\mathcal{Y}_u + \mathcal{Y}_v\right) = 0\,, \tag{4.55}$$

with the subscripts on $\mathcal{Y}$ denoting partial differentiation, and (4.51) becomes

$$\left(\frac{\partial}{\partial v} - \frac{\partial}{\partial u}\right)k^3 + (1 - k^3)\left(\mathcal{Y}\frac{\partial}{\partial \zeta} + \bar{\mathcal{Y}}\frac{\partial}{\partial \bar{\zeta}}\right)k^3 - k^3\left(\frac{\partial}{\partial u} + \frac{\partial}{\partial v}\right)k^3 = 0\,. \tag{4.56}$$

Substituting from (4.52) for k^3 these equations simplify to

$$\mathcal{Y}_v + \mathcal{Y}\,\mathcal{Y}_\zeta + \bar{\mathcal{Y}}\left(\mathcal{Y}_{\bar{\zeta}} - \mathcal{Y}\,\mathcal{Y}_u\right) = 0\,, \tag{4.57}$$

and

$$(\mathcal{Y}\,\bar{\mathcal{Y}})_v - \mathcal{Y}\,\bar{\mathcal{Y}}\,(\mathcal{Y}\,\bar{\mathcal{Y}})_u + \mathcal{Y}\,\{(\mathcal{Y}\,\bar{\mathcal{Y}})_\zeta + \bar{\mathcal{Y}}\,(\mathcal{Y}\,\bar{\mathcal{Y}})_{\bar{\zeta}}\} = 0 \,. \tag{4.58}$$

It is straightforward to check that (4.58) is a consequence of (4.57) and its complex conjugate. Hence the geodesic condition (4.50) is equivalent to the complex equation (4.57).

We now wish to express the shear-free property of k^i in terms of the function $\mathcal{Y}$ and its partial derivatives. We have noted above that, in addition to the geodesic property of k^i, the shear-free property relies on the two real Eqs. (4.35) or equivalently the complex equation

$$k_{\gamma,\beta}\,(e^\gamma + i\,b^\gamma)(e^\beta + i\,b^\beta) = 0 \,. \tag{4.59}$$

Writing $k^\alpha = -k_\alpha$ in terms of $\mathcal{Y}$ from (4.52) we have

$$k^\alpha = (\mathcal{Y}\,\bar{\mathcal{Y}} + 1)^{-1}\hat{k}^\alpha \quad \text{with} \quad \hat{k}^\alpha = \left(\mathcal{Y} + \bar{\mathcal{Y}}, i\,(\mathcal{Y} - \bar{\mathcal{Y}}), \mathcal{Y}\,\bar{\mathcal{Y}} - 1\right) = -\hat{k}_\alpha \,. \tag{4.60}$$

Since **k** is orthogonal to **e** and **b** we can express (4.59) as

$$\hat{k}_{\gamma,\beta}\,(e^\gamma + i\,b^\gamma)(e^\beta + i\,b^\beta) = 0 \,. \tag{4.61}$$

Using the fact that **e**, **b**, **k** form a right handed triad (cf. Eq. (4.15)) we can derive the useful formulas:

$$\frac{e^3 + i\,b^3}{e^1 + i\,b^1 + i\,(e^2 + i\,b^2)} = \frac{k^1 - i\,k^2}{1 - k^3} = \mathcal{Y} \quad \text{and} \quad \frac{e^1 + i\,b^1 - i\,(e^2 + i\,b^2)}{e^1 + i\,b^1 + i\,(e^2 + i\,b^2)} = -\mathcal{Y}^2 \,. \tag{4.62}$$

From these we have

$$e^1 + i\,b^1 = i\left(\frac{1 - \mathcal{Y}^2}{1 + \mathcal{Y}^2}\right)(e^2 + i\,b^2) \quad \text{and} \quad e^3 + i\,b^3 = \frac{2\,i\,\mathcal{Y}}{1 + \mathcal{Y}^2}\,(e^2 + i\,b^2) \,. \tag{4.63}$$

When these are substituted into (4.60), $(e^1 + i\,b^1)^2$ is a common factor in all terms on the left hand side and so, dropping this factor, we are left with

$$-2\,\mathcal{Y}\,(1 - \mathcal{Y}^2)(\hat{k}_{1,3} + \hat{k}_{3,1}) + 2\,i\,\mathcal{Y}\,(1 + \mathcal{Y}^2)\,(\hat{k}_{2,3} + \hat{k}_{3,2}) + i\,(1 - \mathcal{Y}^4)\,(\hat{k}_{1,2} + \hat{k}_{2,1})$$
$$-(1 - \mathcal{Y}^2)^2\hat{k}_{1,1} + (1 + \mathcal{Y}^2)^2\hat{k}_{2,2} - 4\,\mathcal{Y}^2\hat{k}_{3,3} = 0 \,. \tag{4.64}$$

Substituting for $\hat{k}_\alpha = -\hat{k}^\alpha$ from (4.60) we arrive at

$$-2\,\mathcal{Y}\,(1-\mathcal{Y}^2)(\hat{k}_{1,3}+\hat{k}_{3,1}) + 2\,i\,\mathcal{Y}\,(1+\mathcal{Y}^2)\,(\hat{k}_{2,3}+\hat{k}_{3,2}) = \\ -4\,\mathcal{Y}\,\{\mathcal{Y}_u + \mathcal{Y}_v - (\mathcal{Y}\,\bar{\mathcal{Y}})_{\bar{\zeta}}\} + 4\,\mathcal{Y}^3\{\bar{\mathcal{Y}}_u + \bar{\mathcal{Y}}_v - (\mathcal{Y}\,\bar{\mathcal{Y}})_\zeta\}\,, \qquad (4.65)$$

and

$$i\,(1-\mathcal{Y}^4)\,(\hat{k}_{1,2}+\hat{k}_{2,1}) = 2\,(1-\mathcal{Y}^4)(Y_{\bar{\zeta}} - \bar{Y}_\zeta)\,, \qquad (4.66)$$

and finally

$$-(1-\mathcal{Y}^2)\hat{k}_{1,1} + (1+\mathcal{Y}^2)^2\hat{k}_{2,2} - 4\,Y^2\hat{k}_{3,3} = 2\,(\mathcal{Y}_{\bar{\zeta}} + \bar{\mathcal{Y}}_\zeta) - 4\,\mathcal{Y}^2(\mathcal{Y}_\zeta + \bar{\mathcal{Y}}_{\bar{\zeta}}) \\ + 2\,\mathcal{Y}^4(\mathcal{Y}_{\bar{\zeta}} + \bar{\mathcal{Y}}_\zeta) - 4\,\mathcal{Y}^2\left((\mathcal{Y}\,\bar{\mathcal{Y}})_u + (\mathcal{Y}\,\bar{\mathcal{Y}})_v\right)\,. \qquad (4.67)$$

Entering (4.65)–(4.67) into (4.64) results in a substantial simplification to read

$$\mathcal{Y}\,(\mathcal{Y}_v + \mathcal{Y}\,\mathcal{Y}_\zeta) - (\mathcal{Y}_{\bar{\zeta}} - \mathcal{Y}\,\mathcal{Y}_u) = 0\,. \qquad (4.68)$$

This is the shear-free condition in terms of $\mathcal{Y}$ satisfied by k^i. Putting it together with the geodesic condition (4.57) results in $\mathcal{Y}(\zeta, \bar{\zeta}, u, v)$ satisfying the differential equations

$$\mathcal{Y}_v + \mathcal{Y}\,\mathcal{Y}_\zeta = 0 \quad \text{and} \quad \mathcal{Y}_{\bar{\zeta}} - \mathcal{Y}\,\mathcal{Y}_u = 0\,. \qquad (4.69)$$

We turn now to the Maxwell tensor (4.11) which we write as the complex bivector

$$\mathfrak{F}_{ij} = F_{ij} + i\,{}^*F_{ij} = (q_i + i\,w_i)\,k_j - (q_j + i\,w_j)\,k_i\,, \qquad (4.70)$$

or equivalently as the complex 2-form

$$\mathfrak{F} = \frac{1}{2}\mathfrak{F}_{ij}\,dx^i \wedge dx^j = -|\mathbf{E}|\,(e^\alpha + i\,b^\alpha)\,dx^\alpha \wedge (dt - k^\beta\,dx^\beta)\,. \qquad (4.71)$$

Using the fact that $\mathbf{b} \times \mathbf{k} = \mathbf{e}$ and $\mathbf{k} \times \mathbf{e} = \mathbf{b}$ we can write this 2-form as

$$\mathfrak{F} = -|\mathbf{E}|\left\{(e^1 + ib^1)\,(dx \wedge dt - i\,dy \wedge dz) + (e^2 + i\,b^2)\,(dy \wedge dt \right. \\ \left. + i\,dx \wedge dz) + (e^3 + i\,b^3)\,(dz \wedge dt - i\,dx \wedge dy)\right\}\,. \qquad (4.72)$$

Transforming to the coordinates $\zeta, \bar{\zeta}, u, v$ via (4.53) this becomes

$$\mathfrak{F} = -\frac{1}{2}|\mathbf{E}|\Big\{ \{e^1 + i\,b^1 + i\,(e^2 + i\,b^2)\}\,du \wedge d\zeta - (e^3 + i\,b^3)\,du \wedge dv$$

$$+ (e^3 + i\,b^3)\,d\bar{\zeta} \wedge d\zeta + \{e^1 + i\,b^1 - i\,(e^2 + i\,b^2)\}\,d\bar{\zeta} \wedge dv \Big\}. \qquad (4.73)$$

Using (4.62) this simplifies to

$$\mathfrak{F} = -\frac{1}{2}|\mathbf{E}|\,\{e^1 + i\,b^1 + i\,(e^2 + i\,b^2)\}\,\Big\{ du \wedge d\zeta - \mathcal{Y}\,du \wedge dv$$

$$+ \mathcal{Y}\,d\bar{\zeta} \wedge d\zeta - \mathcal{Y}^2\,d\bar{\zeta} \wedge dv \Big\}. \qquad (4.74)$$

It thus follows that we can write

$$\mathfrak{F} = g(\zeta, \bar{\zeta}, u, v)\,(du + Y\,d\bar{\zeta}) \wedge (d\zeta - Y\,dv)\,, \qquad (4.75)$$

for some complex valued function g to be determined from Maxwell's vacuum field equations. These equations correspond to the vanishing of the exterior derivative of this complex 2-form. This then leads to the differential equations (4.69) for $\mathcal{Y}$ (confirming that Maxwell's equations require k^i to be geodesic *and* shear-free [2]) and the following differential equations for g:

$$\frac{\partial g}{\partial v} + \frac{\partial}{\partial \zeta}(g\,\mathcal{Y}) = 0 \quad\text{and}\quad \frac{\partial g}{\partial \bar{\zeta}} - \frac{\partial}{\partial u}(g\,\mathcal{Y}) = 0\,. \qquad (4.76)$$

To solve these we first define $G(\zeta, \bar{\zeta}, u, v)$ by putting

$$g = (1 + \bar{\zeta}\,\mathcal{Y}_u - v\,\mathcal{Y}_\zeta)\,G\,, \qquad (4.77)$$

and then (4.76) become

$$G_v + \mathcal{Y}\,G_\zeta = 0 \quad\text{and}\quad G_{\bar{\zeta}} - \mathcal{Y}\,G_u = 0\,, \qquad (4.78)$$

with the subscripts again denoting partial derivatives. We note the operator equation

$$\left(\frac{\partial}{\partial \bar{\zeta}} - \mathcal{Y}\frac{\partial}{\partial u}\right)\left(\frac{\partial}{\partial v} + \mathcal{Y}\frac{\partial}{\partial \zeta}\right) - \left(\frac{\partial}{\partial v} + \mathcal{Y}\frac{\partial}{\partial \zeta}\right)\left(\frac{\partial}{\partial \bar{\zeta}} - \mathcal{Y}\frac{\partial}{\partial u}\right)$$

$$= (\mathcal{Y}_{\bar{\zeta}} - \mathcal{Y}\,\mathcal{Y}_u)\frac{\partial}{\partial \zeta} + (\mathcal{Y}_v + \mathcal{Y}\,\mathcal{Y}_\zeta)\frac{\partial}{\partial u}\,, \qquad (4.79)$$

from which we conclude that the geodesic and shear-free conditions (4.69) are the integrability conditions for the differential equations (4.78). The general solution of (4.78) is $G = G(u + \bar{\zeta}\,\mathcal{Y}, \zeta - v\,\mathcal{Y})$, where G is an arbitrary analytic function of its arguments. If we write $X_1 = u + \bar{\zeta}\,\mathcal{Y}$, $X_2 = \zeta - v\,\mathcal{Y}$ then

$$dX_1 \wedge dX_2 = (1 + \bar{\zeta}\,\mathcal{Y}_u - v\,\mathcal{Y}_\zeta)\,(du + \mathcal{Y}\,d\bar{\zeta}) \wedge (d\zeta - \mathcal{Y}\,dv)\,, \qquad (4.80)$$

and now we can write the Maxwell 2-form (4.75) as

$$\mathfrak{F} = \frac{1}{2}\mathfrak{F}_{ij}\,dx^i \wedge dx^j = \frac{1}{2}(F_{ij} + i^*F_{ij})\,dx^i \wedge dx^j = G(X_1, X_2)\,dX_1 \wedge dX_2\,.$$
$$(4.81)$$

Using (4.6) and (4.7), and the standard notation for Jacobian determinants, we find from (4.81) that the electric and magnetic 3-vectors $\mathbf{E}$, $\mathbf{B}$ are given by

$$E^1 + i\,B^1 = \mathfrak{F}_{01} = G(X_1, X_2)\,\frac{\partial(X_1, X_2)}{\partial(t, x)}\,, \qquad (4.82)$$

$$E^2 + i\,B^2 = \mathfrak{F}_{02} = G(X_1, X_2)\,\frac{\partial(X_1, X_2)}{\partial(t, y)}\,, \qquad (4.83)$$

$$E^3 + i\,B^3 = \mathfrak{F}_{03} = G(X_1, X_2)\,\frac{\partial(X_1, X_2)}{\partial(t, z)}\,. \qquad (4.84)$$

We also have

$$i\,(E^3 + i\,B^3) = \mathfrak{F}_{12} = G(X_1, X_2)\,\frac{\partial(X_1, X_2)}{\partial(x, y)}\,, \qquad (4.85)$$

$$-i\,(E^2 + i\,B^2) = \mathfrak{F}_{13} = G(X_1, X_2)\,\frac{\partial(X_1, X_2)}{\partial(x, z)}\,, \qquad (4.86)$$

$$i\,(E^1 + i\,B^1) = \mathfrak{F}_{23} = G(X_1, X_2)\,\frac{\partial(X_1, X_2)}{\partial(y, z)}\,. \qquad (4.87)$$

However (4.85)–(4.87) are consistent with (4.82)–(4.84) because, by direct calculation, we have

$$\frac{\partial(X_1, X_2)}{\partial(x, y)} = 2i\,(-\mathcal{Y} + v\,\mathcal{Y}\,\mathcal{Y}_\zeta - \bar{\zeta}\,\mathcal{Y}_{\bar{\zeta}}) = i\,\frac{\partial(X_1, X_2)}{\partial(t, z)}\,, \qquad (4.88)$$

$$\frac{\partial(X_1, X_2)}{\partial(t, y)} = i\,\left\{\mathcal{Y}^2 + 1 - v\,(\mathcal{Y}_\zeta - \mathcal{Y}\,\mathcal{Y}_v) + \zeta\,(\mathcal{Y}\,\mathcal{Y}_{\bar{\zeta}} + \mathcal{Y}_u)\right\} = i\,\frac{\partial(X_1, X_2)}{\partial(x, z)}\,, \qquad (4.89)$$

$$\frac{\partial(X_1, X_2)}{\partial(y, z)} = i\,\left\{\mathcal{Y}^2 - 1 + v\,(\mathcal{Y}_\zeta + \mathcal{Y}\,\mathcal{Y}_v) + \bar{\zeta}\,(\mathcal{Y}\,\mathcal{Y}_{\bar{\zeta}} - \mathcal{Y}_u)\right\} = i\,\frac{\partial(X_1, X_2)}{\partial(t, x)}\,. \qquad (4.90)$$

The electromagnetic waves described by (4.82)–(4.84) with (4.88)–(4.90) are Bateman waves [3]. As a result we can say that *every electromagnetic wave is a Bateman wave* [4]. Finally we observe that since $G(X_1, X_2)$ is the general solution of (4.78) then the general solution of (4.69) is clearly any $\mathcal{Y}(\zeta, \bar{\zeta}, u, v)$ given implicitly by an equation of the form

$$W(\mathcal{Y}, X_1, X_2) = 0, \tag{4.91}$$

where W is an arbitrary complex analytic function of its arguments. This important result in the study of geodesic and shear-free null congruences is originally due to R. P. Kerr (unpublished) and is referred to in the literature as the Kerr theorem [5].

4.3   Some 'Spherical' Electromagnetic Waves

We now look for spherical electromagnetic waves propagating in the radial direction in R^3 with respect to the origin $r = 0$ of R^3. Thus the propagation direction in R^3 is

$$\mathbf{k} = \left(\frac{x}{r}, \frac{y}{r}, \frac{z}{r} \right) \quad \text{with} \quad r = \sqrt{x^2 + y^2 + z^2}. \tag{4.92}$$

Thus from (4.52) and (4.53) we have

$$\mathcal{Y} = \frac{k^1 - ik^2}{1 - k^3} = \frac{x - iy}{r - z} = \frac{2\zeta}{2r + u + v} \quad \text{with} \quad r^2 = \zeta\bar{\zeta} + \frac{1}{4}(u + v)^2. \tag{4.93}$$

Using these we can easily check that (4.69) are satisfied. Turning now to (4.81) we first note that

$$X_1 = r - t \quad \text{and} \quad X_2 = \left(\frac{x - iy}{r - z} \right) (r - t). \tag{4.94}$$

Then the Bateman waves (4.81) are given by the complex 2-form

$$\begin{aligned}
\mathfrak{F} &= \frac{1}{2}\mathfrak{F}_{ij}\, dx^i \wedge dx^j = \frac{1}{2}(F_{ij} + i{}^*F_{ij})\, dx^i \wedge dx^j \\
&= \hat{G}\left(r - t, \frac{x - iy}{r - z} \right) (dr - dt) \wedge d\left(\frac{x - iy}{r - z} \right).
\end{aligned} \tag{4.95}$$

for some complex analytic function $\hat{G}$ of its arguments. We will consider arguably the simplest example of such waves by specialising to the case

$$\hat{G} = \hat{G}(r - t) = A(r - t) + i\,C(r - t), \tag{4.96}$$

where A, C are real-valued functions of $r - t$. In this case the histories of the wave fronts in Minkowskian space-time are the null cones $r - t = $ constant. These are the space-time histories of spheres, centred on $r = 0$, expanding with the speed of light. It is for this reason that we call these waves spherical waves. However we shall see that they have a property of having a point on each wave front where the electromagnetic field is singular and therefore we have referred to the waves in the title of this section as 'spherical'. Calculating the components of the electric and magnetic 3-vectors **E** and **B** from

$$\mathfrak{F} = (A + i\,C)(dr - dt) \wedge \left(\frac{x - i\,y}{r - z} \right) , \qquad (4.97)$$

we arrive at

$$E^1 = \frac{1}{r(r - z)^2} \left[C\,x\,y - A\{r(r - z) - x^2\} \right] , \qquad (4.98)$$

$$E^2 = \frac{1}{r(r - z)^2} \left[A\,x\,y - C\{r(r - z) - y^2\} \right] , \qquad (4.99)$$

$$E^3 = -\frac{1}{r(r - z)}(A\,x + C\,y) , \qquad (4.100)$$

and

$$B^1 = \frac{1}{r(r - z)^2} \left[-A\,x\,y - C\{r(r - z) - x^2\} \right] , \qquad (4.101)$$

$$B^2 = \frac{1}{r(r - z)^2} \left[C\,x\,y - A\{r(r - z) - y^2\} \right] , \qquad (4.102)$$

$$B^3 = \frac{1}{r(r - z)}(A\,y - C\,x) . \qquad (4.103)$$

Direct calculation with **E** and **B** reveals that

$$|\mathbf{E}|^2 = |\mathbf{B}|^2 = \frac{A^2 + C^2}{(r - z)^2} \quad \text{and} \quad \mathbf{E} \cdot \mathbf{B} = 0 . \qquad (4.104)$$

This confirms that the electromagnetic field is pure radiation. The first of these also demonstrates that **E** and **B** are not only singular at the spatial origin $r = 0$ in R^3 but also along the positive z-axis $x = y = 0, z > 0$. This *line singularity* is responsible for the singular point on each wave front mentioned above. We also find, from (4.98)–(4.102), that

$$\mathbf{E} \times \mathbf{B} = |\mathbf{E}|^2 \mathbf{k} , \qquad (4.105)$$

demonstrating the $\mathbf{E}, \mathbf{B}, \mathbf{k}$ form a right handed triad as in the general case (4.15). If $w = r - t$ and $A(w) = da(w)/dw, C(w) = dc(w)/dw$ then $\partial a/\partial t = -A, \partial c/\partial t = -C$ and $\partial a/\partial r = A, \partial c/\partial r = C$. Hence we see that we can write

$$\mathbf{E} = -\frac{\partial \boldsymbol{\sigma}}{\partial t} \quad \text{and} \quad \mathbf{B} = \nabla \times \boldsymbol{\sigma} , \tag{4.106}$$

with $\boldsymbol{\sigma}$ having components

$$\sigma^1 = \frac{1}{r(r-z)^2} \left[c\, x\, y - a\{r(r-z) - x^2\} \right] , \tag{4.107}$$

$$\sigma^2 = \frac{1}{r(r-z)^2} \left[a\, x\, y - c\{r(r-z) - y^2\} \right] , \tag{4.108}$$

$$\sigma^3 = -\frac{1}{r(r-z)}(a\, x + c\, y) . \tag{4.109}$$

This 3-potential satisfies

$$\nabla \cdot \boldsymbol{\sigma} = 0 . \tag{4.110}$$

Defining the 3-vector $\boldsymbol{\omega} = (\omega^1, \omega^2, \omega^3)$ by

$$\omega^1 = \frac{1}{r(r-z)^2} \left[-a\, x\, y - c\{r(r-z) - x^2\} \right] , \tag{4.111}$$

$$\omega^2 = \frac{1}{r(r-z)^2} \left[c\, x\, y + a\{r(r-z) - y^2\} \right] , \tag{4.112}$$

$$\omega^3 = -\frac{1}{r(r-z)}(-c\, x + a\, y) , \tag{4.113}$$

we find that

$$(\sigma^\alpha + i\, \omega^\alpha)\, dx^\alpha = -(a + i\, c)\, d\left(\frac{x - i\, y}{r - z} \right) . \tag{4.114}$$

Thus the 3-potential $\boldsymbol{\sigma}$ is given neatly by the 1-form

$$\sigma^\alpha\, dx^\alpha = \mathrm{Re} \left\{ -(a + i\, c)\, d\left(\frac{x - i\, y}{r - z} \right) \right\} . \tag{4.115}$$

Taking the exterior derivative of (4.114) we have

$$d\left\{ (\sigma^\alpha + i\, \omega^\alpha)\, dx^\alpha \right\} = -(A + i\, C)\, (dr - dt) \wedge d\left(\frac{x - i\, y}{r - z} \right)$$

$$= -\mathfrak{F} , \tag{4.116}$$

so that $-\sigma^\alpha - i\,\omega^\alpha$ acts as a complex 3-potential for the complex 2-form $\mathfrak{F}$. This example is particularly interesting when extended below to the case of gravitational waves in the linear approximation.

4.4   Gravitational Radiation

With points of Minkowskian space-time labelled by rectangular Cartesian coordinates x, y, z and time t, the t-lines are the integral curves of the unit time-like vector field $u^i = \delta_0^i$ with $\eta_{ij}\,u^i\,u^j = u_j\,u^j = +1$. These curves are time-like geodesics with vanishing expansion, twist and shear. In a vacuum space-time chosen to be a small perturbation of Minkowskian space-time due to the presence of gravitational radiation, the perturbed t-lines acquire shear, described by the shear tensor σ_{ij} which is small of first order, but otherwise the perturbed t-lines are geodesic, expansion-free and twist-free. Since $\sigma_{ij}\,u^j = 0$ and $\eta^{ij}\,\sigma_{ij} = 0$ we have $\sigma_{0i} = 0$ and $\sigma_{\alpha\beta} = \sigma_{\beta\alpha} \neq 0$ with $\sigma_{\alpha\alpha} = 0$. The electric and magnetic parts of the perturbed Weyl tensor, $\mathcal{E}_{ij}$ and $\mathcal{H}_{ij}$ respectively, both small of first order, satisfy $\mathcal{E}_{0i} = 0 = \mathcal{H}_{0i}$ and $\mathcal{E}_{\alpha\beta} = \mathcal{E}_{\beta\alpha} \neq 0$, $\mathcal{H}_{\alpha\beta} = \mathcal{H}_{\beta\alpha} \neq 0$ with $\mathcal{E}_{\alpha\alpha} = 0 = \mathcal{H}_{\alpha\alpha}$. The vacuum Ricci identities and Bianchi identities in first approximation provide us with the partial differential equations to be satisfied by $\sigma_{\alpha\beta}$, $\mathcal{E}_{\alpha\beta}$ and $\mathcal{H}_{\alpha\beta}$ (see Appendix B):

$$\mathcal{E}_{\alpha\beta} = -\frac{\partial\sigma_{\alpha\beta}}{\partial t}\,, \quad \mathcal{H}_{\alpha\beta} = -\epsilon_{\gamma\lambda(\alpha}\,\sigma_{\beta)\gamma,\lambda}\,, \quad \sigma_{\alpha\beta,\beta} = 0\,, \tag{4.117}$$

and

$$\frac{\partial\mathcal{E}_{\alpha\beta}}{\partial t} = \epsilon_{\lambda\sigma(\alpha}\,\mathcal{H}_{\beta)\sigma,\lambda}\,, \tag{4.118}$$

$$\frac{\partial\mathcal{H}_{\alpha\beta}}{\partial t} = -\epsilon_{\lambda\sigma(\alpha}\,\mathcal{E}_{\beta)\sigma,\lambda}\,. \tag{4.119}$$

Since the Riemann tensor is purely radiative (remembering that the Ricci tensor vanishes) it is given algebraically by

$$R_{ijkl} + i\,{}^*R_{ijkl} = N_{ij}\,N_{kl}\,, \tag{4.120}$$

with

$$N_{ij} = (q_i + i\,w_i)\,k_j - (q_j + i\,w_j)\,k_i = -N_{ji}\,, \tag{4.121}$$

and

$$k^i\,k_i = 0\,, \quad k^i\,(q_i + i\,w_i) = 0\,. \tag{4.122}$$

The null vector field is the degenerate principal null direction of the Riemann tensor. We now choose the time-like u^i to be orthogonal to each of the space-like vectors q^i and w^i and we normalise k^i with $k_i \, u^i = +1$. Then with respect to this unit time-like vector field the electric and magnetic parts of the Riemann tensor are given by

$$\mathcal{E}_{ik} + i \, \mathcal{H}_{ik} = (R_{ijkl} + i \, {}^*R_{ijkl})u^j \, u^l = (q_i + i \, w_i)(q_k + i \, w_k) \, . \tag{4.123}$$

Hence

$$\mathcal{E}_{ik} = q_i \, q_k - w_i \, w_k \quad \text{and} \quad \mathcal{H}_{ik} = q_i \, w_k + q_k \, w_i \, , \tag{4.124}$$

and since both of these quantities must be trace-free we have

$$q_i \, q^i = w_i \, w^i \quad \text{and} \quad q_i \, w^i = 0 \, . \tag{4.125}$$

The vector fields u^i, k^i, q^i, w^i are vector fields on Minkowskian space-time. With $u^i = (1, 0, 0, 0)$ we have

$$q^i = (0, \mathbf{q}) \, , \quad w^i = (0, \mathbf{w}) \, , \quad k^i = (1, \mathbf{k}) \, , \tag{4.126}$$

with

$$\mathbf{q} = (q^\alpha) = (-q_\alpha) \, , \quad \mathbf{w} = (w^\alpha) = (-w_\alpha) \, , \quad \mathbf{k} = (k^\alpha) = (-k_\alpha) \, , \tag{4.127}$$

and

$$|\mathbf{q}|^2 = q^\alpha \, q^\alpha = |\mathbf{w}|^2 \, , \quad |\mathbf{k}|^2 = 1 \, , \tag{4.128}$$

and the 3-vectors $\mathbf{q}, \mathbf{w}, \mathbf{k}$ are mutually orthogonal. We shall write

$$\mathbf{q} = |\mathbf{q}| \, \hat{q} \, , \quad \mathbf{w} = |\mathbf{q}| \, \hat{w} \quad \text{with} \quad \hat{q} \cdot \hat{q} = \hat{w} \cdot \hat{w} = 1 \quad \text{and} \quad \hat{q} \cdot \hat{w} = 0 \, . \tag{4.129}$$

Hence, from (4.124),

$$\mathcal{E}_{\alpha\beta} = |\mathbf{q}|^2(\hat{q}_\alpha \, \hat{q}_\beta - \hat{w}_\alpha \, \hat{w}_\beta) = \mathcal{E}^{\alpha\beta} \, , \quad \mathcal{H}_{\alpha\beta} = |\mathbf{q}|^2(\hat{q}_\alpha \, \hat{w}_\beta + \hat{q}_\beta \, \hat{w}_\alpha) = \mathcal{H}^{\alpha\beta} \, . \tag{4.130}$$

Using (4.120) and (4.124) we have

$$R_{ijkl} \, u^l = \mathcal{E}_{ki} \, k_j - \mathcal{E}_{kj} \, k_i \, . \tag{4.131}$$

However in general [6]

$$R_{ijkl} \, u^l = \mathcal{E}_{ki} \, u_j - \mathcal{E}_{kj} \, u_i - \eta_{pijq} \, \mathcal{H}^p{}_k \, u^q \, . \tag{4.132}$$

Putting (4.131) and (4.132) together we can solve for k_i to find that

$$k_j = u_j - \frac{1}{\mathcal{E}^{pq}\,\mathcal{E}_{pq}}\,\eta_{pijq}\,\mathcal{H}^P{}_k\,u^q\,\mathcal{E}^{ki}\,. \tag{4.133}$$

But

$$\mathcal{E}^{pq}\,\mathcal{E}_{pq} = 2|\mathbf{q}|^4\,,\ \ \mathcal{H}^P{}_k\,\mathcal{E}^{ki} = |\mathbf{q}|^2(q^p\,w^i - q^i\,w^p)\,, \tag{4.134}$$

and so

$$k_j = u_j + \frac{\eta_{0jkl}\,q^k\,w^l}{|\mathbf{q}|^2} = u_j + \frac{\epsilon_{0j\beta\gamma}\,q^\beta\,w^\gamma}{|\mathbf{q}|^2}\,. \tag{4.135}$$

With $\epsilon_{0123} = -1$ this gives us

$$k_j = \left(1, -\frac{\mathbf{q}\times\mathbf{w}}{|\mathbf{q}|^2}\right) \Leftrightarrow\ k^j = (1,\mathbf{k})\ \text{ with } \mathbf{k} = \hat{q}\times\hat{w} = \hat{k}\ (\text{say})\,. \tag{4.136}$$

Hence we see that $\hat{q},\hat{w},\hat{k}$ form a right-handed orthonormal triad and it follows that

$$\delta_{\alpha\beta} = \hat{q}^\alpha\,\hat{q}^\beta + \hat{w}^\alpha\,\hat{w}^\beta + \hat{k}^\alpha\,\hat{k}^\beta\,. \tag{4.137}$$

With (4.130) and the scalar products (4.129) we have

$$\mathcal{E}_{\alpha\beta}\,\mathcal{E}_{\beta\sigma} = |\mathbf{q}|^4(\hat{q}_\alpha\,\hat{q}_\sigma + \hat{w}_\alpha\,\hat{w}_\sigma) = \mathcal{H}_{\alpha\beta}\,\mathcal{H}_{\beta\sigma}\,, \tag{4.138}$$

and

$$\mathcal{E}_{\alpha\beta}\,\mathcal{H}_{\beta\sigma} = |\mathbf{q}|^4(\hat{q}_\alpha\,\hat{w}_\sigma - \hat{q}_\sigma\,\hat{w}_\alpha) = -\mathcal{H}_{\alpha\beta}\,\mathcal{E}_{\beta\sigma}\,. \tag{4.139}$$

Using the time evolution equations (4.118) and (4.119) we find that

$$\frac{\partial}{\partial t}\left(\mathcal{E}_{\alpha\beta}\,\mathcal{E}_{\beta\sigma} - \mathcal{H}_{\alpha\beta}\,\mathcal{H}_{\beta\sigma}\right) = |\mathbf{q}|^2\mathcal{E}_{\alpha\sigma}\left\{\hat{q}\cdot\nabla\times\hat{w}) + \hat{w}\cdot(\nabla\times\hat{q})\right\}$$

$$+|\mathbf{q}|^2\mathcal{H}_{\alpha\sigma}\left\{\hat{q}\cdot(\nabla\times\hat{q}) - \hat{w}\cdot(\nabla\times\hat{w})\right\}\,, \tag{4.140}$$

and

$$\frac{\partial}{\partial t}\left(\mathcal{E}_{\alpha\beta}\,\mathcal{H}_{\beta\sigma} + \mathcal{H}_{\alpha\beta}\,\mathcal{E}_{\beta\sigma}\right) = |\mathbf{q}|^2\mathcal{E}_{\alpha\sigma}\left\{\hat{w}\cdot(\nabla\times\hat{w}) - \hat{q}\cdot(\nabla\times\hat{q})\right\}$$

$$+|\mathbf{q}|^2\mathcal{H}_{\alpha\sigma}\left\{\hat{q}\cdot(\nabla\times\hat{w}) + \hat{w}\cdot(\nabla\times\hat{q})\right\}\,, \tag{4.141}$$

and so the algebraic conditions (4.138) and (4.139) are preserved under this time evolution provided $\hat{q}$, $\hat{w}$ satisfy

$$\hat{q} \cdot (\nabla \times \hat{w}) + \hat{w} \cdot (\nabla \times \hat{q}) = 0 , \tag{4.142}$$

and

$$\hat{w} \cdot (\nabla \times \hat{w}) - \hat{q} \cdot (\nabla \times \hat{q}) = 0 . \tag{4.143}$$

Also we see from (4.117) that $\mathcal{E}_{\alpha\beta} \mathcal{E}_{\beta\alpha} = 2 |\mathbf{q}|^4$ and thus we find from (4.118) that

$$\frac{\partial}{\partial t} |\mathbf{q}|^4 + \nabla \cdot (|\mathbf{q}|^4 \hat{k}) = 0 . \tag{4.144}$$

To interpret the Eqs. (4.142)–(4.144) we first note that the Bel–Robinson tensor (with vanishing Ricci tensor) is given by Penrose and Rindler [7]

$$T_{ijkl} = \frac{1}{4} \left(R_i{}^p{}_j{}^q R_{kplq} + {}^*R_i{}^p{}_j{}^q {}^*R_{kplq} \right) . \tag{4.145}$$

This symmetric tensor is, as a result of the Bianchi identities, divergence-free which, in the linear approximation, reads

$$T^{ijkl}{}_{,i} = 0 . \tag{4.146}$$

For the case of (4.120) the Bel–Robinson tensor is given by

$$T^{ijkl} = |\mathbf{q}|^4 k^i k^j k^k k^l . \tag{4.147}$$

Since (4.144) is equivalent to

$$\left(|\mathbf{q}|^4 k^i \right)_{,i} = 0 , \tag{4.148}$$

we have from (4.146) and (4.147) that

$$(k^i k^j{}_{,i}) k^k k^l + k^j (k^i k^k{}_{,i}) k^l + k^j k^k (k^i k^l{}_{,i}) = 0 \quad \Rightarrow \quad k^j{}_{,i} k^i = 0 , \tag{4.149}$$

and so *the integral curves of k^i are null geodesics* on account of (4.148). We note that with k^i given by (4.136) we can write (4.149) in the form

$$\frac{\partial \hat{k}^\alpha}{\partial t} = -\hat{k}^\alpha{}_{,\beta} \hat{k}^\beta . \tag{4.150}$$

Just as in the electromagnetic case the Eqs. (4.142) and (4.143) can be rewritten using (4.137) in the forms

$$\hat{k}_{\gamma,\beta}\,\hat{w}^{\gamma}\,\hat{w}^{\beta} = \hat{k}_{\gamma,\beta}\,\hat{q}^{\gamma}\,\hat{q}^{\beta}\,, \tag{4.151}$$

and

$$\hat{k}_{\gamma,\beta}\,\hat{q}^{\gamma}\,\hat{w}^{\beta} + \hat{k}_{\gamma,\beta}\,\hat{w}^{\gamma}\,\hat{q}^{\beta} = 0\,, \tag{4.152}$$

respectively. From these it follows that

$$\hat{k}_{\alpha,\beta} + \hat{k}_{\beta,\alpha} = -\hat{k}^{\gamma}{}_{,\gamma}\,\delta_{\alpha\beta} + \xi_{\alpha}\,\hat{k}_{\beta} + \xi_{\beta}\,\hat{k}_{\alpha}\,, \tag{4.153}$$

with

$$\xi_{\alpha} = (\hat{k}_{\gamma,\sigma}\,\hat{q}^{\gamma}\,\hat{k}^{\sigma})\,\hat{q}_{\alpha} + (\hat{k}_{\gamma,\sigma}\,\hat{w}^{\gamma}\,\hat{k}^{\sigma})\,\hat{w}_{\alpha} + \frac{1}{2}\,\hat{k}^{\gamma}{}_{,\gamma}\,\hat{k}_{\alpha}\,. \tag{4.154}$$

It now follows from (4.150) and (4.154) that, as in the electromagnetic case,

$$\xi_{\alpha}\,\hat{k}_{\alpha} = \frac{1}{2}\hat{k}^{\alpha}{}_{,\alpha} \quad \text{and} \quad \xi^{\alpha}\,\xi^{\alpha} = \frac{1}{4}(\hat{k}^{\alpha}{}_{,\alpha})^2 + \left|\frac{\partial\hat{k}}{\partial t}\right|^2. \tag{4.155}$$

These equations then lead again to Robinson's *shear-free condition* on k^i:

$$k_{(i,j)}\,k^{i,j} - \frac{1}{2}(k^i{}_{,i})^2 = 0\,. \tag{4.156}$$

4.5 Bateman Gravitational Waves

From the Riemann tensor components given by (4.120) and (4.121) we form the complex 2-form

$$\begin{aligned}
{}^+\Omega_{ij} &= \frac{1}{2}(R_{ijkl} + i\,{}^*R_{ijkl})\,dx^k \wedge dx^l \\
&= (\mathcal{E}_{ik}\,k_j - \mathcal{E}_{jk}\,k_i)k_l\,dx^k \wedge dx^l + i\,(\mathcal{H}_{ik}\,k_j - \mathcal{H}_{jk}\,k_i)k_l\,dx^k \wedge dx^l \\
&= -{}^+\Omega_{ji}\,.
\end{aligned} \tag{4.157}$$

With k^i in (4.136) we can write these more explicitly as

$${}^+\Omega_{0\alpha} = -(\mathcal{E}^{\alpha\beta} + i\,\mathcal{H}^{\alpha\beta})\,dx^{\beta} \wedge (dt - \hat{k}^{\gamma}\,dx^{\gamma})\,, \tag{4.158}$$

$${}^+\Omega_{\alpha\beta} = \left\{\hat{k}^{\alpha}(\mathcal{E}^{\beta\sigma} + i\,\mathcal{H}^{\beta\sigma}) - \hat{k}^{\beta}(\mathcal{E}^{\alpha\sigma} + i\,\mathcal{H}^{\alpha\sigma})\right\}\,dx^{\sigma} \wedge (dt - \hat{k}^{\gamma}\,dx^{\gamma})\,. \tag{4.159}$$

From (4.130),

$$\mathcal{E}^{\alpha\beta} + i\,\mathcal{H}^{\alpha\beta} = |\mathbf{q}|^2(\hat{q}^\alpha + i\,\hat{w}^\alpha)(\hat{q}^\beta + i\,\hat{w}^\beta)\,, \tag{4.160}$$

and using this in (4.158) gives

$$^+\Omega_{0\alpha} = -|\mathbf{q}|^2(\hat{q}^\alpha + i\,w^\alpha)\Big\{(\hat{q}^1 + i\,\hat{w}^1)\,dx \wedge dt + (\hat{q}^2 + i\,\hat{w}^2)\,dy \wedge dt$$

$$+(\hat{q}^3 + i\,\hat{w}^3)\,dz \wedge dt - (\hat{q}^\beta + i\,\hat{w}^\beta)\,\hat{k}^\gamma\,dx^\beta \wedge dx^\gamma\Big\}\,. \tag{4.161}$$

Since $\hat{q} = \hat{w} \times \hat{k}$ and $\hat{w} = \hat{k} \times \hat{q}$ it follows that

$$(\hat{q}^\beta + i\,\hat{w}^\beta)\,\hat{k}^\gamma\,dx^\beta \wedge dx^\gamma = i\,(\hat{q}^3 + i\,\hat{w}^3)\,dx \wedge dy - i\,(\hat{q}^2 + i\,\hat{w}^2)\,dx \wedge dz$$

$$+i\,(\hat{q}^1 + i\,\hat{w}^1)\,dy \wedge dz\,. \tag{4.162}$$

As a result (4.161) becomes

$$^+\Omega_{0\alpha} = -|\mathbf{q}|^2(\hat{q}^\alpha + i\,\hat{w}^\alpha)\Big\{(\hat{q}^1 + i\,\hat{w}^1)(dx \wedge dt - i\,dy \wedge dz)$$

$$+(\hat{q}^2 + i\,\hat{w}^2)(dy \wedge dt + i\,dx \wedge dz)$$

$$+(\hat{q}^3 + i\,\hat{w}^3)(dz \wedge dt - i\,dx \wedge dy)\Big\}\,. \tag{4.163}$$

With (4.160) in (4.159) we have

$$^+\Omega_{\alpha\beta} = |\mathbf{q}|^2\{\hat{k}^\alpha(\hat{q}^\beta + i\,\hat{w}^\beta) - \hat{k}^\beta(\hat{q}^\alpha + i\,\hat{w}^\alpha)\}(\hat{q}^\sigma + i\,\hat{w}^\sigma)\,dx^\sigma \wedge (dt - \hat{k}^\gamma\,dx^\gamma)\,. \tag{4.164}$$

Making use of (4.162) again and the simplifications

$$-i\,(\hat{q}^3 + i\,\hat{w}^3) = \hat{k}^1\,\hat{q}^2 - \hat{k}^2\,\hat{q}^1 + i\,(\hat{k}^1\,\hat{w}^2 - \hat{k}^2\,\hat{w}^1)\,, \tag{4.165}$$

$$i\,(\hat{q}^2 + i\,\hat{w}^2) = \hat{k}^1\,\hat{q}^3 - \hat{k}^3\,\hat{q}^1 + i\,(\hat{k}^1\,\hat{w}^3 - \hat{k}^3\,\hat{w}^1)\,, \tag{4.166}$$

$$-i\,(\hat{q}^1 + i\,\hat{w}^1) = \hat{k}^2\,\hat{q}^3 - \hat{k}^3\,\hat{q}^2 + i\,(\hat{k}^2\,\hat{w}^3 - \hat{k}^3\,\hat{w}^2)\,, \tag{4.167}$$

(4.164) is given by

$$^+\Omega_{12} = -i|\mathbf{q}|^2(\hat{q}^3 + i\,\hat{w}^3)\Big\{(\hat{q}^1 + i\,\hat{w}^2)(dx \wedge dt - i\,dy \wedge dz)$$

$$+(\hat{q}^2 + i\,\hat{w}^2)(dy \wedge dt + i\,dx \wedge dz)$$

$$+(\hat{q}^3 + i\,\hat{w}^3)(dz \wedge dt - i\,dx \wedge dz)\bigg\}\,, \tag{4.168}$$

$$^+\Omega_{13} = i|\mathbf{q}|^2(\hat{q}^2 + i\,\hat{w}^2)\bigg\{(\hat{q}^1 + i\,\hat{w}^2)(dx \wedge dt - i\,dy \wedge dz)$$

$$+(\hat{q}^2 + i\,\hat{w}^2)(dy \wedge dt + i\,dx \wedge dz)$$

$$+(\hat{q}^3 + i\,\hat{w}^3)(dz \wedge dt - i\,dx \wedge dz)\bigg\}\,, \tag{4.169}$$

$$^+\Omega_{23} = -i|\mathbf{q}|^2(\hat{q}^1 + i\,\hat{w}^1)\bigg\{(\hat{q}^1 + i\,\hat{w}^2)(dx \wedge dt - i\,dy \wedge dz)$$

$$+(\hat{q}^2 + i\,\hat{w}^2)(dy \wedge dt + i\,dx \wedge dz)$$

$$+(\hat{q}^3 + i\,\hat{w}^3)(dz \wedge dt - i\,dx \wedge dz)\bigg\}\,. \tag{4.170}$$

Next we make the coordinate transformations used already in the electromagnetic case:

$$\zeta = x - i\,y\,, \ \bar{\zeta} = x + i\,y\,, \ u = -t - z\,, \ v = t - z\,. \tag{4.171}$$

At the same time it is useful to define

$$\frac{\hat{k}^1 - i\,\hat{k}^2}{1 - \hat{k}^3} = \mathcal{Y}\,, \tag{4.172}$$

from which it follows that

$$\frac{\hat{q}^3 + i\,\hat{w}^3}{\hat{q}^1 + i\,\hat{w}^1 + i(\hat{q}^2 + i\,\hat{w}^2)} = \mathcal{Y} \ \text{ and } \ \frac{\hat{q}^1 + i\,\hat{w}^1 - i\,(\hat{q}^2 + i\,\hat{w}^2)}{\hat{q}^1 + i\,\hat{w}^1 + i\,(\hat{q}^2 + i\,\hat{w}^2)} = -\mathcal{Y}^2\,. \tag{4.173}$$

Finally putting

$$|\mathbf{q}|^2(\hat{q}^3 + i\,\hat{w}^3)^2 = \mathcal{Y}\,g(\zeta, \bar{\zeta}, u, v)\,, \tag{4.174}$$

for some complex valued function g, we can write $^+\Omega_{ij}$ as

$$^+\Omega_{12} = -\frac{i}{2} g\,(du + \mathcal{Y}\,d\bar{\zeta}) \wedge (d\zeta - \mathcal{Y}\,dv)\,, \tag{4.175}$$

$$^+\Omega_{13} = \frac{1}{4}(\mathcal{Y}^{-1} + \mathcal{Y})\,g\,(du + \mathcal{Y}\,d\bar{\zeta}) \wedge (d\zeta - \mathcal{Y}\,dv)\,, \tag{4.176}$$

$$^+\Omega_{23} = -\frac{i}{4}(\mathcal{Y}^{-1} - \mathcal{Y})\, g\, (du + \mathcal{Y}\, d\bar{\zeta}) \wedge (d\zeta - \mathcal{Y}\, dv)\,, \qquad (4.177)$$

$$^+\Omega_{01} = -\frac{1}{4}(\mathcal{Y}^{-1} - \mathcal{Y})\, g\, (du + \mathcal{Y}\, d\bar{\zeta}) \wedge (d\zeta - \mathcal{Y}\, dv)\,, \qquad (4.178)$$

$$^+\Omega_{02} = \frac{i}{4}(\mathcal{Y}^{-1} + \mathcal{Y})\, g\, (du + \mathcal{Y}\, d\bar{\zeta}) \wedge (d\zeta - \mathcal{Y}\, dv)\,, \qquad (4.179)$$

$$^+\Omega_{03} = -\frac{1}{2}\, g\, (du + \mathcal{Y}\, d\bar{\zeta}) \wedge (d\zeta - \mathcal{Y}\, dv)\,. \qquad (4.180)$$

The vacuum field equations (in the linear approximation) imply that the Bianchi identities take the form $d^+\Omega = 0$. When this condition is imposed on (4.175) or (4.180) we find that $\mathcal{Y}$ must satisfy the geodesic and shear-free conditions

$$\mathcal{Y}_{\bar{\zeta}} - \mathcal{Y}\, \mathcal{Y}_u = 0 \quad\text{and}\quad \mathcal{Y}_v + \mathcal{Y}\, \mathcal{Y}_\zeta = 0\,, \qquad (4.181)$$

with the subscripts denoting partial derivatives, and g must satisfy

$$g_{\bar{\zeta}} - (g\, \mathcal{Y})_u = 0 \quad\text{and}\quad g_v + (g\, \mathcal{Y})_\zeta = 0\,. \qquad (4.182)$$

With (4.181) and (4.182) satisfied the exterior derivatives of (4.176)–(4.179) automatically vanish. As in the electromagnetic case we put

$$g = (1 + \bar{\zeta}\, \mathcal{Y}_u - v\, \mathcal{Y}_\zeta)\, G\,, \qquad (4.183)$$

and have $G = G(X_1, X_2)$ with $X_1 = u + \bar{\zeta}\, \mathcal{Y}$, $X_2 = \zeta - v\, \mathcal{Y}$. We can then simplify (4.175)–(4.180) to describe the Bateman gravitational waves by

$$^+\Omega_{12} = -\frac{i}{2}\, G(X_1, X_2)\, dX_1 \wedge dX_2\,, \qquad (4.184)$$

$$^+\Omega_{13} = \frac{1}{4}(\mathcal{Y}^{-1} + \mathcal{Y})\, G(X_1, X_2)\, dX_1 \wedge dX_2\,, \qquad (4.185)$$

$$^+\Omega_{23} = -\frac{i}{4}(\mathcal{Y}^{-1} - \mathcal{Y})\, G(X_1, X_2)\, dX_1 \wedge dX_2\,, \qquad (4.186)$$

$$^+\Omega_{01} = -\frac{1}{4}(\mathcal{Y}^{-1} - \mathcal{Y})\, G(X_1, X_2)\, dX_1 \wedge dX_2\,, \qquad (4.187)$$

$$^+\Omega_{02} = \frac{i}{4}(\mathcal{Y}^{-1} + \mathcal{Y})\, G(X_1, X_2)\, dX_1 \wedge dX_2\,, \qquad (4.188)$$

$$^+\Omega_{03} = -\frac{1}{2}\, G(X_1, X_2)\, dX_1 \wedge dX_2\,. \qquad (4.189)$$

4.6   Some 'Spherical' Gravitational Waves

We now set out to construct the gravitational analogue of the spherical waves described in electromagnetic theory. The propagation direction in R^3 is

$$\hat{k} = \left(\frac{x}{r}, \frac{y}{r}, \frac{z}{r}\right) \quad \text{with} \quad r = \sqrt{x^2 + y^2 + z^2} \,. \tag{4.190}$$

The function $\mathcal{Y}$, which plays a key role in the construction of the Bateman waves, is

$$\mathcal{Y} = \frac{\hat{k}^1 - i\,\hat{k}^2}{1 - \hat{k}^3} = \frac{x - i\,y}{r - z} \,, \tag{4.191}$$

and the functions X_1, X_2 are given by

$$X_1 = r - t \,, \quad X_2 = \left(\frac{x - i\,y}{r - z}\right)(r - t) \,, \tag{4.192}$$

resulting in

$$G(X_1, X_2)\,dX_1 \wedge dX_2 = \hat{G}\left(r - t, \frac{x - i\,y}{r - z}\right)(dr - dt) \wedge d\left(\frac{x - i\,y}{r - z}\right) , \tag{4.193}$$

for some function $\hat{G}$ of its arguments. For arguably the simplest 'spherical' waves we take

$$\hat{G} = A(r - t) + i\,C(r - t) \,, \tag{4.194}$$

where A, C are arbitrary real-valued functions of $r - t$. Now the 2-forms (4.184)–(4.189) read

$$^{+}\Omega_{12} = -\frac{i}{2}(A + i\,C)\,(dr - dt) \wedge d\left(\frac{x - i\,y}{r - z}\right) , \tag{4.195}$$

$$^{+}\Omega_{13} = \frac{1}{4}\left(\frac{1}{Y} + Y\right)(A + i\,C)\,(dr - dt) \wedge d\left(\frac{x - i\,y}{r - z}\right) , \tag{4.196}$$

$$^{+}\Omega_{23} = -\frac{i}{4}\left(\frac{1}{Y} - Y\right)(A + i\,C)\,(dr - dt) \wedge d\left(\frac{x - i\,y}{r - z}\right) , \tag{4.197}$$

$$^{+}\Omega_{01} = -\frac{1}{4}\left(\frac{1}{Y} - Y\right)(A + i\,C)\,(dr - dt) \wedge d\left(\frac{x - i\,y}{r - z}\right) , \tag{4.198}$$

$$^+\Omega_{02} = \frac{i}{4}\left(\frac{1}{Y} + Y\right)(A + i\,C)\,(dr - dt) \wedge d\left(\frac{x - i\,y}{r - z}\right), \qquad (4.199)$$

$$^+\Omega_{03} = -\frac{1}{2}(A + i\,C)\,(dr - dt) \wedge d\left(\frac{x - i\,y}{r - z}\right). \qquad (4.200)$$

When these are equated to (4.158) and (4.159) we find the following components of the electric and magnetic parts of the perturbed Weyl tensor:

$$\mathcal{E}^{11} = \frac{-A\left[x\,z\,\{r(r - z) - x^2\} + r\,x\,y^2\right] + C\left[-r\,y\,\{r(r - z) - x^2\} + x^2y\,z\right]}{2\,r\,(r - z)^2(r^2 - z^2)},$$

$$(4.201)$$

$$\mathcal{E}^{12} = \frac{A\left[r\,y\,\{r(r - z) - y^2\} + x^2y\,z\right] + C\left[-x\,z\,\{r(r - z) - y^2\} + r\,x\,y^2\right]}{2\,r\,(r - z)^2(r^2 - z^2)},$$

$$(4.202)$$

$$\mathcal{E}^{13} = \frac{A\left[-x^2z + r\,y^2\right] + C\left[-x\,y\,z - r\,x\,y\right]}{2\,r\,(r - z)(r^2 - z^2)}, \qquad (4.203)$$

$$\mathcal{E}^{22} = \frac{-A\left[r\,x\,\{r(r - z) - y^2\} - x\,y^2z\right] - C\left[y\,z\,\{r(r - z) - y^2\} + r\,x^2y\right]}{2\,r\,(r - z)^2(r^2 - z^2)},$$

$$(4.204)$$

$$\mathcal{E}^{23} = \frac{-A\left[x\,y\,z + r\,x\,y\right] + C\left[-y^2z + r\,x^2\right]}{2\,r\,(r - z)(r^2 - z^2)}, \qquad (4.205)$$

$$\mathcal{E}^{33} = \frac{A\,x + C\,y}{2\,r\,(r - z)}, \qquad (4.206)$$

and

$$\mathcal{H}^{11} = \frac{A\left[r\,y\,\{r(r - z) - x^2\} - x^2y\,z\right] - C\left[x\,z\,\{r(r - z) - x^2\} + r\,x\,y^2\right]}{2\,r\,(r - z)^2(r^2 - z^2)},$$

$$(4.207)$$

$$\mathcal{H}^{12} = \frac{A\left[x\,z\,\{r(r - z) - y^2\} - r\,x\,y^2\right] + C\left[r\,y\,\{r(r - z) - y^2\} + x^2y\,z\right]}{2\,r\,(r - z)^2(r^2 - z^2)},$$

$$(4.208)$$

$$\mathcal{H}^{13} = \frac{A\left[x\,y\,z + r\,x\,y\right] + C\left[-x^2z + r\,y^2\right]}{2\,r\,(r - z)(r^2 - z^2)}, \qquad (4.209)$$

$$\mathcal{H}^{22} = \frac{A\left[y\,z\,\{r(r - z) - y^2\} + r\,x^2y\right] - C\left[r\,x\,\{r(r - z) - y^2\} - x\,y^2z\right]}{2\,r\,(r - z)^2(r^2 - z^2)},$$

$$(4.210)$$

$$\mathcal{H}^{23} = \frac{A\,[y^2 z - r\,x^2] - C\,[x\,y\,z + r\,x\,y]}{2\,r\,(r-z)(r^2 - z^2)}\,,\tag{4.211}$$

$$\mathcal{H}^{33} = \frac{-A\,y + C\,x}{2\,r\,(r-z)}\,.\tag{4.212}$$

Next with $w = r - t$ we can write $A(w) = da/dw$ and $C(w) = dc/dw$ and then $-\partial a/\partial t = \partial a/\partial r = A$ and $-\partial c/\partial t = \partial c/\partial r = C$. Consequently we can write

$$\mathcal{E}^{\alpha\beta} = -\frac{\partial \sigma_{\alpha\beta}}{\partial t} \quad \text{and} \quad \mathcal{H}^{\alpha\beta} = -\frac{\partial \lambda_{\alpha\beta}}{\partial t}\,,\tag{4.213}$$

with $\sigma_{\alpha\beta}$, $\lambda_{\alpha\beta}$, satisfying $\sigma_{\alpha\beta} = \sigma_{\beta\alpha}$, $\sigma_{\alpha\alpha} = 0$ and $\lambda_{\alpha\beta} = \lambda_{\beta\alpha}$, $\lambda_{\alpha\alpha} = 0$, given by

$$\sigma_{11} + i\,\lambda_{11} = -\frac{i\,(a+i\,c)(r\,y+i\,xz)}{2\,(r-z)^2(r+z)}\left(\frac{x\,(x-i\,y)}{r\,(r-z)} - 1\right),\tag{4.214}$$

$$\sigma_{22} + i\,\lambda_{22} = \frac{(a+i\,c)\,(r\,x - i\,yz)}{2\,(r-z)^2(r+z)}\left(\frac{i\,y(x-i\,y)}{r\,(r-z)} - 1\right),\tag{4.215}$$

$$\sigma_{33} + i\,\lambda_{33} = \frac{(a+i\,c)}{2\,r}\left(\frac{x - i\,y}{r-z}\right),\tag{4.216}$$

$$\sigma_{12} + i\,\lambda_{12} = -\frac{(a+i\,c)\,(r\,y+i\,xz)}{2\,(r-z)^2(r+z)}\left(\frac{i\,y(x-i\,y)}{r\,(r-z)} - 1\right),\tag{4.217}$$

$$\sigma_{13} + i\,\lambda_{13} = \frac{i\,(a+i\,c)\,(r\,y+i\,xz)}{2\,r\,(r^2 - z^2)}\left(\frac{x - i\,y}{r-z}\right),\tag{4.218}$$

$$\sigma_{23} + i\,\lambda_{23} = -\frac{i\,(a+i\,c)\,(r\,x - i\,yz)}{2\,r\,(r^2 - z^2)}\left(\frac{x - i\,y}{r-z}\right).\tag{4.219}$$

We emphasise that $\sigma_{\alpha\beta}$ has the geometrical interpretation of the perturbed shear of the t-lines in the perturbed Minkowskian space-time due to the presence of the gravitational waves. In addition $\sigma_{\alpha\beta}$, $\lambda_{\alpha\beta}$ satisfy the differential equations

$$\sigma_{\alpha\beta,\beta} = 0 \quad \text{and} \quad \lambda_{\alpha\beta,\beta} = 0\,.\tag{4.220}$$

To verify these it is useful to first express (4.214)–(4.219) in terms of the function $\mathcal{Y}$ in (4.191). This results in

$$\sigma_{11} + i\,\lambda_{11} = \frac{(a+i\,c)}{8\,r}\,\mathcal{Y}\left(\mathcal{Y} - \frac{1}{\mathcal{Y}}\right)^2,\tag{4.221}$$

$$\sigma_{22} + i\,\lambda_{22} = -\frac{(a+i\,c)}{8\,r}\,\mathcal{Y}\left(\mathcal{Y} + \frac{1}{\mathcal{Y}}\right)^2,\tag{4.222}$$

$$\sigma_{33} + i\,\lambda_{33} = \frac{(a+ic)}{2\,r}\,\mathcal{Y}\,, \tag{4.223}$$

$$\sigma_{12} + i\,\lambda_{12} = \frac{i\,(a+ic)}{8\,r}\,\mathcal{Y}\left(\mathcal{Y}^2 - \frac{1}{\mathcal{Y}^2}\right)\,, \tag{4.224}$$

$$\sigma_{13} + i\,\lambda_{13} = -\frac{(a+ic)}{4\,r}\,\mathcal{Y}\left(\mathcal{Y} - \frac{1}{\mathcal{Y}}\right)\,, \tag{4.225}$$

$$\sigma_{23} + i\,\lambda_{23} = -\frac{i\,(a+ic)}{4\,r}\,\mathcal{Y}\left(\mathcal{Y} + \frac{1}{\mathcal{Y}}\right)\,. \tag{4.226}$$

The following formulas, with partial derivatives denoted by subscripts and a bar denoting complex conjugation, are very useful in verifying (4.220):

$$\mathcal{Y}_x = -\frac{\mathcal{Y}}{2\,r}\left(\mathcal{Y} - \frac{1}{\mathcal{Y}}\right)\,,\ \mathcal{Y}_y = -\frac{i\,\mathcal{Y}}{2\,r}\left(\mathcal{Y} + \frac{1}{\mathcal{Y}}\right)\,,\ \mathcal{Y}_z = \frac{\mathcal{Y}}{r}\,, \tag{4.227}$$

and

$$r_x = \frac{\mathcal{Y} + \bar{\mathcal{Y}}}{\mathcal{Y}\bar{\mathcal{Y}} + 1}\,,\ r_y = \frac{i\,(\mathcal{Y} - \bar{\mathcal{Y}})}{\mathcal{Y}\bar{\mathcal{Y}} + 1}\,,\ r_z = \frac{\mathcal{Y}\bar{\mathcal{Y}} - 1}{\mathcal{Y}\bar{\mathcal{Y}} + 1}\,. \tag{4.228}$$

Finally we wish to verify the second of (4.117). We see from (4.213) that

$$\mathcal{E}_{\alpha\beta} + i\,\mathcal{H}_{\alpha\beta} = -\frac{\partial}{\partial t}(\sigma_{\alpha\beta} + i\,\lambda_{\alpha\beta})\,, \tag{4.229}$$

and this implies that

$$\mathcal{E}_{11} + i\,\mathcal{H}_{11} = \frac{(A+iC)}{8\,r}\,\mathcal{Y}\left(\mathcal{Y} - \frac{1}{\mathcal{Y}}\right)^2\,, \tag{4.230}$$

$$\mathcal{E}_{22} + i\,\mathcal{H}_{22} = -\frac{(A+iC)}{8\,r}\,\mathcal{Y}\left(\mathcal{Y} + \frac{1}{\mathcal{Y}}\right)^2\,, \tag{4.231}$$

$$\mathcal{E}_{33} + i\,\mathcal{H}_{33} = \frac{(A+iC)}{2\,r}\,\mathcal{Y}\,, \tag{4.232}$$

$$\mathcal{E}_{12} + i\,\mathcal{H}_{12} = \frac{i\,(A+iC)}{8\,r}\,\mathcal{Y}\left(\mathcal{Y}^2 - \frac{1}{\mathcal{Y}^2}\right)\,, \tag{4.233}$$

$$\mathcal{E}_{13} + i\,\mathcal{H}_{13} = -\frac{(A+iC)}{4\,r}\,\mathcal{Y}\left(\mathcal{Y} - \frac{1}{\mathcal{Y}}\right)\,, \tag{4.234}$$

$$\mathcal{E}_{23} + i\,\mathcal{H}_{23} = -\frac{i\,(A+iC)}{4\,r}\,\mathcal{Y}\left(\mathcal{Y} + \frac{1}{\mathcal{Y}}\right)\,. \tag{4.235}$$

Also (4.117) is equivalent to the equations:

$$\mathcal{H}_{11} = \sigma_{13,2} - \sigma_{12,3} \, , \quad \mathcal{H}_{12} = \sigma_{23,2} - \sigma_{22,3} \, , \quad \mathcal{H}_{13} = \sigma_{33,2} - \sigma_{23,3} \, ,$$
$$\mathcal{H}_{22} = \sigma_{12,3} - \sigma_{23,1} \, , \quad \mathcal{H}_{23} = \sigma_{13,3} - \sigma_{33,1} \, , \quad \mathcal{H}_{33} = \sigma_{23,1} - \sigma_{13,2} \, .$$

$$(4.236)$$

Now using (4.221)–(4.226), and the very useful Eqs. (4.227) and (4.228), we find that

$$(\sigma_{13} + i\,\lambda_{13})_{,2} - (\sigma_{12} + i\,\lambda_{12})_{,3} = -\frac{i\,(A + i\,C)}{8\,r}\,y\left(y - \frac{1}{y}\right)^2 = \mathcal{H}_{11} - i\,\mathcal{E}_{11} \, ,$$

$$(4.237)$$

$$(\sigma_{23} + i\,\lambda_{23})_{,2} - (\sigma_{22} + i\,\lambda_{22})_{,3} = \frac{(A + i\,C)}{8\,r}\,y\left(y^2 - \frac{1}{y^2}\right) = \mathcal{H}_{12} - i\,\mathcal{E}_{12} \, ,$$

$$(4.238)$$

$$(\sigma_{33} + i\,\lambda_{33})_{,2} - (\sigma_{23} + i\,\lambda_{23})_{,3} = \frac{i\,(A + i\,C)}{4\,r}\,y\left(y - \frac{1}{y}\right) = \mathcal{H}_{13} - i\,\mathcal{E}_{13} \, ,$$

$$(4.239)$$

$$(\sigma_{12} + i\,\lambda_{12})_{,3} - (\sigma_{23} + i\,\lambda_{23})_{,1} = \frac{i\,(A + i\,C)}{8\,r}\,y\left(y + \frac{1}{y}\right)^2 = \mathcal{H}_{22} - i\,\mathcal{E}_{22} \, ,$$

$$(4.240)$$

$$(\sigma_{13} + i\,\lambda_{13})_{,3} - (\sigma_{33} + i\,\lambda_{33})_{,1} = -\frac{(A + i\,C)}{4\,r}\,y\left(y + \frac{1}{y}\right) = \mathcal{H}_{23} - i\,\mathcal{E}_{23} \, ,$$

$$(4.241)$$

$$(\sigma_{23} + i\,\lambda_{23})_{,1} - (\sigma_{13} + i\,\lambda_{13})_{,2} = -\frac{i\,(A + i\,C)}{2\,r}\,y = \mathcal{H}_{33} - i\,\mathcal{E}_{33} \, . \qquad (4.242)$$

Comparing the real parts on either side of these equations reveals that (4.236) are satisfied.

References

1. L. Mariot, C. R. Acad. Sci **238**, 2055 (1954)
2. I. Robinson, J. Math. Phys. **2**, 290 (1961)
3. H. Bateman, *The Mathematical Analysis of Electrical and Optical Wave-Motion* (Dover, New York, 1955)
4. P.A. Hogan, Proc. R. Soc. Lond. A **396**, 199 (1984)
5. D. Cox, E.J. Flaherty, Communs. Math. Phys. **47**, 75 (1976)
6. G.F.R. Ellis, *Relativistic Cosmology* (Gordon and Breach, London, 1971)
7. R. Penrose, W. Rindler, *Spinors and Space-Time*, vol. 1 (Cambridge University Press, Cambridge, 1984)

Gravitational (Clock) Compass

5

Abstract

A central question in general relativity is how the components of the Riemann curvature tensor (the gravitational field) can be determined in an operational way. Here we review two different methods which allow for a complete determination of the Riemann curvature tensor. The first method relies on measuring the accelerations of a suitably prepared set of test bodies relative to an observer. The second method utilizes a set of suitably prepared clocks.

5.1 Determination of the Gravitational Field by Means of Test Bodies

Historically, Felix Pirani [1] was the first to point out that one could determine the full Riemann tensor with the help of a (sufficiently large) number of test bodies in the vicinity of an observer's world line. Pirani's suggestion to measure the curvature was based on the equation which describes the dynamics of a vector connecting two adjacent geodesics in spacetime. In the literature this equation is known as the Jacobi equation or the geodesic deviation equation. Its early derivations in a Riemannian context can be found in [2–4]. See Sect. 1.3 for a derivation and extension based on Synge's world function.

In the following we explicitly show how a suitably prepared set of test bodies can be used to determine all components of the curvature of spacetime (and thereby to measure the gravitational field) with the help of an exact solution for the components of the Riemann tensor in terms of the mutual accelerations between the constituents of a cloud of test bodies and the observer. The analysis follows the presentation given in [5].

This can be viewed as an explicit realization of Szekeres' "gravitational compass" [6], or Synge's "curvature detector" [7]. The operational procedure, see Fig. 5.1, is to monitor the accelerations of a set of test bodies w.r.t. to an observer

P. A. Hogan, D. Puetzfeld, *Frontiers in General Relativity*, Lecture Notes
in Physics 984, https://doi.org/10.1007/978-3-030-69370-1_5

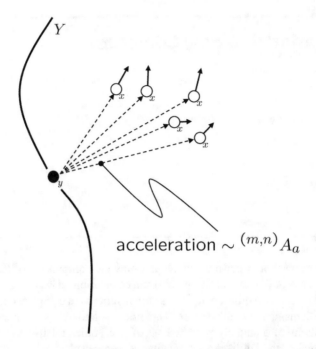

Fig. 5.1 Sketch of the operational procedure to measure the curvature of spacetime. An observer moving along a world line Y monitors the accelerations $^{(m,n)}A_a$ of a set of suitably prepared test bodies (hollow circles). The number of test bodies required for the determination of all curvature components depends on the type of the underlying spacetime

moving along a reference world line Y. A mechanical analogue would be to measure the forces between the test bodies and the reference body via a spring connecting them.

5.2   Gravitational Compass

The curvature tensor in Einstein's theory has twenty (20) independent components for the most general field configurations produced by nontrivial matter sources, whereas in vacuum the number of independent components reduces to ten (10). As compared to Newton's theory, the gravitational field thus has more degrees of freedom in the relativistic framework. The explicit determination of the curvature of spacetime in the context of deviation equations has been discussed in [6–8]. In particular, Szekeres coined in [6] the notion of a gravitational compass, which we will adopt from here on for a set of suitably prepared test bodies which allow for the measurement of the curvature and, thereby, the gravitational field.

The starting point for setting up a compass is the standard geodesic deviation equation, which can be obtained from the generalized deviation equation (1.90) by considering only the first order and non-accelerated curves, resulting in

$$\frac{D^2}{ds^2}\eta^a = R^a{}_{bcd}u^b\eta^c u^d. \tag{5.1}$$

The goal is to express the curvature in terms of measured quantities, i.e. the velocities and the forces (accelerations) of the test bodies in the compass configuration. Hence, we rewrite (5.1) in terms of the standard (non-covariant) derivative w.r.t. the proper time and employ normal coordinates, i.e. we have on the world line of the reference test body

$$\Gamma_{ab}{}^c|_Y = 0, \qquad \partial_a\Gamma_{bc}{}^d|_Y = \frac{2}{3}R_{a(bc)}{}^d. \tag{5.2}$$

In terms of the standard total derivative w.r.t. to the proper time s, the deviation equation (5.1) takes the form:

$$\frac{d^2}{ds^2}\eta_a \overset{|_Y}{=} \frac{4}{3}R_{abcd}u^b\eta^c u^d. \tag{5.3}$$

The observer at the position of the reference test body will use 3 additional test bodies at locations:

$$^{(1)}\eta^a = \begin{pmatrix} 0 \\ 1 \\ 0 \\ 0 \end{pmatrix}, \; ^{(2)}\eta^a = \begin{pmatrix} 0 \\ 0 \\ 1 \\ 0 \end{pmatrix}, \; ^{(3)}\eta^a = \begin{pmatrix} 0 \\ 0 \\ 0 \\ 1 \end{pmatrix}. \tag{5.4}$$

In addition to the positions of the compass constituents, one also has to make a choice for their relative velocities, i.e.

$$^{(1)}u^a = \begin{pmatrix} c_{10} \\ 0 \\ 0 \\ 0 \end{pmatrix}, \; ^{(2)}u^a = \begin{pmatrix} c_{20} \\ c_{21} \\ 0 \\ 0 \end{pmatrix}, \; ^{(3)}u^a = \begin{pmatrix} c_{30} \\ 0 \\ c_{32} \\ 0 \end{pmatrix},$$

$$^{(4)}u^a = \begin{pmatrix} c_{40} \\ 0 \\ 0 \\ c_{43} \end{pmatrix}, \; ^{(5)}u^a = \begin{pmatrix} c_{50} \\ c_{51} \\ c_{52} \\ 0 \end{pmatrix}, \; ^{(6)}u^a = \begin{pmatrix} c_{60} \\ 0 \\ c_{62} \\ c_{63} \end{pmatrix}. \tag{5.5}$$

Here $c_{(m)a}$ are just constants, chosen appropriately to ensure the normalization of the 4-velocity of each compass.

For (n) bodies at locations $^{(n)}\eta^a$ relative to the reference body, moving with relative (m) velocities $^{(m)}u^a$ one ends up with the system

$$^{(m,n)}A_a \overset{\text{ly}}{=} \frac{4}{3}R_{abcd}{}^{(m)}u^b{}^{(n)}\eta^c{}^{(m)}u^d. \tag{5.6}$$

Here the $^{(m,n)}A_a$ denote the measured accelerations relative to the reference point Y. Physically, these A's correspond to the springs in the mechanical compass picture of Szekeres [6].

General Solution

As was shown in [5], the 20 independent components of the curvature tensor can be explicitly determined in terms of the accelerations $^{(m,n)}A_a$ and velocities $^{(m)}u^a$. The algebraic system (5.6) allows to express the Riemann curvature tensor components as follows:

$$01 : R_{1010} = \frac{3}{4}{}^{(1,1)}A_1 c_{10}^{-2}, \tag{5.7}$$

$$02 : R_{2010} = \frac{3}{4}{}^{(1,1)}A_2 c_{10}^{-2}, \tag{5.8}$$

$$03 : R_{3010} = \frac{3}{4}{}^{(1,1)}A_3 c_{10}^{-2}, \tag{5.9}$$

$$04 : R_{2020} = \frac{3}{4}{}^{(1,2)}A_2 c_{10}^{-2}, \tag{5.10}$$

$$05 : R_{3020} = \frac{3}{4}{}^{(1,2)}A_3 c_{10}^{-2}, \tag{5.11}$$

$$06 : R_{3030} = \frac{3}{4}{}^{(1,3)}A_3 c_{10}^{-2}, \tag{5.12}$$

$$07 : R_{2110} = \frac{3}{4}{}^{(2,1)}A_2 c_{21}^{-1} c_{20}^{-1} - R_{2010} c_{21}^{-1} c_{20}, \tag{5.13}$$

$$08 : R_{3110} = \frac{3}{4}{}^{(2,1)}A_3 c_{21}^{-1} c_{20}^{-1} - R_{3010} c_{21}^{-1} c_{20}, \tag{5.14}$$

$$09 : R_{0212} = \frac{3}{4}{}^{(3,1)}A_0 c_{32}^{-2} + R_{2010} c_{32}^{-1} c_{30}, \tag{5.15}$$

$$10 : R_{1212} = \frac{3}{4}{}^{(2,2)}A_2 c_{21}^{-2} - R_{2020} c_{20}^{2} c_{21}^{-2} - 2 R_{0212} c_{21}^{-1} c_{20}, \tag{5.16}$$

$$11 : R_{3220} = \frac{3}{4}{}^{(3,2)}A_3 c_{32}^{-1} c_{30}^{-1} - R_{3020} c_{32}^{-1} c_{30}, \tag{5.17}$$

$$12 : R_{0313} = \frac{3}{4} {}^{(4,1)}A_0 c_{43}^{-2} + R_{3010} c_{43}^{-1} c_{40}, \tag{5.18}$$

$$13 : R_{1313} = \frac{3}{4} {}^{(2,3)}A_3 c_{21}^{-2} - R_{3030} c_{20}^2 c_{21}^{-2} - 2 R_{0313} c_{21}^{-1} c_{20}, \tag{5.19}$$

$$14 : R_{0323} = \frac{3}{4} {}^{(4,2)}A_0 c_{43}^{-2} + R_{3020} c_{43}^{-1} c_{40}, \tag{5.20}$$

$$15 : R_{2323} = \frac{3}{4} {}^{(4,2)}A_2 c_{43}^{-2} - R_{2020} c_{43}^{-2} c_{40}^2 + 2 R_{3220} c_{43}^{-1} c_{40}, \tag{5.21}$$

$$16 : R_{3132} = \frac{3}{8} {}^{(5,3)}A_3 c_{52}^{-1} c_{51}^{-1} - \frac{1}{2} R_{3030} c_{52}^{-1} c_{51}^{-1} c_{50}^2$$
$$- R_{0313} c_{52}^{-1} c_{50} - R_{0323} c_{51}^{-1} c_{50} - \frac{1}{2} R_{1313} c_{52}^{-1} c_{51}$$
$$- \frac{1}{2} R_{2323} c_{52} c_{51}^{-1}, \tag{5.22}$$

$$17 : R_{1213} = \frac{3}{8} {}^{(6,1)}A_1 c_{63}^{-1} c_{62}^{-1} - \frac{1}{2} R_{1010} c_{63}^{-1} c_{62}^{-1} c_{60}^2$$
$$+ R_{2110} c_{63}^{-1} c_{60} + R_{3110} c_{62}^{-1} c_{60} - \frac{1}{2} R_{1212} c_{63}^{-1} c_{62}$$
$$- \frac{1}{2} R_{1313} c_{63} c_{62}^{-1}, \tag{5.23}$$

$$18 : R_{0231} = \frac{1}{4} {}^{(4,1)}A_2 c_{40}^{-1} c_{43}^{-1} - \frac{1}{4} {}^{(2,2)}A_3 c_{20}^{-1} c_{21}^{-1}$$
$$+ \frac{1}{3} \big(R_{3020} c_{20} c_{21}^{-1} + R_{3121} c_{21} c_{20}^{-1}$$
$$- R_{2010} c_{40} c_{43}^{-1} - R_{2313} c_{43} c_{40}^{-1} \big), \tag{5.24}$$

$$19 : R_{0312} = \frac{1}{4} {}^{(4,1)}A_2 c_{40}^{-1} c_{42}^{-1} + \frac{1}{2} {}^{(2,2)}A_3 c_{20}^{-1} c_{21}^{-1}$$
$$- \frac{1}{3} \big(2 R_{3020} c_{20} c_{21}^{-1} + 2 R_{3121} c_{21} c_{20}^{-1}$$
$$+ R_{2010} c_{40} c_{43}^{-1} + R_{2313} c_{43} c_{40}^{-1} \big), \tag{5.25}$$

$$20 : R_{3212} = \frac{3}{4} {}^{(4,1)}A_3 c_{20}^{-1} c_{21}^{-1} c_{50} c_{52}^{-1} - \frac{3}{4} {}^{(5,2)}A_3 c_{51}^{-1} c_{52}^{-1}$$
$$+ R_{3121} c_{52}^{-1} \big(c_{51} - c_{50} c_{21} c_{20}^{-1} \big) + R_{3220} c_{50} c_{51}^{-1}$$
$$+ R_{3020} c_{50} c_{52}^{-1} \big(c_{50} c_{51}^{-1} - c_{20} c_{21}^{-1} \big). \tag{5.26}$$

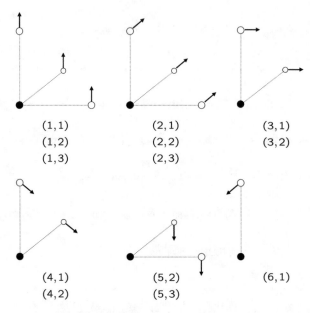

Fig. 5.2 Symbolical sketch of the explicit compass solution in (5.7)–(5.26). In total 13 suitably prepared test bodies (hollow circles) are needed to determine all curvature components. The reference body is denoted by the black circle

This implies that 13 test bodies are needed to measure the gravitational field completely. In Fig. 5.2 a graphical form of the solution in (5.7)-(5.26) is given.

Vacuum Solution

In vacuum the number of independent components of the curvature is reduced to the 10 components of the Weyl tensor C_{abcd}. Replacing R_{abcd} in the compass solution (5.7)-(5.26), and taking into account the symmetries of Weyl (in particular the double-self-duality property $C_{abcd} = -\frac{1}{4}\epsilon_{abef}\epsilon_{cdgh}C^{efgh}$, where ϵ_{abcd} is the totally antisymmetric Levi-Civita tensor with $\epsilon_{0123} = 1$), the gravitational field may be explicitly expressed as follows:

$$01 : C_{1010} = \frac{3}{4}{}^{(1,1)}A_1 c_{10}^{-2}, \tag{5.27}$$

$$02 : C_{2010} = \frac{3}{4}{}^{(1,1)}A_2 c_{10}^{-2}, \tag{5.28}$$

$$03 : C_{3010} = \frac{3}{4}{}^{(1,1)}A_3 c_{10}^{-2}, \tag{5.29}$$

$$04 : C_{2020} = \frac{3}{4} {}^{(1,2)} A_2 c_{10}^{-2}, \tag{5.30}$$

$$05 : C_{3020} = \frac{3}{4} {}^{(1,2)} A_3 c_{10}^{-2}, \tag{5.31}$$

$$06 : C_{2110} = \frac{3}{4} {}^{(2,1)} A_2 c_{21}^{-1} c_{20}^{-1} - C_{2010} c_{21}^{-1} c_{20}, \tag{5.32}$$

$$07 : C_{3110} = \frac{3}{4} {}^{(2,1)} A_3 c_{21}^{-1} c_{20}^{-1} - C_{3010} c_{21}^{-1} c_{20}, \tag{5.33}$$

$$08 : C_{0212} = \frac{3}{4} {}^{(3,1)} A_0 c_{32}^{-2} + C_{2010} c_{32}^{-1} c_{30}, \tag{5.34}$$

$$09 : C_{0231} = \frac{1}{4} {}^{(4,1)} A_2 c_{40}^{-1} c_{43}^{-1} - \frac{1}{4} {}^{(2,2)} A_3 c_{20}^{-1} c_{21}^{-1}$$

$$+ \frac{1}{3} C_{3020} \left(c_{20} c_{21}^{-1} + c_{21} c_{20}^{-1} \right)$$

$$- \frac{1}{3} C_{2010} \left(c_{40} c_{43}^{-1} + c_{43} c_{40}^{-1} \right), \tag{5.35}$$

$$10 : C_{0312} = \frac{1}{4} {}^{(4,1)} A_2 c_{40}^{-1} c_{42}^{-1} + \frac{1}{2} {}^{(2,2)} A_3 c_{20}^{-1} c_{21}^{-1}$$

$$- \frac{2}{3} C_{3020} \left(c_{20} c_{21}^{-1} + c_{21} c_{20}^{-1} \right)$$

$$+ \frac{1}{3} C_{2010} \left(c_{40} c_{43}^{-1} + c_{43} c_{40}^{-1} \right). \tag{5.36}$$

This implies that 6 test bodies are required to measure the gravitational field in vacuum. See Fig. 5.3 for a graphical representation of the solution.

Summary

We have reviewed how the standard geodesic deviation equation can be used to determine the curvature of space-time, and the solutions given here can be viewed as an explicit realization of Szekeres' gravitational compass [6] and of Synge's curvature detector [7]. With the standard geodesic deviation equation one needs at least 13 test bodies to determine all curvature components in a general spacetime, and 6 test bodies in vacuum.

It is interesting to note that the use of generalized deviation equations for the curvature determination has been discussed in the literature. Depending on the underlying generalization a reduction of the number of required test bodies has been reported [5, 8, 9].

Fig. 5.3 Symbolical sketch
of the explicit compass
solution in (5.27)–(5.36) for
the vacuum case. In total 6
suitably prepared test bodies
(hollow circles) are needed to
determine all components of
the Weyl tensor. The
reference body is denoted by
the black circle

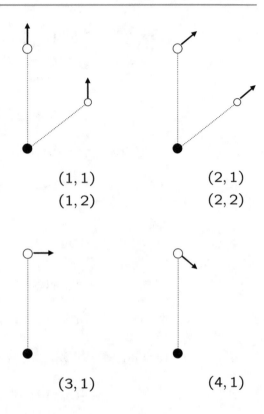

$$(1, 1)$$
$$(1, 2)$$

$$(2, 1)$$
$$(2, 2)$$

$$(3, 1)$$

$$(4, 1)$$

5.3 Determination of the Gravitational Field by Means of Clocks

Another method for the determination of the gravitational field relies on the mutual
frequency comparison of an ensemble of clocks. Clock based methods for the
field determination are particularly attractive due to their unprecedented level of
accuracy and stability [10–16] in recent years. Following [17, 18] we call such
an experimental setup a *gravitational clock compass*, in analogy to the usual
gravitational compass. A sketch of the procedure is depicted in Fig. 5.4.

We first show how the frequency ratio of two clocks moving on two general
curves within an arbitrary space-time manifold can be derived. Such a derivation
is naturally based on a suitable choice of coordinates, and there have been several
suggestions for the construction and realization of coordinates in the literature [7,
17, 19–44].

Here we follow the construction from [45], which was motivated by earlier work
on radiation from isolated systems [46] and on work on the equations of motion in
general relativity [47, 48]. It offers a different perspective on the derivation of the
measurable frequency ratio between the clocks and is not, like [17], based on [49]
as a starting point.

Fig. 5.4 Sketch of the operational procedure to measure the curvature of spacetime by means of clocks. An observer with a clock moving along a world line Y compares his clock readings C to a set of suitably prepared clocks in the vicinity of Y. The number of clocks required for the determination of all curvature components depends on the type of the underlying spacetime

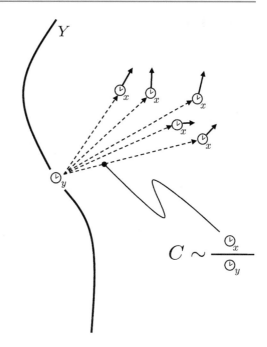

Frequency Ratio in Flat Space-Time

Here we review the frequency ratio between a clock carried by the observer and a clock in the vicinity of his world line is given in a flat background. Readers should also consult Appendix C.1, in which a suitable form of the flat space-time metric in the neighbourhood of a time-like world line, in a reference frame carried by a general observer, is constructed.

With $X^i = (X, Y, Z, T)$ and $x^i = (x, y, r, u)$ the parametric equations of an arbitrary time-like world line, with arc length s as parameter along it, are

$$X^i = X^i(s) \quad \Leftrightarrow \quad x^i = x^i(s) \,, \tag{5.37}$$

with

$$\eta_{ij} \frac{dX^i}{ds} \frac{dX^j}{ds} = +1 \,. \tag{5.38}$$

Using (C.27) and writing

$$\frac{dx^i}{ds} = \frac{dx^i}{du} \frac{du}{ds} \,, \tag{5.39}$$

in effect changing from the parameter s to the parameter u along the arbitrary world line, we can write

$$\frac{dX^i}{ds} = \frac{du}{ds} \left\{ [1 + (\mathbf{a} \cdot \mathbf{p}) r] v^i + W^{(\alpha)} \lambda^i_{(\alpha)} \right\} , \tag{5.40}$$

with

$$W^{(\alpha)} = p^{(\alpha)} \frac{dr}{du} + r \frac{\partial p^{(\alpha)}}{\partial x} \frac{dx}{du} + r \frac{\partial p^{(\alpha)}}{\partial y} \frac{dy}{du} + r \, \omega^{(\alpha)(\beta)} \, p_{(\beta)} . \tag{5.41}$$

Substituting (5.40) into (5.38) we obtain

$$\left(\frac{ds}{du} \right)^2 = (1 + (\mathbf{a} \cdot \mathbf{p}) r)^2 - |\mathbf{W}|^2 , \tag{5.42}$$

with

$$\mathbf{W} = \mathbf{u} + r (\boldsymbol{\omega} \times \mathbf{p}) , \tag{5.43}$$

and

$$\mathbf{u} = \frac{dr}{du} \mathbf{p} + r \frac{dx}{du} \frac{\partial \mathbf{p}}{\partial x} + r \frac{dy}{du} \frac{\partial \mathbf{p}}{\partial y} . \tag{5.44}$$

We see that, since u is proper-time along the world line $r = 0$, $\mathbf{u}$ is the 3-velocity of the observer with world line (5.37) relative to the observer with world line $r = 0$. Hence (5.42) can finally be written

$$\left(\frac{ds}{du} \right)^2 = 1 - |\mathbf{u}|^2 + 2 \{ \mathbf{a} \cdot \mathbf{p} - \mathbf{u} \cdot (\boldsymbol{\omega} \times \mathbf{p}) \} r$$

$$+ \{ (\mathbf{a} \cdot \mathbf{p})^2 - |\boldsymbol{\omega} \times \mathbf{p}|^2 \} r^2 . \tag{5.45}$$

This formula is, of course, exact (in particular it does not have any restriction on r). Equation (5.42), or equivalently the equation (5.45), can be compared directly to the results in [49] and with eq. (22) in [17].

Frequency Ratio in Curved Space-Time

We now consider a general curved space-time, guided by Eqs. (C.42) and (C.45) we choose a family of time-like hypersurfaces $r(x^i) = $ constant in this space-time with unit space-like normal

$$p^i = \frac{dx^i}{dr} = -g^{ij} r_{,j} \quad \text{and} \quad g_{ij} \, p^i \, p^j = -1 . \tag{5.46}$$

Taking $x^3 = r$ as a coordinate and labelling the remaining coordinates $x^i = (x, y, r, u)$ we have from (5.46)

$$g_{ij} \frac{dx^j}{dr} = -r_{,i} \quad \Leftrightarrow \quad g_{i3} = -\delta_i^3 . \tag{5.47}$$

Hence four components of the metric tensor of the space-time are fixed. Using straightforward algebra the remaining six components can be expressed in terms of six functions $P, \alpha, \beta, a, b, c$ of the four coordinates x, y, r, u in such a way that the line element of the space-time is given by

$$ds^2 = -(\vartheta^{(1)})^2 - (\vartheta^{(2)})^2 - (\vartheta^{(3)})^2 + (\vartheta^{(4)})^2 , \tag{5.48}$$

with

$$\vartheta^{(1)} = -\vartheta_{(1)i} \, dx^i$$
$$= r \, P^{-1}(e^\alpha \cosh \beta \, dx + e^{-\alpha} \sinh \beta \, dy + a \, du), \tag{5.49}$$
$$\vartheta^{(2)} = -\vartheta_{(2)i} \, dx^i$$
$$= r \, P^{-1}(e^\alpha \sinh \beta \, dx + e^{-\alpha} \cosh \beta \, dy + b \, du), \tag{5.50}$$
$$\vartheta^{(3)} = -\vartheta_{(3)i} \, dx^i = dr, \tag{5.51}$$
$$\vartheta^{(4)} = \vartheta_{(4)i} \, dx^i = c \, du. \tag{5.52}$$

We take $r = 0$ in this space-time to be a time-like world line with u as proper time along it. Then in the neighbourhood of $r = 0$ (i.e. for small values of r) we expand the functions of x, y, r, u in (5.49)–(5.52) in positive powers of r, with coefficients functions of x, y, u, in such a way that the line element (5.48) is a perturbation of the Minkowskian line element (C.30). Clearly this involves taking

$$P = P_0 + O(r) , \quad \alpha = O(r) , \quad \beta = O(r) , \quad a = a_0 + O(r) ,$$
$$b = b_0 + O(r) , \quad c = 1 - h_0 r + O(r^2) , \tag{5.53}$$

but we need to know the leading powers of r in the $O(r)$-terms here. To find these we make use of (C.57)–(C.60). With p^i given by (5.46), and using (5.49)–(5.52), and denoting by a semicolon covariant differentiation with respect to the Riemannian connection associated with the metric tensor g_{ij} given by the line element (5.48) we start by recording that

$$\vartheta^i_{(1)} = r^{-1} P \, (e^{-\alpha} \cosh \beta, -e^\alpha \sinh \beta, 0, 0) , \tag{5.54}$$
$$\vartheta^i_{(2)} = r^{-1} P \, (-e^{-\alpha} \sinh \beta, e^\alpha \cosh \beta, 0, 0) , \tag{5.55}$$
$$\vartheta^i_{(3)} = (0, 0, 1, 0) , \tag{5.56}$$

$$\vartheta^i_{(4)} = c^{-1}\Big(- e^{-\alpha}\{a \cosh \beta - b \sinh \beta\},$$

$$e^{\alpha}\{a \sinh \beta - b \cosh \beta\}, 0, 1\Big), \tag{5.57}$$

and thus

$$P_{i;j}\,\vartheta^i_{(1)}\,\vartheta^j_{(1)} - P_{i;j}\,\vartheta^i_{(2)}\,\vartheta^j_{(2)} = -2\,\frac{\partial \alpha}{\partial r}\,\cosh 2\beta\,, \tag{5.58}$$

$$P_{i;j}\,\vartheta^i_{(1)}\,\vartheta^j_{(2)} = -\frac{\partial \beta}{\partial r}\,, \tag{5.59}$$

$$P_{i;j}\,\vartheta^i_{(1)}\,\vartheta^j_{(1)} + P_{i;j}\,\vartheta^i_{(2)}\,\vartheta^j_{(2)} = -2\,\frac{\partial}{\partial r}\,\log(r\,P^{-1})\,, \tag{5.60}$$

$$P_{i;j}\,\vartheta^i_{(4)}\,\vartheta^j_{(4)} = \frac{\partial}{\partial r}\,\log c\,, \tag{5.61}$$

$$P_{i;j}\,\vartheta^i_{(1)}\,\vartheta^j_{(4)} = \frac{1}{2}r\,(P\,c)^{-1}\Big\{-\frac{\partial a}{\partial r} + (a \cosh 2\beta$$

$$-b \sinh 2\beta)\frac{\partial \alpha}{\partial r} + b\,\frac{\partial \beta}{\partial r}\Big\}\,, \tag{5.62}$$

$$P_{i;j}\,\vartheta^i_{(2)}\,\vartheta^j_{(4)} = \frac{1}{2}r\,(P\,c)^{-1}\Big\{-\frac{\partial b}{\partial r} + (a \sinh 2\beta$$

$$-b \cosh 2\beta)\frac{\partial \alpha}{\partial r} + a\,\frac{\partial \beta}{\partial r}\Big\}\,. \tag{5.63}$$

From (C.57) perturbed for small values of r we require the left hand sides of (5.58) and (5.59) to be small of order r and this is achieved with

$$\alpha = \alpha_2(x, y, u)\,r^2 + O(r^3)\,,$$

$$\beta = \beta_2(x, y, u)\,r^2 + O(r^3)\,. \tag{5.64}$$

From (C.57) we require the left hand side of (5.60) to have the form $-2\,r^{-1} + O(r)$ and this is achieved with

$$P = P_0\{1 + q_2(x, y, u)\,r^2 + O(r^3)\}\,. \tag{5.65}$$

Next from (C.59) the left hand side of (5.61) should have the form $-h_0 + O(r)$ and this occurs if

$$c = 1 - h_0\,r + c_2(x, y, u)\,r^2 + O(r^3)\,. \tag{5.66}$$

Finally from (C.60) with $a = a_0 + O(r)$ and $b = b_0 + O(r)$ we now have from (5.62) and (5.63):

$$p_{i;j}\,\vartheta^i_{(1)}\,\vartheta^j_{(4)} = -\frac{1}{2}r\,P_0^{-1}\frac{\partial a}{\partial r} + O(r^2)\,, \tag{5.67}$$

$$p_{i;j}\,\vartheta^i_{(2)}\,\vartheta^j_{(4)} = -\frac{1}{2}r\,P_0^{-1}\frac{\partial b}{\partial r} + O(r^2)\,, \tag{5.68}$$

and to have the right hand sides of these $O(r)$ we take

$$a = a_0 + a_1(x, y, u)\,r + O(r^2)\,,$$
$$b = b_0 + b_1(x, y, u)\,r + O(r^2)\,. \tag{5.69}$$

The components of the Riemann curvature tensor calculated on the world line $r = 0$, and expressed on the orthonormal tetrad $\lambda^i_{(a)}$ with $a = 1, 2, 3, 4$ defined by (C.7)–(C.11), are denoted

$$R_{(a)(b)(c)(d)} = R_{ijkl}\,\lambda^i_{(a)}\,\lambda^j_{(b)}\,\lambda^k_{(c)}\,\lambda^l_{(d)}. \tag{5.70}$$

Calculating the Riemann tensor of the space-time evaluated on $r = 0$ allows us to determine the functions $\alpha_2, \beta_2, q_2, c_2, a_1, b_1$ appearing in (5.64), (5.65), (5.66) and (5.69) in terms of the tetrad components (5.70). We find the following expressions for these functions of x, y, u:

$$\alpha_2 = \frac{1}{6}P_0^2\,R_{(\alpha)(\beta)(\gamma)(\sigma)}\,\frac{\partial p^{(\alpha)}}{\partial x}\,p^{(\beta)}\,\frac{\partial p^{(\gamma)}}{\partial x}\,p^{(\sigma)}$$
$$-\frac{1}{12}R_{(\alpha)(\beta)(\alpha)(\sigma)}\,p^{(\beta)}\,p^{(\sigma)}\,,$$

$$= -\frac{1}{6}P_0^2\,R_{(\alpha)(\beta)(\gamma)(\sigma)}\,\frac{\partial p^{(\alpha)}}{\partial y}\,p^{(\beta)}\,\frac{\partial p^{(\gamma)}}{\partial y}\,p^{(\sigma)}$$
$$+\frac{1}{12}R_{(\alpha)(\beta)(\alpha)(\sigma)}\,p^{(\beta)}\,p^{(\sigma)}\,, \tag{5.71}$$

with the second equality following from the use of (C.23),

$$\beta_2 = \frac{1}{6}P_0^2\,R_{(\alpha)(\beta)(\gamma)(\sigma)}\,\frac{\partial p^{(\alpha)}}{\partial x}\,p^{(\beta)}\,\frac{\partial p^{(\gamma)}}{\partial y}\,p^{(\sigma)}\,, \tag{5.72}$$

$$q_2 = -\frac{1}{12}R_{(\alpha)(\beta)(\alpha)(\sigma)}\,p^{(\beta)}\,p^{(\sigma)}\,, \tag{5.73}$$

$$c_2 = -\frac{1}{2}R_{(\alpha)(4)(\beta)(4)}\,p^{(\alpha)}\,p^{(\beta)}\,, \tag{5.74}$$

$$a_1 = -\frac{2}{3}\, P_0^2\, R_{(\alpha)(4)(\beta)(\gamma)}\, p^{(\alpha)}\, \frac{\partial p^{(\beta)}}{\partial x}\, p^{(\gamma)}\,, \tag{5.75}$$

$$b_1 = -\frac{2}{3}\, P_0^2\, R_{(\alpha)(4)(\beta)(\gamma)}\, p^{(\alpha)}\, \frac{\partial p^{(\beta)}}{\partial y}\, p^{(\gamma)}\,. \tag{5.76}$$

When the functions are substituted into the 1-forms (5.49)–(5.52) the line element (5.48) is given, in the coordinates $x^i = (x, y, r, u)$ with $i = 1, 2, 3, 4$ as $ds^2 = g_{ij}\, dx^i\, dx^j$ with

$$g_{11} = -r^2 P_0^{-2}\{1 + 2\,(\alpha_2 - q_2)\, r^2 + O(r^3)\}\,, \tag{5.77}$$

$$g_{22} = -r^2 P_0^{-2}\{1 - 2\,(\alpha_2 + q_2)\, r^2 + O(r^3)\}\,, \tag{5.78}$$

$$g_{12} = -r^2 P_0^{-2}\{2\,\beta_2\, r^2 + O(r^3)\}\,, \tag{5.79}$$

$$g_{33} = -1\ \ (g_{31} = g_{32} = g_{34} = 0)\,, \tag{5.80}$$

$$g_{14} = -r^2\, P_0^{-2}\{a_0 + a_1\, r + O(r^2)\}\,, \tag{5.81}$$

$$g_{24} = -r^2\, P_0^{-2}\{b_0 + b_1\, r + O(r^2)\}\,, \tag{5.82}$$

$$g_{44} = \{1 + (\mathbf{a} \cdot \mathbf{p})\, r\}^2 - |\boldsymbol{\omega} \times \mathbf{p}|^2 r^2$$
$$+2\, c_2\, r^2 + O(r^3)\,, \tag{5.83}$$

with P_0 given by (C.17).

If $x^i = x^i(s)$ with $x^i = (x, y, r, u)$ is an arbitrary time-like world line in the neighbourhood of $r = 0$ with s proper time along it then, for small values of r and using the line element (5.48) the formula (5.45) is modified to read

$$\left(\frac{ds}{du}\right)^2 = 1 - |\mathbf{u}|^2 + 2\,\{\mathbf{a} \cdot \mathbf{p} - \mathbf{u} \cdot (\boldsymbol{\omega} \times \mathbf{p})\}\, r + \{(\mathbf{a} \cdot \mathbf{p})^2 - |\boldsymbol{\omega} \times \mathbf{p}|^2\}\, r^2$$

$$+2\, c_2\, r^2 - 2\, a_1\, r^3\, P_0^{-2}\, \frac{dx}{du} - 2\, b_1\, r^3\, P_0^{-2}\, \frac{dy}{du} - 4\, \beta_2\, r^4 P_0^{-2}\, \frac{dx}{du}\frac{dy}{du}$$

$$-2\,(\alpha_2 - q_2)\, r^4 P_0^{-2}\left(\frac{dx}{du}\right)^2 + 2\,(\alpha_2 + q_2)\, r^4 P_0^{-2}\left(\frac{dy}{du}\right)^2 + O(r^3)\,. \tag{5.84}$$

Here c_2 is given by (5.74). Using (5.75) and (5.76) we have

$$a_1\, r^3 P_0^{-2}\frac{dx}{du} + b_1\, r^3 P_0^{-2}\frac{dy}{du} = -\frac{2}{3}\, r^2 R_{(\alpha)(4)(\beta)(\gamma)}\, p^{(\alpha)}\, u^{(\beta)}\, p^{(\gamma)}\,, \tag{5.85}$$

using (5.44). Next using (5.71)–(5.73) and (5.44) again we have

$$-4\,\beta_2\,r^4 P_0^{-2}\frac{dx}{du}\frac{dy}{du} - 2\,(\alpha_2 - q_2)\,r^4 P_0^{-2}\left(\frac{dx}{du}\right)^2 + 2\,(\alpha_2 + q_2)r^4 P_0^{-2}\left(\frac{dy}{du}\right)^2$$

$$= -\frac{1}{3}r^2 R_{(\alpha)(\beta)(\gamma)(\sigma)}\,u^{(\alpha)}\,p^{(\beta)}\,u^{(\gamma)}\,p^{(\sigma)}\,. \tag{5.86}$$

Substituting (5.85) and (5.86) into (5.84) results in

$$\left(\frac{ds}{du}\right)^2 = 1 - |\mathbf{u}|^2 + 2\{(\mathbf{a}\cdot\mathbf{p}) - \mathbf{u}\cdot(\boldsymbol{\omega}\times\mathbf{p})\}\,r + \Big\{(\mathbf{a}\cdot\mathbf{p})^2 - |\boldsymbol{\omega}\times\mathbf{p}|^2$$

$$- R_{(\alpha)(4)(\beta)(4)}\,p^{(\alpha)}\,p^{(\beta)} + \frac{4}{3}R_{(\alpha)(4)(\beta)(\gamma)}\,p^{(\alpha)}\,u^{(\beta)}\,p^{(\gamma)}$$

$$- \frac{1}{3}R_{(\alpha)(\beta)(\gamma)(\sigma)}\,u^{(\alpha)}\,p^{(\beta)}\,u^{(\gamma)}\,p^{(\sigma)}\Big\}\,r^2 + O(r^3)\,. \tag{5.87}$$

5.4 Gravitational Clock Compass

Now we turn to the determination of the curvature in a general space-time by means of a clock compass. Here we consider non-accelerated and non-rotating configurations, i.e. we consider a rearranged version of the system (5.87) in which the dependence on the acceleration and the rotation is assumed to be known. In analogy to the analysis of the gravitational compass in Sect. 5.2 we now have:

$$B(r, p^\alpha, u^\alpha, a^\alpha, \omega^{\alpha\beta}) = R_{(\alpha)(4)(\beta)(4)}\,p^{(\alpha)}\,p^{(\beta)} - \frac{4}{3}R_{(\alpha)(4)(\beta)(\gamma)}\,p^{(\alpha)}\,u^{(\beta)}\,p^{(\gamma)}$$

$$+ \frac{1}{3}R_{(\alpha)(\beta)(\gamma)(\sigma)}\,u^{(\alpha)}\,p^{(\beta)}\,u^{(\gamma)}\,p^{(\sigma)}, \tag{5.88}$$

where

$$B(r, p^\alpha, u^\alpha, a^\alpha, \omega^{\alpha\beta}) := (\mathbf{a}\cdot\mathbf{p})^2 - |\boldsymbol{\omega}\times\mathbf{p}|^2 + \frac{2}{r}\Big\{(\mathbf{a}\cdot\mathbf{p}) - \mathbf{u}\cdot(\boldsymbol{\omega}\times\mathbf{p})\Big\}$$

$$+ \frac{1}{r^2}\Big(1 - C - |\mathbf{u}|^2\Big). \tag{5.89}$$

Here we introduced C for the frequency ratio.

General Solution

Considering the non-vacuum case first one needs to measure 20 independent components of the Riemann curvature tensor R_{abcd}. Analogously to the analysis of the gravitational compass in Sect. 5.2, we may now consider different setups of clocks to measure as many curvature components as possible. Introducing different initial values for the clocks:

$$^{(1)}p^\alpha = \begin{pmatrix} 1 \\ 0 \\ 0 \end{pmatrix}, \,^{(2)}p^\alpha = \begin{pmatrix} 0 \\ 1 \\ 0 \end{pmatrix}, \,^{(3)}p^\alpha = \begin{pmatrix} 0 \\ 0 \\ 1 \end{pmatrix},$$

$$^{(4)}p^\alpha = \begin{pmatrix} 1 \\ 1 \\ 0 \end{pmatrix}, \,^{(5)}p^\alpha = \begin{pmatrix} 0 \\ 1 \\ 1 \end{pmatrix}, \,^{(6)}p^\alpha = \begin{pmatrix} 1 \\ 0 \\ 1 \end{pmatrix},$$

(5.90)

and

$$^{(1)}u^\alpha = \begin{pmatrix} c_{11} \\ 0 \\ 0 \end{pmatrix}, \,^{(2)}u^\alpha = \begin{pmatrix} 0 \\ c_{22} \\ 0 \end{pmatrix}, \,^{(3)}u^\alpha = \begin{pmatrix} 0 \\ 0 \\ c_{33} \end{pmatrix},$$

$$^{(4)}u^\alpha = \begin{pmatrix} c_{41} \\ c_{42} \\ 0 \end{pmatrix}, \,^{(5)}u^\alpha = \begin{pmatrix} 0 \\ c_{52} \\ c_{53} \end{pmatrix}, \,^{(6)}u^\alpha = \begin{pmatrix} c_{61} \\ 0 \\ c_{63} \end{pmatrix},$$

(5.91)

we have again an algebraic system. It was shown in [17], that this system can be used to determine all gravitational field components as follows:

$$01: R_{(1)(0)(1)(0)} = \,^{(1,1)}B,$$

(5.92)

$$02: R_{(2)(1)(1)(0)} = \frac{3}{4}c_{22}^{-1}c_{42}^{-1}(c_{22} - c_{42})^{-1}\left(^{(1,1)}Bc_{22}^2 - \,^{(1,1)}Bc_{42}^2 \right.$$

$$\left. +^{(1,2)}Bc_{42}^2 - \,^{(1,4)}Bc_{22}^2\right),$$

(5.93)

$$03: R_{(1)(2)(1)(2)} = -3c_{22}^{-1}c_{42}^{-1}(c_{22} - c_{42})^{-1}\left(^{(1,1)}Bc_{22} - \,^{(1,1)}Bc_{42} \right.$$

$$\left. +^{(1,2)}Bc_{42} - \,^{(1,4)}Bc_{22}\right),$$

(5.94)

$$04: R_{(3)(1)(1)(0)} = \frac{3}{4}c_{33}^{-1}c_{63}^{-1}\left(c_{33}-c_{63}\right)^{-1}\left({}^{(1,1)}Bc_{33}^2 - {}^{(1,1)}Bc_{63}^2\right.$$

$$\left. + {}^{(1,3)}Bc_{63}^2 - {}^{(1,6)}Bc_{33}^2\right), \tag{5.95}$$

$$05: R_{(1)(3)(1)(3)} = -3c_{33}^{-1}c_{63}^{-1}\left(c_{33}-c_{63}\right)^{-1}\left({}^{(1,1)}Bc_{33} - {}^{(1,1)}Bc_{63}\right.$$

$$\left. + {}^{(1,3)}Bc_{63} - {}^{(1,6)}Bc_{33}\right), \tag{5.96}$$

$$06: R_{(1)(2)(1)(3)} = \frac{3}{2}c_{52}^{-1}c_{53}^{-1}\left(- {}^{(1,5)}B + R_{(1)(0)(1)(0)} - \frac{4}{3}R_{(2)(1)(1)(0)}c_{52}\right.$$

$$\left. -\frac{4}{3}R_{(3)(1)(1)(0)}c_{53} - \frac{1}{3}R_{(1)(2)(1)(2)}c_{52}^2 - \frac{1}{3}R_{(1)(3)(1)(3)}c_{53}^2\right), \tag{5.97}$$

$$07: R_{(2)(0)(2)(0)} = {}^{(2,2)}B, \tag{5.98}$$

$$08: R_{(0)(2)(1)(2)} = \frac{3}{4}c_{11}^{-1}\left({}^{(2,1)}B - R_{(2)(0)(2)(0)} + \frac{1}{3}R_{(1)(2)(1)(2)}c_{11}^2\right), \tag{5.99}$$

$$09: R_{(3)(2)(2)(0)} = \frac{3}{4}c_{33}^{-1}c_{53}^{-1}\left(c_{33}-c_{53}\right)^{-1}\left({}^{(2,2)}Bc_{33}^2 - {}^{(2,2)}Bc_{53}^2\right.$$

$$\left. + {}^{(2,3)}Bc_{53}^2 - {}^{(2,5)}Bc_{33}^2\right), \tag{5.100}$$

$$10: R_{(2)(3)(2)(3)} = -3c_{33}^{-1}c_{53}^{-1}\left(c_{33}-c_{53}\right)^{-1}\left({}^{(2,2)}Bc_{33} - {}^{(2,2)}Bc_{53}\right.$$

$$\left. + {}^{(2,3)}Bc_{53} - {}^{(5,2)}Bc_{33}\right), \tag{5.101}$$

$$11: R_{(3)(2)(1)(2)} = \frac{3}{2}c_{61}^{-1}c_{63}^{-1}\left(- {}^{(2,6)}B + R_{(2)(0)(2)(0)} + \frac{4}{3}R_{(0)(2)(1)(2)}c_{61}\right.$$

$$\left. -\frac{4}{3}R_{(3)(2)(2)(0)}c_{63} - \frac{1}{3}R_{(1)(2)(1)(2)}c_{61}^2 - \frac{1}{3}R_{(2)(3)(2)(3)}c_{63}^2\right), \tag{5.102}$$

$$12: R_{(3)(0)(3)(0)} = {}^{(3,3)}B, \tag{5.103}$$

$$13: R_{(0)(3)(1)(3)} = \frac{3}{4}c_{11}^{-1}\left({}^{(3,1)}B - R_{(3)(0)(3)(0)} + \frac{1}{3}R_{(1)(3)(1)(3)}c_{11}^2\right), \tag{5.104}$$

$$14: R_{(0)(3)(2)(3)} = \frac{3}{4}c_{22}^{-1}\left({}^{(3,2)}B - R_{(3)(0)(3)(0)} + \frac{1}{3}R_{(2)(3)(2)(3)}c_{22}^2\right), \tag{5.105}$$

$$15: R_{(3)(1)(3)(2)} = \frac{3}{2}c_{41}^{-1}c_{42}^{-1}\left(-{}^{(3,4)}B + R_{(3)(0)(3)(0)} + \frac{4}{3}R_{(0)(3)(1)(3)}c_{41}\right.$$

$$\left. + \frac{4}{3}R_{(0)(3)(2)(3)}c_{42} - \frac{1}{3}R_{(1)(3)(1)(3)}c_{41}^2 - \frac{1}{3}R_{(2)(3)(2)(3)}c_{42}^2\right), \tag{5.106}$$

$$16: R_{(2)(0)(1)(0)} = \frac{1}{2}\left({}^{(4,1)}B - R_{(1)(0)(1)(0)} - R_{(2)(0)(2)(0)} - \frac{4}{3}R_{(0)(2)(1)(2)}c_{11}\right.$$

$$\left. - \frac{4}{3}R_{(2)(1)(1)(0)}c_{11} + \frac{1}{3}R_{(1)(2)(1)(2)}c_{11}^2\right), \tag{5.107}$$

$$17: R_{(3)(0)(2)(0)} = \frac{1}{2}\left({}^{(5,2)}B - R_{(2)(0)(2)(0)} - R_{(3)(0)(3)(0)} - \frac{4}{3}R_{(0)(3)(2)(3)}c_{22}\right.$$

$$\left. - \frac{4}{3}R_{(3)(2)(2)(0)}c_{22} + \frac{1}{3}R_{(2)(3)(2)(3)}c_{22}^2\right), \tag{5.108}$$

$$18: R_{(3)(0)(1)(0)} = \frac{1}{2}\left({}^{(6,1)}B - R_{(1)(0)(1)(0)} - R_{(3)(0)(3)(0)} - \frac{4}{3}R_{(0)(3)(1)(3)}c_{11}\right.$$

$$\left. - \frac{4}{3}R_{(3)(1)(1)(0)}c_{11} + \frac{1}{3}R_{(1)(3)(1)(3)}c_{11}^2\right). \tag{5.109}$$

Introducing abbreviations

$$K_1 := \frac{3}{4}c_{33}^{-1}\left[-{}^{(4,3)}B + R_{(1)(0)(1)(0)} + 2R_{(2)(0)(1)(0)} + R_{(2)(0)(2)(0)}\right.$$

$$-\frac{1}{3}(R_{(1)(3)(1)(3)} + 2R_{(3)(1)(3)(2)} + R_{(2)(3)(2)(3)})c_{33}^2$$

$$\left. -\frac{4}{3}(R_{(3)(1)(1)(0)} + R_{(3)(2)(2)(0)})c_{33}\right], \tag{5.110}$$

$$K_2 := \frac{3}{4}c_{11}^{-1}\left[- {}^{(5,1)}B + R_{(2)(0)(2)(0)} + 2R_{3020} + R_{3030}\right.$$

$$-\frac{1}{3}(R_{(1)(2)(1)(2)} + 2R_{(1)(2)(1)(3)} + R_{(1)(3)(1)(3)})c_{11}^2$$

$$\left. +\frac{4}{3}(R_{(0)(2)(1)(2)} + R_{(0)(3)(1)(3)})c_{11}\right], \tag{5.111}$$

$$K_3 := \frac{3}{4}c_{22}^{-1}\left[- {}^{(6,2)}B + R_{(1)(0)(1)(0)} + 2R_{3010} + R_{(3)(0)(3)(0)}\right.$$

$$-\frac{1}{3}(R_{(1)(2)(1)(2)} + 2R_{(3)(2)(1)(2)} + R_{(2)(3)(2)(3)})c_{22}^2$$

$$\left. -\frac{4}{3}(R_{(2)(1)(1)(0)} + R_{(0)(3)(2)(3)})c_{22}\right], \tag{5.112}$$

the three remaining curvature components can be expressed as:

$$19 : R_{(1)(0)(2)(3)} = \frac{1}{3}(K_3 - K_1), \tag{5.113}$$

$$20 : R_{(2)(0)(1)(3)} = \frac{1}{3}(K_2 - K_1), \tag{5.114}$$

$$21 : R_{(3)(0)(2)(1)} = \frac{1}{3}(K_3 - K_2). \tag{5.115}$$

See Fig. 5.5 for a symbolical sketch of the solution. Note that the sketches of the clock configurations make use of a notation analogous to the one in [17]. The observer is indicated by a black circle, the prepared clocks are indicated by hollow circles. Furthermore, we note that the sketches were introduced in [17] to give a two dimensional visual representation of the solution. In particular they are designed for counting the number of clocks/measurements at a glance, they do not directly represent the three dimensional geometry of the measurement (we order hollow circles, corresponding to different positions (n), starting at the three o'clock position, advancing counter clockwise in 45 degree angles depending on the position index n). Note that we order arrows, corresponding to different velocities (m), starting at the twelve o'clock position, advancing clockwise in 45 degree angles depending on the velocity index m.

Vacuum Solution

In vacuum the number of independent components of the curvature is reduced to the 10 components of the Weyl tensor C_{abcd}. Replacing R_{abcd} in the compass solution

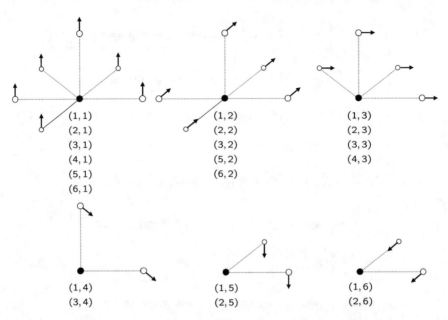

Fig. 5.5 Symbolical sketch of the explicit solution for the curvature (5.92)–(5.115). In total 21 suitably prepared clocks (hollow circles) are needed to determine all curvature components. The observer is denoted by the black circle

(5.92)–(5.109), and taking into account the symmetries of the Weyl tensor, we may use a reduced clock setup to completely determine the gravitational field. All other components may be obtained from the double self-duality property $C_{abcd} = -\frac{1}{4}\varepsilon_{abef}\varepsilon_{cdgh}C^{efgh}$.

$$01 : C_{(2)(3)(2)(3)} = -^{(1,1)}B, \tag{5.116}$$

$$02 : C_{(0)(3)(2)(3)} = \frac{3}{4}c_{22}^{-1}c_{42}^{-1}(c_{22} - c_{42})^{-1}\left(^{(1,1)}Bc_{22}^2 - ^{(1,1)}Bc_{42}^2\right.$$

$$\left. +^{(1,2)}Bc_{42}^2 - ^{(1,4)}Bc_{22}^2\right), \tag{5.117}$$

$$03 : C_{(3)(0)(3)(0)} = 3c_{22}^{-1}c_{42}^{-1}(c_{22} - c_{42})^{-1}\left(^{(1,1)}Bc_{22} - ^{(1,1)}Bc_{42}\right.$$

$$\left. +^{(1,2)}Bc_{42} - ^{(1,4)}Bc_{22}\right), \tag{5.118}$$

$$04 : C_{(2)(0)(2)(0)} = ^{(2,2)}B, \tag{5.119}$$

$$05 : C_{(3)(2)(2)(0)} = \frac{3}{4}c_{33}^{-1}\left(^{(1,3)}B + C_{(2)(3)(2)(3)} - \frac{1}{3}C_{(2)(0)(2)(0)}c_{33}^2\right), \tag{5.120}$$

$$06 : C_{(0)(3)(1)(3)} = -\frac{3}{4}c_{11}^{-1}\left({}^{(2,1)}B - C_{(2)(0)(2)(0)} - \frac{1}{3}C_{(3)(0)(3)(0)}c_{11}^2\right),$$

(5.121)

$$07 : C_{(3)(0)(2)(0)} = -\frac{3}{2}c_{52}^{-1}c_{53}^{-1}\left({}^{(1,5)}B + C_{(2)(3)(2)(3)} + \frac{4}{3}C_{(0)(3)(2)(3)}c_{52}\right.$$
$$\left. -\frac{4}{3}C_{(3)(2)(2)(0)}c_{53} - \frac{1}{3}C_{(3)(0)(3)(0)}c_{52}^2 - \frac{1}{3}C_{(2)(0)(2)(0)}c_{53}^2\right),$$

(5.122)

$$08 : C_{(3)(2)(1)(2)} = -\frac{3}{2}c_{61}^{-1}c_{63}^{-1}\left({}^{(2,6)}B - C_{(2)(0)(2)(0)} + \frac{4}{3}C_{(0)(3)(1)(3)}c_{61}\right.$$
$$\left. +\frac{4}{3}C_{(3)(2)(2)(0)}c_{63} - \frac{1}{3}C_{(3)(0)(3)(0)}c_{61}^2 + \frac{1}{3}C_{(2)(3)(2)(3)}c_{63}^2\right),$$

(5.123)

$$09 : C_{(3)(1)(3)(2)} = -\frac{3}{2}c_{41}^{-1}c_{42}^{-1}\left({}^{(3,4)}B - C_{(3)(0)(3)(0)} - \frac{4}{3}C_{(0)(3)(1)(3)}c_{41}\right.$$
$$\left. -\frac{4}{3}C_{(0)(3)(2)(3)}c_{42} - \frac{1}{3}C_{(2)(0)(2)(0)}c_{41}^2 + \frac{1}{3}C_{(2)(3)(2)(3)}c_{42}^2\right).$$

(5.124)

With the abbreviations

$$K_1 := \frac{3}{4}c_{33}^{-1}\left[-{}^{(4,3)}B - C_{(2)(3)(2)(3)} + 2C_{(3)(1)(3)(2)} + C_{(2)(0)(2)(0)}\right.$$
$$\left. +\frac{1}{3}(C_{(2)(0)(2)(0)} - 2C_{(3)(1)(3)(2)} - C_{(2)(3)(2)(3)})c_{33}^2\right],$$

(5.125)

$$K_2 := \frac{3}{4}c_{11}^{-1}\left[-{}^{(5,1)}B + C_{(2)(0)(2)(0)} + 2C_{(3)(0)(2)(0)} + C_{(3)(0)(3)(0)}\right.$$
$$\left. +\frac{1}{3}(C_{(3)(0)(3)(0)} - 2C_{(3)(0)(2)(0)} + C_{(2)(0)(2)(0)})c_{11}^2\right],$$

(5.126)

$$K_3 := -\frac{3}{4}c_{22}^{-1}\left[{}^{(6,2)}B + C_{(2)(3)(2)(3)} - 2C_{(3)(2)(1)(2)} - C_{(3)(0)(3)(0)}\right.$$
$$\left. -\frac{1}{3}(C_{(3)(0)(3)(0)} - 2C_{(3)(2)(1)(2)} - C_{(2)(3)(2)(3)})c_{22}^2\right],$$

(5.127)

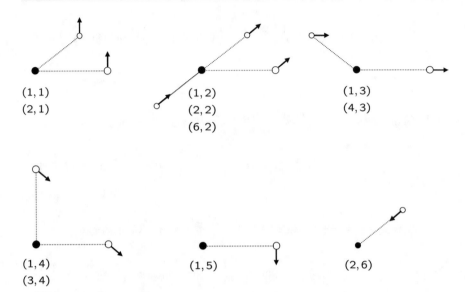

Fig. 5.6 Symbolical sketch of the explicit vacuum solution for the curvature (5.116)–(5.130). In total 11 suitably prepared clocks (hollow circles) are needed to determine all curvature components. The observer is denoted by the black circle

the remaining three curvature components read

$$10 : C_{(1)(0)(2)(3)} = \frac{1}{3} (K_3 - K_1),$$
(5.128)

$$11 : C_{(2)(0)(1)(3)} = \frac{1}{3} (K_2 - K_1),$$
(5.129)

$$12 : C_{(3)(0)(2)(1)} = \frac{1}{3} (K_3 - K_2).$$
(5.130)

A symbolical sketch of the solution is given in Fig. 5.6.

Special Spacetimes

In the following we consider setups of clocks which allow for a determination of the field in special space-times. As in the previous section on general and vacuum spacetimes we look for arrangements of n clocks, at positions $^{(n)}p^\alpha$, with velocities $^{(m)}u^\alpha$ w.r.t. the reference world line of the observer. While possible in principle, and in particular covered by our general formalism, we are again not going to allow for situations with additional acceleration or rotations. The explicit derivation of the frequency ratios is given in Appendices C.2–C.4.

Plane Gravitational Waves

The starting point is (C.82), which is the measurable frequency ratio as a function of the quantities characterizing the state of motion as well as the space-time.

Assuming that all quantities but the gravitational field can be prescribed by the experimentalist, we can rearrange (C.82) as follows:

$$B(r, p^\alpha, u^\alpha, a^\alpha, \omega^{\alpha\beta}) = R_{(\alpha)(4)(\beta)(4)} \, p^{(\alpha)} \, p^{(\beta)} \,,$$

$$(5.131)$$

where

$$B(r, p^\alpha, u^\alpha, a^\alpha, \omega^{\alpha\beta}) := (\mathbf{a} \cdot \mathbf{p})^2 - |\boldsymbol{\omega} \times \mathbf{p}|^2$$

$$+ \frac{2}{r}\left\{(\mathbf{a} \cdot \mathbf{p}) - \mathbf{u} \cdot (\boldsymbol{\omega} \times \mathbf{p})\right\} + \frac{1}{r^2}\left(1 - C - |\mathbf{u}|^2\right). \qquad (5.132)$$

Employing the same strategy as before, we are now looking for a configuration of clocks, which allows for a determination of all components of the gravitational field in terms of the measured quantities B. By labelling different positions of the clocks by an additional index (n) Eq. (5.131) turns into the system

$$^{(n)}B = R_{(\alpha)(4)(\beta)(4)} \, ^{(n)}p^{(\alpha)} \, ^{(n)}p^{(\beta)} \,, \qquad (5.133)$$

in which we suppressed all indices of quantities entering $^{(n)}B$ which are directly controlled by the experimentalist. Considering different choices for the positions $^{(n)}p^\alpha$, we notice that we end up with the constrained vacuum clock compass solution given in [17, (114)–(119)]:

$$01 : R_{(1)(4)(1)(4)} = {}^{(1)}B, \qquad (5.134)$$

$$02 : R_{(2)(4)(2)(4)} = {}^{(2)}B, \qquad (5.135)$$

$$03 : R_{(3)(4)(3)(4)} = {}^{(3)}B, \qquad (5.136)$$

$$04 : R_{(2)(4)(1)(4)} = \frac{1}{2}\left({}^{(4)}B - {}^{(1)}B - {}^{(2)}B\right), \qquad (5.137)$$

$$05 : R_{(3)(4)(2)(4)} = \frac{1}{2}\left({}^{(5)}B - {}^{(2)}B - {}^{(3)}B\right), \qquad (5.138)$$

$$06 : R_{(3)(4)(1)(4)} = \frac{1}{2}\left({}^{(6)}B - {}^{(1)}B - {}^{(3)}B\right). \qquad (5.139)$$

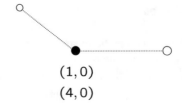

$$(1,0)$$
$$(4,0)$$

Fig. 5.7   Symbolical sketch of the explicit clock configuration which allows for a complete determination of the gravitational field (5.141). In total 2 suitably prepared clocks (hollow circles) are needed to determine all curvature components. The observer is denoted by the black circle

Of course in our case the situation is simplified even further due to (C.65) and (C.66). From the constrained system we can infer that two clocks at positions

$$
{}^{(1)}p^\alpha = \begin{pmatrix} 1 \\ 0 \\ 0 \end{pmatrix}, \qquad {}^{(4)}p^\alpha = \begin{pmatrix} 1 \\ 1 \\ 0 \end{pmatrix}, \tag{5.140}
$$

allow for a complete determination of the gravitational field, i.e. the functions a and b are given by

$$
a = -\frac{1}{2}\,{}^{(1)}B, \qquad b = -\frac{1}{4}\,{}^{(4)}B. \tag{5.141}
$$

See Fig. 5.7 for a symbolical sketch of the solution. In contrast to the notation in (5.141)—in which all indices but the relevant position index (n) are suppressed— the second (velocity) index (m) is explicitly given in Fig. 5.7 and set to $m = 0$, indicating that the clocks in this configuration do not move w.r.t. to the observer.

Waves Radial Relative to $r = 0$

Following the same line of reasoning as in the case of plane gravitational waves, we use the definition for B as given in (5.132), however now we have a system of clocks at positions ${}^{(n)}p^{(\alpha)}$ moving with velocities ${}^{(m)}u^{(\alpha)}$, and we are left with the system

$$
\begin{aligned}
{}^{(n,m)}B &= {}^{(n)}p^{(\alpha)}\,{}^{(n)}p^{(\beta)}\left(R_{(\alpha)(4)(\beta)(4)} - \frac{4}{3} R_{(\alpha)(4)(\gamma)(\beta)} \right. \\
&\quad \left. \times {}^{(m)}u^{(\gamma)} + \frac{1}{3} R_{(\gamma)(\alpha)(\delta)(\beta)}\,{}^{(m)}u^{(\gamma)}\,{}^{(m)}u^{(\delta)} \right).
\end{aligned} \tag{5.142}
$$

In vacuum, the general clock compass solution on the basis of (5.142) was given in [17]. Taking into account the non-vanishing curvature components in the radial case as indicated in (C.93) and (C.94), one may infer several clock configurations which allow for a determination of the curvature components.

One configuration coincides with the one already given in the plane gravitational wave case, c.f. Eq. (5.141) and Fig. 5.7. However, due to the more general nature of the compass equation (5.142) one may now also construct configurations in which the clocks are in motion. We briefly mention here two possible configurations, i.e.

$$R_{(1)(4)(1)(4)} = 3c_{11}^{-2} \,^{(3,1)}B, \tag{5.143}$$

$$R_{(1)(4)(2)(4)} = \left(2 - \frac{8}{3}c_{33} + \frac{2}{3}c_{33}^2\right)^{-1} \,^{(4,3)}B. \tag{5.144}$$

An alternative solution for the second curvature component is given by

$$R_{(1)(4)(2)(4)} = \frac{3}{2c_{41}c_{42}} \left\{ {}^{(3,4)}B \right.$$
$$\left. - \left[\left(\frac{c_{41}}{c_{11}}\right)^2 - \left(\frac{c_{42}}{c_{11}}\right)^2\right] \,^{(3,1)}B \right\}. $$

$$\tag{5.145}$$

Here we used the same nomenclature for the positions and velocities as before, i.e.

$$^{(3)}p^\alpha = \begin{pmatrix} 0 \\ 0 \\ 1 \end{pmatrix}, \quad ^{(1)}u^\alpha = \begin{pmatrix} c_{11} \\ 0 \\ 0 \end{pmatrix},$$

$$^{(3)}u^\alpha = \begin{pmatrix} 0 \\ 0 \\ c_{33} \end{pmatrix}, \quad ^{(4)}u^\alpha = \begin{pmatrix} c_{41} \\ c_{42} \\ 0 \end{pmatrix}. \tag{5.146}$$

Symbolical sketches of the solutions (5.143)-(5.145) are given in Fig. 5.8.

Summary

We have shown that a suitably prepared set of clocks can be used to determine all components of the gravitational field, i.e. the curvature, in general relativity, as well as to describe the state of motion of a noninertial observer. One needs 21 and 11 clocks, respectively, to determine all curvature components in a general curved space-time and in vacuum. Building upon this result, we were able to specialize the general compass setup to two special types of space-times, describing plane gravitational waves and waves moving radially with respect to an observer. It should be stressed that the measurement by means of the gravitational clock compass differs somewhat from other works in the gravitational wave context. Our main focus was

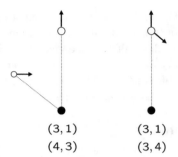

$$(3,1) \qquad (3,1)$$
$$(4,3) \qquad (3,4)$$

Fig. 5.8 Symbolical sketch of the two explicit clock configurations which allow for a complete determination of the gravitational field (5.143)–(5.145). In both cases two suitably prepared clocks (hollow circles) are needed to determine all curvature components. Again the observer is denoted by the black circle

on the general geometry of the clock configuration required for a complete field determination, and not the possible measurement of the wave character (profile). In contrast to classical works on (indirect) timing experiments like [50, 51], a clock compass relies on the direct frequency comparison of a suitably prepared set of local clocks.

It is clear that the highly idealized situations of plane and spherically gravitational waves should be generalized. Still they demonstrate the direct operational relevance of a clock compass. We hope this motivates future works on an approximate description of more general radiative space-times, an interesting future application being the realization of an omnidirectional (tensorial) [52–54] gravitational wave detector based on clocks.

References

1. F.A.E. Pirani, Acta Phys. Pol. **15**, 389 (1956)
2. T. Levi-Civita, Math. Ann. **97**, 291 (1927)
3. J.L. Synge, Proc. Lond. Math. Soc. **25**, 247 (1926)
4. J.L. Synge, Phil. Trans. R. Soc. Lond. A **226**, 31 (1927)
5. D. Puetzfeld, Y.N. Obukhov, Phys. Rev. D **93**, 044073 (2016)
6. P. Szekeres, J. Math. Phys. **6**, 1387 (1965)
7. J.L. Synge, *Relativity: The General Theory* (North-Holland, Amsterdam, 1960)
8. I. Ciufolini, M. Demianski, Phys. Rev. D **34**, 1018 (1986)
9. I. Ciufolini, Phys. Rev. D **34**, 1014 (1986)
10. C.W. Chou, et al., Phys. Rev. Lett. **104**, 070802 (2010)
11. N. Huntemann, et al., Phys. Rev. Lett. **108**, 090801 (2012)
12. J. Guéna, et al., IEEE Trans. Ultrason. Ferroelectr. Freq. Control **59**, 391 (2012)
13. S. Falke, et al., New J. Phys. **16**, 073023 (2014)
14. B.J. Bloom, et al., Nature **506**, 71 (2014)
15. M. Schioppo, et al., Nat. Photonics **11**, 48 (2017)
16. A. Bauch, *Relativistic Geodesy: Foundations and Application*, vol. 196, ed. by D. Puetzfeld, et al., Fundamental Theories of Physics (Springer, Cham, 2019), p. 1
17. D. Puetzfeld, Y.N. Obukhov, C. Lämmerzahl, Phys. Rev. D **98**, 024032 (2018)

18. Y.N. Obukhov, D. Puetzfeld, *Relativistic Geodesy: Foundations and Application*, vol. 196, ed. by D. Puetzfeld et al. Fundamental Theories of Physics (Springer, Cham, 2019), p. 87
19. E. Fermi, Atti. Accad. Naz. Lincei Cl, Sci. Fis. Mat. Nat. Rend. **31**, 2151101 (1922)
20. E. Fermi, *Collected Papers*, vol. 1, ed. by E. Amaldi, E. Persico, F. Rasetti, E. Segrè (University of Chicago Press, Chicago, 1962)
21. O. Veblen, Proc. Nat. Acad. Sci. **8**, 192 (1922)
22. O. Veblen, T.Y. Thomas, Trans. Amer. Math. Soc. **25**, 551 (1923)
23. J.L. Synge, Proc. London Math. Soc. **32**, 241 (1931)
24. A.G. Walker, Proc. Roy. Soc. Edinb. **52**, 345 (1932)
25. F.K. Manasse, C.W. Misner, J. Math. Phys. **4**, 735 (1963)
26. C.W. Misner, K.S. Thorne, J.A. Wheeler, *Gravitation* (Freeman, San Francisco, 1973)
27. W.T. Ni, Chin. J. Phys. **15**, 51 (1977)
28. B. Mashhoon, Astrophys. J. **216**, 591 (1977)
29. W.T. Ni, M. Zimmermann, Phys. Rev. D **17**, 1473 (1978)
30. W.Q. Li, W.T. Ni, Chin. J. Phys. **16**, 214 (1978)
31. W.T. Ni, Chin. J. Phys. **16**, 223 (1978)
32. W.Q. Li, W.T. Ni, J. Math. Phys. **20**, 1473 (1979)
33. W.Q. Li, W.T. Ni, J. Math. Phys. **20**, 1925 (1979)
34. N. Ashby, B. Bertotti, Phys. Rev. D **34**, 2246 (1986)
35. A.M. Eisele, Helv. Phys. Acta **60**, 1024 (1987)
36. T. Fukushima, Cel. Mech. **44**, 1024 (1988)
37. O. Semerák, Gen. Rel. Grav. **25**, 1041 (1993)
38. K.P. Marzlin, Phys. Rev. D **50**, 888 (1994)
39. D. Bini, A. Geralico, R.T. Jantzen, J. Math. Phys. **22**, 4729 (2005)
40. C. Chicone, B. Mashhoon, Phys. Rev. D **74**, 064019 (2006)
41. D. Klein, P. Collas, Class. Quant. Grav. **25**, 145019 (2008)
42. D. Klein, P. Collas, J. Math. Phys. **51**, 022501 (2010)
43. P. Delva, M.C. Angonin, Gen. Rel. Grav. **44**, 1 (2012)
44. S.G. Turyshev, O.L. Minazzoli, V.T. Toth, J. Math. Phys. **53**, 032501 (2012)
45. P.A. Hogan, D. Puetzfeld, Phys. Rev. D **101**, 044012 (2020)
46. P.A. Hogan, A. Trautman, *Gravitation and Geometry* (Bibliopolis, Naples, 1987), p. 215
47. P.A. Hogan, T. Futamase, Y. Itoh, Phys. Rev. D **78**, 104014 (2008)
48. P.A. Hogan, H. Asada, T. Futamase, *Equations of motion in General Relativity* (Oxford University Press, Oxford, 2010)
49. F.W. Hehl, W.T. Ni, Phys. Rev. D **42**, 2045 (1990)
50. S. Detweiler, Astrophys. J. **234**, 1100 (1979)
51. F.B. Estabrook, H.D. Wahlquist, Gen. Rel. Grav. **6**, 439 (1975)
52. R.L. Forward, Gen. Rel. Grav. **2**, 149 (1971)
53. R.V. Wagoner, H.J. Paik, *Accademia Nazionale dei Lincei International Symposium on Experimental Gravitation* (Accademia Nazionale, Roma, 1976), p. 257
54. H.J. Paik, et al., Class. Q. Grav. **33**, 075003 (2016)

de Sitter Cosmology

6

Abstract

The equation $u \equiv t - z = $ constant in Minkowskian space-time with line element

$$ds^2 = dt^2 - dx^2 - dy^2 - dz^2 = \eta_{ij}\, dx^i\, dx^j\,,$$

is an example of a null hyperplane. It represents the history of a 2-plane, parallel to the x, y-plane in three dimensional Euclidean space, moving with the speed of light in the positive z-direction. Thus the family of null hyperplanes $u = $ constant could be the histories of the wave fronts of plane electromagnetic waves travelling in the positive z-direction in Euclidean space. The vector field normal to the hyperplanes is $k_i = u_{,i}$ and with $k^i = \eta^{ij}\, k_j$ this has the properties that $k^i\, k_i = 0$ and $k_{i,j} = 0$. Thus k^i is a null vector field (and therefore tangent to $u = $ constant) and covariantly constant (i.e. a constant vector field in the coordinates $x^i = (t, x, y, z)$ with $i = 0, 1, 2, 3$). To generalise this notion of null hyperplanes to space-times of non-zero constant curvature, the first obstacle one encounters is the non-existence in such a space-time of covariantly constant vector fields.

6.1 Null Hyperplanes in Space-Times of Constant Curvature

In a general space-time with metric tensor components g_{ab} the components of the Riemann curvature tensor are given in terms of the metric tensor and its derivatives by

$$R_{ijkl} = \frac{1}{2}(g_{ik,jl} + g_{jl,ik} - g_{il,jk} - g_{jk,il}) + g_{ab}(\Gamma^a_{ik}\, \Gamma^b_{jl} - \Gamma^a_{il}\, \Gamma^b_{jk})\,, \qquad (6.1)$$

where Γ^a_{ik} are the components of the Riemannian connection and, as always, a comma denotes partial differentiation with respect to the coordinates x^a (with a comma followed by two indices denoting second partial differentiation). In a space-time of constant curvature the components R_{ijkl} of the Riemann tensor have the algebraic form

$$R_{ijkl} = K (g_{ik} g_{jl} - g_{il} g_{jk}) \quad \Rightarrow R_{jk} = -3 K g_{jk} = -\Lambda g_{jk} , \tag{6.2}$$

where K denotes the constant curvature, Λ is the cosmological constant and $R_{jk} = g^{il} R_{ijkl}$ are the components of the Ricci tensor. In such a space-time any covariant vector field with components v_i satisfies the Ricci identities

$$v_{j;kl} - v_{j;lk} = K (v_k g_{jl} - v_l g_{jk}) . \tag{6.3}$$

It thus follows that if $K \neq 0$ then it is impossible to have a vector field v_i satisfying $v_{i;j} = 0$. Thus in a space-time of constant non-zero curvature it is impossible to have a covariantly constant vector field. In particular it is impossible to have a covariantly constant *null* vector field. In wishing to generalise plane waves with a covariantly constant null propagation direction in space-time to the case of space-times of constant non-zero curvature we seek null hyper*planes*, which are the histories of the wave fronts of such waves. A null hyperplane is a null hypersurface generated by null geodesics which are shear-free and expansion-free (and, of course, twist-free). The properties of being *null*, *geodesic* and *shear-free* are conformally invariant properties and thus help to construct null hyperplanes in space-times of constant curvature since such space-times are conformally flat. In addition all the shear-free, null hypersurfaces in flat space-time are known. They are either *null hyperplanes* or *null cones* or portions thereof [1]. Let $x^i = (t, x, y, z) = (x^0, x^1, x^2, x^3)$ be rectangular Cartesian coordinates and time in Minkowskian space-time with line element

$$ds^2 = dt^2 - dx^2 - dy^2 - dz^2 = \eta_{ij} dx^i dx^j , \tag{6.4}$$

with $\eta_{ij} = \text{diag}(-1, 1, 1, 1)$ the components of the Minkowskian metric tensor in coordinates x^i. Latin indices take values 0, 1, 2, 3 and indices are raised and lowered using η^{ij} and η_{ij} respectively with η^{ij} defined by $\eta^{ij} \eta_{jk} = \delta^i_k$. Shear-free null hypersurfaces in Minkowskian space-time [2] are given by $u(t, x, y, z) = $ constant, with $u(t, x, y, z)$ defined implicitly by the equation of a *null hyperplane*:

$$\eta_{ij} a^i(u) x^j + b(u) = 0 \quad \text{with} \quad \eta_{ij} a^i a^j = 0 , \tag{6.5}$$

or the equation of *null cones* with vertices on an arbitrary world line $x^i = w^i(u)$:

$$\eta_{ij} (x^i - w^i(u))(x^j - w^j(u)) = 0 . \tag{6.6}$$

We first verify that $u = $ constant given by (6.5) are null hyperplanes. Differentiating (6.5) with respect to x^k yields

$$u_{,k} = -\varphi^{-1} a_k \quad \text{with} \quad \varphi = \dot{b} + \dot{a}_i x^i , \tag{6.7}$$

and the partial derivative is, as always, denoted by a comma. The dot denotes differentiation with respect to u. Since a^i is a null vector field this confirms that $u = $ constant are null hypersurfaces. Differentiating (6.7) with respect to x^l results in

$$u_{,kl} = -\varphi^{-1}(\dot{a}_k u_{,l} + \dot{a}_l u_{,k}) - \varphi^{-1}\dot{\varphi} u_{,k} u_{,l} . \tag{6.8}$$

From the algebraic form of the right hand side of this equation [3] it follows that the covariant null vector field $u_{,k}$ is geodesic and shear-free (it is obviously twist-free). From (6.8) we deduce that

$$\eta^{kl} u_{,kl} = -2\varphi^{-1}\eta^{kl}\dot{a}_k u_{,l} = 2\varphi^{-2}\eta^{kl}\dot{a}_k a_l = 0 , \tag{6.9}$$

since a^i is null. Hence, in addition to being *null*, *geodesic*, *shear-free* and *twist-free*, $u_{,k}$ is also *expansion-free*. Thus the null hypersurfaces $u = $ constant are null hyper*planes*.

The space-times of constant (non-zero) curvature K are de Sitter space-time (positive curvature) or anti-de Sitter space-time (negative curvature) depending upon the sign of the cosmological constant $\Lambda = 3 K$. These space-times are conformally flat and the line element can be written in the conformally flat form:

$$ds^2 = \lambda^2 \eta_{ij} dx^i dx^j , \tag{6.10}$$

with

$$\lambda = \left(1 - \frac{\Lambda}{12}\eta_{ij} x^i x^j\right)^{-1} . \tag{6.11}$$

From the conformal invariance of the null, geodesic and shear-free properties we know that $u = $ constant given by (6.5) are null, geodesic and shear-free in the space-time with line element (6.10). We now look for the condition that $u = $ constant are expansion-free in the space-time with line element (6.10). For this we must calculate $u_{,k;l}$ with the semicolon indicating covariant differentiation with respect to the Riemannian connection associated with the metric tensor $g_{ij} = \lambda^2 \eta_{ij}$. The components of this Riemannian connection are

$$\Gamma^i_{jk} = \lambda^{-1}(\lambda_{,j}\delta^i_k + \lambda_{,k}\delta^i_j - \eta^{ip}\lambda_{,p}\eta_{jk}) . \tag{6.12}$$

Thus we find, using (6.8), that

$$u_{,k;l} = -(\varphi^{-1}\dot{a}_k + \lambda^{-1}\lambda_{,k})\,u_{,l} - (\varphi^{-1}\dot{a}_l + \lambda^{-1}\lambda_{,l})\,u_{,k}$$
$$-\varphi^{-1}\dot{\varphi}\,u_{,k}\,u_{,l} + \lambda^{-1}\eta^{pq}\lambda_{,p}\,u_{,q}\,\eta_{kl}\,. \tag{6.13}$$

This again has the correct algebraic form [3] for $u_{,k}$ to be geodesic and shear-free in the space-time with line element (6.10). From (6.13) we see that

$$\eta^{kl}u_{,k;l} = 2\,\lambda^{-1}\eta^{pq}\lambda_{,p}\,u_{,q} = -2\,\lambda^{-1}\varphi^{-1}\eta^{pq}\lambda_{,p}\,a_q\,. \tag{6.14}$$

But $\lambda_{,p} = \lambda^2\Lambda\,\eta_{pq}x^q/6$ and so

$$\eta^{pq}\lambda_{,p}\,a_q = \frac{\Lambda}{6}\lambda^2 a_p\,x^p = -\frac{\Lambda}{6}\lambda^2 b\,, \tag{6.15}$$

using (6.5). Hence the null hyperplanes (6.5) in Minkowskian space-time are null hyper*planes* in the space-time with line element (6.10) provided $b = 0$ (which, by (6.14) and (6.15), is necessary in order to have $u_{,k}$ expansion-free in the space-time with line element (6.10)). Thus the null hyperplanes $u =$ constant in Minkowskian space-time given by

$$a_i(u)\,x^i = 0\,, \tag{6.16}$$

correspond to null hyperplanes in the space-time of constant curvature with line element (6.10). We note, for later consideration, that the null hyperplanes (6.16) pass through the origin $x^i = 0$, are tangent to the null cone with vertex $x^i = 0$ and therefore intersect each other. Only the *direction* of the null vector field a^i is significant in (6.16) and this is determined by two real-valued functions of u or equivalently by one complex-valued function $l(u)$ with complex conjugate $\bar{l}(u)$. Thus we can write

$$a^0 - a^3 = 2\,,\quad a^0 + a^3 = 4l\bar{l}\,,\quad a^1 + ia^2 = 2\sqrt{2}l\,. \tag{6.17}$$

Now (6.16) reads:

$$z + t = \sqrt{2}\bar{l}(x + iy) + \sqrt{2}l(x - iy) + 2l\bar{l}(z - t)\,. \tag{6.18}$$

Using this we find that

$$\eta_{ij}x^i\,x^j = -\left|x + iy + \sqrt{2}l\,(z - t)\right|^2\,, \tag{6.19}$$

which suggests that we introduce a complex coordinate ζ (with complex conjugate denoted by a bar) via

$$\zeta = \frac{1}{\sqrt{2}}(x + iy) + l(z - t) . \tag{6.20}$$

Now instead of using the coordinates t, x, y, z we may use coordinates $\zeta, \bar{\zeta}, u$ and $z - t$ satisfying

$$x + iy = \sqrt{2}\zeta - \sqrt{2}l(z - t) , \tag{6.21}$$

$$z + t = 2(\bar{l}\zeta + l\bar{\zeta}) - 2l\bar{l}(z - t) , \tag{6.22}$$

while the conformal factor λ is given, using (6.11) and (6.19), by

$$\lambda^{-1} = 1 - \frac{\Lambda}{12}\eta_{ij}x^i x^j = 1 + \frac{\Lambda}{6}\zeta\bar{\zeta} = p \text{ (say)} . \tag{6.23}$$

Now the line element (6.10) reads

$$-ds^2 = 2\,p^{-2}d\zeta\,d\bar{\zeta} + 2\,p^{-2}d\Sigma\,du, \tag{6.24}$$

where $d\Sigma$ (which is not necessarily an exact differential) is given by

$$d\Sigma = -(z - t)(\beta\,d\bar{\zeta} + \bar{\beta}\,d\zeta) + (\beta\,\bar{\zeta} + \bar{\beta}\,\zeta)(dz - dt) + \beta\bar{\beta}(z - t)^2 du , \tag{6.25}$$

where $\beta(u) = dl(u)/du$. If we now define

$$q = \beta\bar{\zeta} + \bar{\beta}\zeta , \tag{6.26}$$

and in place of $Z - T$ use a coordinate r defined by

$$z - t = qr , \tag{6.27}$$

then the line element (6.24) takes the Ozsváth-Robinson-Rózga [4] form

$$-ds^2 = 2\,p^{-2}d\zeta\,d\bar{\zeta} + 2\,p^{-2}q^2 du\{dr + (q^{-1}\dot{q}\,r + \beta\bar{\beta}\,r^2)du\} , \tag{6.28}$$

where the dot, as always, denotes differentiation with respect to u. This is a special case of this form of line element as we shall see below. Next writing

$$\xi^i = x^i - w^i(u) , \tag{6.29}$$

we have from (6.6) that

$$\eta_{ij}\xi^i\,\xi^j = 0\,. \tag{6.30}$$

Differentiating this with respect to x^k results in

$$u_{,k} = \frac{\xi_k}{R} = k_k \ \ \text{(say)}\,, \tag{6.31}$$

with

$$R = \eta_{ij}\dot{w}^i\,\xi^j \ \ \text{and thus} \ \ \eta_{ij}\dot{w}^i\,k^j = +1\,. \tag{6.32}$$

As always a dot indicates differentiation with respect to u. It is clear from (6.30) and (6.31) that the hypersurfaces $u = $ constant are null. Straightforward calculations yield

$$\xi^i_{\,,j} = \delta^i_j - \dot{w}^i\,k_j\,, \tag{6.33}$$

and

$$R_{,i} = \dot{w}_i + A\,k_i\,, \tag{6.34}$$

with $\dot{w}_i = \eta_{ij}\dot{w}^j$ and

$$A = -\dot{w}_i\,\dot{w}^i + R\,\ddot{w}_i\,k^i\,. \tag{6.35}$$

Using (6.31)–(6.35) we arrive at

$$k_{i,j} = \frac{1}{R}(\eta_{ij} - \dot{w}_i\,k_j - \dot{w}_j\,k_i - A\,k_i\,k_j) = k_{j,i}\,, \tag{6.36}$$

which displays the algebraic structure [3] guaranteeing that k_i is geodesic and shear-free with expansion $\vartheta = k^i_{\,,i}/2 = 1/R \neq 0$. With a semicolon, as before, indicating covariant differentiation with respect to the Riemannian connection (6.12) associated with the metric tensor given via the line element (6.10) we find that

$$k_{i;j} = (R^{-1} + \lambda^{-1}\eta^{kl}\lambda_{,k}\,k_l)\eta_{ij} - (R^{-1}\dot{w}_i + \lambda^{-1}\lambda_{,i})k_j$$

$$-\,(R^{-1}\dot{w}_j + \lambda^{-1}\lambda_{,j})k_i - A\,R^{-1}k_i\,k_j\,. \tag{6.37}$$

The algebraic form of this [3] ensures that k_i is geodesic and shear-free in the space-time with line element (6.10). The expansion vanishes if $k^i{}_{;i}$ vanishes. It follows from (6.37) that this condition reduces to

$$R^{-1} + \lambda^{-1}\lambda_{,i}\,k^i = 0 . \tag{6.38}$$

With λ given by (6.11) this becomes

$$\eta_{ij}\,w^i\,w^j = \frac{12}{\Lambda} . \tag{6.39}$$

Since $w^i(u)$ has three independent components, a convenient parametrisation in terms of the real-valued function $m(u)$ and the complex-valued function $l(u)$ (with complex conjugate denoted $\bar{l}(u)$) is given by

$$w^0 - w^3 = \frac{6}{\Lambda\,m} , \;\; w^0 + w^3 = \frac{6}{\Lambda\,m}\left(\frac{1}{3}\Lambda\,m^2 + 2l\bar{l}\right) , \;\; w^1 + iw^2 = \frac{6\sqrt{2}\,l}{\Lambda\,m} . \tag{6.40}$$

Here we assume that $m \neq 0$ but if m is small then (6.30) approximates (6.16) with a^i given by (6.17) and so we can expect that the results we obtain now starting with the null cones (6.6) will include those obtained above starting with the null hypersurfaces (6.5) in the limit of small $m(u)$. Writing out (6.6) with $w^i(u)$ given by (6.40) results in

$$z + t = \sqrt{2}\,\bar{l}\,(x + iy) + \sqrt{2}\,l\,(x - iy) + 2\bar{l}\,(z - t)$$
$$+ 2\,m\left(1 + \frac{\Lambda}{6}\,m\,(z - t)\right) + \frac{\Lambda}{6}\,m\,\eta_{ij}x^i\,x^j , \tag{6.41}$$

which specialises to (6.18) when $m = 0$. Using this we can write

$$\left(1 + \frac{\Lambda}{6}\,m\,(z - t)\right)\eta_{ij}\,x^i x^j = -\left|x + iy + \sqrt{2}\,l\,(z - t)\right|^2$$
$$- 2\,m\,(z - t)\left(1 + \frac{\Lambda}{6}\,m\,(z - t)\right) . \tag{6.42}$$

This specialises to (6.19) when $m = 0$. From this we see that

$$\left(1 + \frac{\Lambda}{6}\,m\,(z - t)\right)\lambda^{-1} = \left(1 + \frac{\Lambda}{6}\,m\,(z - t)\right)^2$$
$$+ \frac{\Lambda}{12}\left|x + iy + \sqrt{2}\,l\,(z - t)\right|^2 , \tag{6.43}$$

with λ given by (6.11). In similar fashion to (6.20) this suggests that we should define the new complex coordinate

$$\zeta = \left(1 + \frac{\Lambda}{6} m (z - t)\right)^{-1} \left\{\frac{1}{\sqrt{2}}(x + iy) + l (z - t)\right\} , \qquad (6.44)$$

so that

$$\lambda^{-1} = \left(1 + \frac{\Lambda}{6} m (z - t)\right) p , \qquad (6.45)$$

with

$$p = 1 + \frac{\Lambda}{6} \zeta \bar{\zeta} . \qquad (6.46)$$

Now instead of using the coordinates t, x, y, z we shall use the coordinates $\zeta, \bar{\zeta}, u, z - t$ with

$$x + iy = \sqrt{2} \left(1 + \frac{\Lambda}{6} m (z - t)\right) \zeta - \sqrt{2} l (z - t) , \qquad (6.47)$$

$$z + t = 2 \left\{\bar{l} \zeta + l \bar{\zeta} + m \left(1 - \frac{\Lambda}{6} \zeta \bar{\zeta}\right)\right\} \left(1 + \frac{\Lambda}{6} m (z - t)\right)$$

$$- \left(\frac{\Lambda}{3} m^2 + 2 l \bar{l}\right) (z - t) . \qquad (6.48)$$

Now the line element (6.10) reads

$$- ds^2 = 2 p^{-2} d\zeta \, d\bar{\zeta} + 2 p^{-2} d\Sigma \, du , \qquad (6.49)$$

with

$$d\Sigma = \left(1 + \frac{\Lambda}{6} m (z - t)\right)^{-2} \left\{\left(\frac{1}{2}\kappa - \frac{\Lambda}{6} \alpha q\right) (z - t)^2 du\right.$$

$$\left. + q (dz - dt) - (z - t) \left(1 + \frac{\Lambda}{6} m (z - t)\right) (dq - \dot{q} \, du)\right\} , \qquad (6.50)$$

where $\alpha(u) = dm(u)/du$, $\beta(u) = dl(u)/du$ and

$$\kappa = \frac{\Lambda}{3} \alpha^2 + 2 \beta \bar{\beta} , \qquad (6.51)$$

$$q(\zeta, \bar{\zeta}, u) = \beta \bar{\zeta} + \bar{\beta} \zeta + \alpha \left(1 - \frac{\Lambda}{6} \zeta \bar{\zeta}\right) , \qquad (6.52)$$

and $\dot{q} = \partial q/\partial u$. If we now introduce the coordinate r via the equation

$$z - t = \left(1 + \frac{\Lambda}{6} m (z - t)\right) q r , \tag{6.53}$$

the line element (6.49) takes the general Ozsváth–Robinson–Rózga [4] form

$$ds^2 = 2 p^{-2} d\zeta \, d\bar{\zeta} + 2 p^{-2} q^2 du\{dr + (q^{-1}\dot{q} r + \frac{1}{2}\kappa r^2) \, du\} , \tag{6.54}$$

with p, q given by (6.46) and (6.52) respectively. When $m = 0 (\Rightarrow \alpha = 0)$ this reduces to (6.28). The construction given here illustrates an origin for the arbitrary functions $\alpha(u), \beta(u)$ appearing in (6.54) via (6.51) and (6.52).

6.2   Intersecting Null Hyperplanes

The equations of the null hyperplanes $u(t, x, y, z) =$ constant, are given implicitly by (6.5) with $b = 0$ and by (6.6) with (6.29) holding. These are easy to visualise in Minkowskian space-time and so it is clear that the null hyperplanes given by (6.5) with $b = 0$ intersect. The null cones (6.6) intersect if the world line $x^i = w^i (u)$ is space-like or time-like and they also intersect if this world line is, in general, null except when this null world line is a common generator of the null cones (see Fig. 6.1). For the latter to happen the world line $x^i = w^i (u)$ must be a null geodesic.

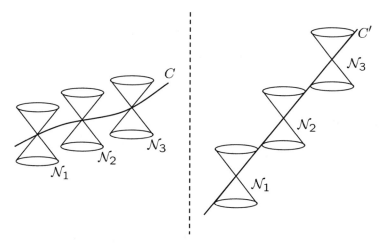

Fig. 6.1   On the lhs we have intersecting null cones N_1, N_2, N_3 with vertices on an arbitrary world line C. On the rhs we have non-intersecting null cones N_1, N_2, N_3 with vertices on a common generator (null geodesic) C'

With $w^i(u)$ given by (6.40) we find that

$$\eta_{ij} \dot{w}^i \dot{w}^j = - \left(\frac{6}{\Lambda m} \right)^2 \kappa ,$$ (6.55)

with κ given by (6.51). Hence the character of the world line $x^i = w^i(u)$ depends upon the sign of κ. For $x^i = w^i(u)$ to be a null geodesic we must have $\kappa = 0$ and

$$\ddot{w}^i = C(u) \, \dot{w}^i ,$$ (6.56)

for some real-valued function $C(u)$. Substituting (6.40) into

$$\ddot{w}^0 - \ddot{w}^3 = C \, (\dot{w}^0 - \dot{w}^3) ,$$ (6.57)

results in

$$C = \frac{1}{\alpha} \frac{d\alpha}{du} - \frac{2\alpha}{m} .$$ (6.58)

Now (6.40) in

$$\ddot{w}^1 + i\ddot{w}^2 = C(\dot{w}^1 + i\dot{w}^2) ,$$ (6.59)

along with (6.58), produces

$$\frac{1}{\beta} \frac{d\beta}{du} = \frac{1}{\alpha} \frac{d\alpha}{du} \quad \Rightarrow \quad \frac{1}{\beta} \frac{d\beta}{du} = \frac{1}{\bar{\beta}} \frac{d\bar{\beta}}{du} .$$ (6.60)

As a consequence of (6.60) the equation

$$\ddot{w}^0 + \ddot{w}^3 = C \, (\dot{w}^0 + \dot{w}^3) ,$$ (6.61)

is automatically satisfied. Finally we note that

$$\kappa = 0 \text{ and } \frac{1}{\beta} \frac{d\beta}{du} = \frac{1}{\bar{\beta}} \frac{d\bar{\beta}}{du} \quad \Rightarrow \quad \frac{1}{\beta} \frac{d\beta}{du} = \frac{1}{\alpha} \frac{d\alpha}{du} ,$$ (6.62)

and

$$\frac{1}{\beta} \frac{d\beta}{du} = \frac{1}{\bar{\beta}} \frac{d\bar{\beta}}{du} \quad \Rightarrow \quad \text{Re } \beta = 0 \text{ or } \text{Im } \beta = 0 \text{ or } \text{Re } \beta = c_0 \text{ Im } \beta ,$$ (6.63)

for some real number c_0. We can summarise the results here in the theorem of Tran and Robinson [5, 6]: (1) $\Lambda > 0 \Rightarrow \kappa > 0 \Rightarrow$ intersecting null hyperplanes; (2) $\Lambda < 0 \Rightarrow \kappa > 0$ or $\kappa < 0$ or $\kappa = 0$ with (i) $\kappa > 0 \Rightarrow$ intersecting null hyperplanes, (ii) $\kappa < 0 \Rightarrow$ intersecting null hyperplanes, (iii) $\kappa = 0 \Rightarrow$ intersecting null hyperplanes *except* when $\mathrm{Re}\,\beta = 0$ or $\mathrm{Im}\,\beta = 0$ or $\mathrm{Re}\,\beta = c_0\,\mathrm{Im}\,\beta$.

6.3 Generalized Kerr–Schild Space-Times and Gravitational Waves

The gravitational field of pure gravitational radiation is described by a Weyl tensor which is algebraically degenerate, as we have seen in Chap. 2. However Trautman [7] has pointed out that "there is another important property of waves, both linear and gravitational: *waves can propagate information*. This means that wave-like solutions depend on arbitrary functions, the shape of which contains the information carried by the wave." A notable aspect of this statement is that it is *a point of view*, which leaves the researcher free as to how it is implemented. Such statements are a powerful stimulus for research. Trautman chose to illustrate this with what is now referred to as a generalised Kerr–Schild [8, 9] metric required to satisfy Einstein's vacuum field equations. Such a metric, satisfying Einstein's field equations with a cosmological constant, is also useful in studying plane fronted gravitational waves in the present context for the reasons we now describe. We assume the metric tensor components, in coordinates x^i, take the form

$$g_{ij} = \underset{(0)}{g}_{ij} + 2\,H\,k_i\,k_j\,, \tag{6.64}$$

with

$$\underset{(0)}{g}{}^{ij}\,k_i\,k_j = k^j\,k_j = 0 \quad\text{and}\quad k^i = \underset{(0)}{g}{}^{ij}\,k_j\,. \tag{6.65}$$

Some useful properties of such a metric tensor are derived in the Appendix D. If a stroke denotes covariant differentiation with respect to the Riemannian connection calculated with the metric tensor g_{ij} and a semicolon denotes covariant differentiation with respect to the Riemannian connection calculated with the metric tensor $\underset{(0)}{g}{}_{ij}$ then

$$k_{i|j} = k_{i;j} + H'\,k_i\,k_j + H\,(k_{i;k}\,k^k\,k_j + k_{j;k}\,k^k\,k_i)\,, \tag{6.66}$$

where $H' = H_{,i}\,k^i$. We will apply this to the case in which $\underset{(0)}{g}{}_{ij}$ is the de Sitter or anti-de Sitter metric tensor given via the line element (6.54) and $k_i = u_{,i}$ ($\Leftrightarrow k_i\,dx^i = du$). Thus k_i satisfies

$$k_{i;j} = k_{j;i}\,, \tag{6.67}$$

$$k^i{}_{;i} = 0\,, \tag{6.68}$$

$$k_{i;j} = \xi_i\,k_j + \xi_j\,k_i\,, \tag{6.69}$$

for some ξ_i such that $\xi_i\, k^i = 0$, so that k_i is twist-free, expansion-free, geodesic and shear-free in the space-time with metric tensor $\underset{(0)}{g}_{ij}$. It follows from (6.66) that

$$k_{i|j} = k_{j|i}\,, \tag{6.70}$$

$$k^i{}_{|i} = 0\,, \tag{6.71}$$

$$k_{i|j} = \hat{\xi}_i\, k_j + \hat{\xi}_j\, k_i\,, \tag{6.72}$$

for some $\hat{\xi}_i$ such that $\hat{\xi}_i\, k^i = 0$, so that k_i is twist-free, expansion-free, geodesic and shear-free in the space-time with metric tensor g_{ij}. In particular we note that (6.71) is a direct consequence of (6.68) as a result of (D.9). The line element corresponding to the metric tensor (6.64) is the Ozsváth–Robinson–Rógza line element

$$- ds^2 = 2\, p^{-2}\, d\zeta\, d\bar{\zeta} + 2\, q^2\, p^{-2}\, du\, (dr + G\, du)\,, \tag{6.73}$$

with

$$G = \frac{1}{2}\, \kappa\, r^2 + q^{-1} \dot{q}\, r + p\, q^{-1}\, H(\zeta, \bar{\zeta}, r, u)\,, \tag{6.74}$$

$$p = 1 + \frac{\Lambda}{6}\, \zeta\, \bar{\zeta}\,, \quad q = \beta(u)\, \bar{\zeta} + \bar{\beta}(u)\, \zeta + \alpha(u)\left(1 - \frac{\Lambda}{6}\, \zeta\, \bar{\zeta}\right)\,, \tag{6.75}$$

and

$$\kappa = 2\, \beta\, \bar{\beta} + \frac{\Lambda}{3}\, \alpha^2\,. \tag{6.76}$$

We have replaced H in (6.64) by $q\, p^{-1}\, H$ for convenience. Now Einstein's field equations $R_{ij} = -\Lambda\, g_{ij}$ are satisfied provided

$$p^2 \frac{\partial^2 H}{\partial\zeta\,\partial\bar{\zeta}} + \frac{\Lambda}{3}\, H = 0\,, \tag{6.77}$$

and

$$\frac{\partial}{\partial\zeta}\left(p\, q^{-1} \frac{\partial H}{\partial r}\right) = 0\,, \quad \frac{\partial}{\partial\bar{\zeta}}\left(p\, q^{-1} \frac{\partial H}{\partial r}\right) = 0 \quad \text{and} \quad \frac{\partial^2 H}{\partial r^2} = 0\,. \tag{6.78}$$

The only non-vanishing Newman–Penrose component of the Weyl tensor is

$$\Psi_0 = q^{-2}\, p^2 \frac{\partial}{\partial\zeta}\left(q^2 \frac{\partial}{\partial\zeta}(p\, q^{-1} H)\right)\,, \tag{6.79}$$

which indicates that the Weyl tensor is Type N (pure gravitational radiation) in the Petrov classification, with degenerate principal null direction k^i. Solving (6.78) we have

$$H = p^{-1} q \, f(u) \, r + \mathcal{H}(\zeta, \bar{\zeta}, u) \, . \tag{6.80}$$

Substituting this into (6.77) and (6.79) we find that

$$p^2 \frac{\partial^2 \mathcal{H}}{\partial \zeta \partial \bar{\zeta}} + \frac{\Lambda}{3} \mathcal{H} = 0 \, , \tag{6.81}$$

and

$$\Psi_0 = q^{-2} p^2 \frac{\partial}{\partial \zeta} \left(q^2 \frac{\partial}{\partial \zeta} (p \, q^{-1} \mathcal{H}) \right) \, , \tag{6.82}$$

which indicates that the arbitrary function $f(u)$ in (6.80) is disposable. It can be removed from the line element (6.73), without affecting the algebraic form of the line element, by the coordinate transformation

$$u = \gamma(u') \, , \quad r = \left(\frac{d\gamma}{du'} \right)^{-1} r' \quad \text{with} \quad \frac{d^2 \gamma}{du'^2} = \left(\frac{d\gamma}{du'} \right)^2 f(\gamma(u')) \, . \tag{6.83}$$

Arguable the simplest geometrical construction is to have the null hyperplanes $u = $ constant *nonintersecting*. This requires $\kappa = 0$ and either $Re\beta = 0$ or $Im\beta = 0$ or $Re\beta = $ const. $\times Im\beta$. We shall take $Im\beta = 0$. With $\kappa = 0$ we must have $\Lambda < 0$ and so, writing

$$n = \sqrt{-\frac{\Lambda}{6}} = \frac{\beta}{\alpha} \, , \tag{6.84}$$

we have

$$p = 1 - n^2 \zeta \bar{\zeta} \quad \text{and} \quad q = \alpha \, |1 + n \, \zeta|^2 \, . \tag{6.85}$$

Now the line element (6.73) reads (noting that $q^{-1} \dot{q} = \alpha^{-1} \dot{\alpha}$)

$$- ds^2 = 2 \, p^{-2} d\zeta \, d\bar{\zeta} + 2 \, p^{-2} |1 + n \, \zeta|^4 \alpha \, du \, \{ d(\alpha \, r) + p \, q^{-1} H \, \alpha \, du \} \, . \tag{6.86}$$

A change of coordinates r, u replaced by $r' = r \alpha$ and $u = u(u')$ given by $du' = \alpha \, du$ effectively reduces α to unity so that we may write (6.86) as

$$- ds^2 = 2 \, p^{-2} d\zeta \, d\bar{\zeta} + 2 \, p^{-2} |1 + n \, \zeta|^4 \, du \, \{ dr + p \, |1 + n \, \zeta|^{-2} H \, du \} \, . \tag{6.87}$$

To exhibit a solution of the field equation (6.77) we note that if $W = (p/q)^2$ then, in general,

$$p^2 \frac{\partial^2 W}{\partial \zeta \partial \bar{\zeta}} + \frac{\Lambda}{3} W = 3 \left(\frac{p}{q}\right)^4 \kappa . \qquad (6.88)$$

Hence when $\kappa = 0$ we have a solution of (6.77) given by

$$H = c_0 \frac{p^2}{q^2} , \qquad (6.89)$$

and, for simplicity, we will take $c_0 = $ constant (although we can have $c_0 = c_0(u)$). Now (6.87) reads

$$- ds^2 = 2 p^{-2} d\zeta \, d\bar{\zeta} + 2 p^{-2} |1 + n \zeta|^4 du \, \{dr + c_0 p^3 |1 + n \zeta|^{-6} du\} , \qquad (6.90)$$

with p given by (6.85). This is the solution of Ozsváth [10]. To put it in a more familiar form make the coordinate transformation:

$$n \zeta = \frac{2\sqrt{2} n + e^{-\sqrt{2} n X} + i \sqrt{2} n Y}{2\sqrt{2} n - e^{-\sqrt{2} n X} - i \sqrt{2} n Y} . \qquad (6.91)$$

Absorbing constants into u and c_0 we arrive at

$$- ds^2 = dX^2 + e^{2\sqrt{2} n X} (dY^2 + 2 \, du \, dr) + f_0 e^{-\sqrt{2} n X} du^2 , \qquad (6.92)$$

with f_0 a constant. This is the form of the solution given in [10]. Its significance is that the maximal group of isometries of this space-time has five parameters and, in this sense, it appears to be the closest one can get to plane gravitational waves when a cosmological constant is present [11].

6.4   de Sitter Space-Time Revisited

When introducing parametrisations, as in (6.40), exceptional cases can arise. For example if $w^0 = 0$ in (6.40) then the parametrisation given is impossible if $\Lambda > 0$. An example of this occurs when the de Sitter line element is given in the quite familiar form of

$$ds^2 = dt^2 - e^{2\sqrt{2} k t} (dx^2 + dy^2 + dz^2) \quad \text{with} \quad k = \sqrt{\Lambda/6} . \qquad (6.93)$$

The only null hyperplanes available in the space-time with this line element are $u(t, x, y, z) = $ constant with $u(t, x, y, z)$ given implicitly by the equation

$$|\mathbf{x} - \mathbf{w}(u)|^2 = \frac{1}{2k^2} e^{-2\sqrt{2}kt} , \tag{6.94}$$

where the components of the 3-vector $\mathbf{w}(u) = (w^\alpha(u))$ are three arbitrary functions of u. To establish this result we begin by writing (6.93) in manifestly conformally flat form by putting

$$T(t) = \frac{1}{\sqrt{2}k} e^{-\sqrt{2}kt} , \tag{6.95}$$

and then rewriting (6.93) as

$$ds^2 = \frac{1}{2k^2 T^2} (dT^2 - dx^2 - dy^2 - dz^2) = \lambda^2 \eta_{a'b'} dx^{a'} dx^{b'} , \tag{6.96}$$

with

$$\lambda^{-1} = \sqrt{2}kT , \tag{6.97}$$

and here $x^{a'} = (T, x, y, z) = (T, x^\alpha)$ for $a' = 0', 1', 2', 3'$. We now utilise the argument of Sect. 6.1 which first notes that the properties of a congruence in space-time being null, geodesic and shear-free (in the optical sense) are conformally invariant and that the only shear-free null hypersurfaces in Minkowskian space-time are null hyperplanes or null cones (or portions thereof). We therefore begin, as in Sect. 6.1, by considering shear-free null hyperplanes in Minkowskian space-time. These have equations $u(T, x, y, z) = $ constant with u given implicitly by the equation

$$\eta_{b'c'} a^{b'}(u) x^{c'} + b(u) = 0 \quad \text{with} \quad \eta_{b'c'} a^{b'} a^{c'} = 0 . \tag{6.98}$$

The covariant vector field $u_{,a'}$ is null, geodesic, shear-free and expansion-free (see Sect. 6.1). In the space-time with line element (6.96), and thus with metric $g_{a'b'} = \lambda^2 \eta_{a'b'}$, we have

$$\lambda_{,a'} = -\frac{1}{\sqrt{2}kT^2} \delta^{0'}_{a'} , \tag{6.99}$$

and

$$u_{,a';b'} = \xi_{a'} u_{,b'} + \xi_{b'} u_{,a'} + \lambda^{-1} \eta^{r's'} u_{,r'} \lambda_{,s'} \eta_{a'b'} , \tag{6.100}$$

with

$$\xi'_c = -f^{-1}\dot{a}_{c'} - \lambda^{-1}\lambda_{,c'} - \frac{1}{2}f^{-1}\dot{f}\,u_{,c'}\,, \tag{6.101}$$

and

$$f = \dot{b} + \dot{a}_{c'}\,x^{c'}\,, \tag{6.102}$$

and the dot denoting differentiation with respect to u. The fact that $u_{,a';b'}$ has the algebraic form of the right hand side of (6.100) confirms that, in the space-time with metric $g_{a'b'}$, $u_{,a'}$ is geodesic and shear-free following the Robinson–Trautman [3] test for these properties. However we also require $u_{,a'}$ to be expansion-free if $u = $ constant are to be null hyper*planes* in the space-time with metric $g_{a'b'}$. This means that we must have

$$g^{a'b'}\,u_{,a';b'} = -2\,\lambda^{-3}f^{-1}\lambda_{,b'}\,a^{b'} = 0 \quad\Rightarrow\quad a^{0'} = 0\,. \tag{6.103}$$

But since $a_{b'}\,a^{b'} = 0$ this means that we must have $a^{b'} = 0$ and so we get no null hyper*planes* in the de Sitter space-time in this case. Next we choose $u(T, x, y, z)$ to be given implicitly by

$$\eta_{a'b'}\,(x^{a'} - w^{a'}(u))(x^{b'} - w^{b'}(u)) = 0\,. \tag{6.104}$$

Thus each $u = $ constant is a null cone in Minkowskian space-time. The vertices of the null cones $u = $ constant lie on an arbitrary world line $x^{a'} = w^{a'}(u)$. From (6.104) it follows that

$$u_{,a'} = \frac{x^{a'} - w^{a'}(u)}{\eta_{c'd'}\,\dot{w}^{c'}(x^{d'} - w^{d'})}\,. \tag{6.105}$$

These null cones $u = $ constant are generated by expanding, shear-free null geodesics in Minkowskian space-time. In the de Sitter space-time $u = $ constant will be null hyper*planes* provided $u_{,a'}$ is expansion-free. This is the case if

$$g^{a'b'}\,u_{,a';b'} = 0 \quad\Rightarrow\quad 1 + \sqrt{2}\,k\,T\,\lambda_{,a'}\,(x^{a'} - w^{a'}) \quad\Rightarrow\quad w^{0'}(u) = 0\,. \tag{6.106}$$

Hence the world line of the vertices of the null cones $u = $ constant in Minkowskian space-time is space-like. Consequently the null cones must intersect each other and so the corresponding null hyperplanes in the de Sitter space-time also intersect each other since they are described by (6.104) too, which now reads

$$(x^{\alpha} - w^{\alpha})(x^{\alpha} - w^{\alpha}) = T^2 = \frac{1}{2\,k^2}\,e^{-2\sqrt{2}kt}\,. \tag{6.107}$$

Using (6.95) we see that (6.107) coincides with (6.94).

To verify that $u = $ constant, with $u(t, x, y, z)$ given by (6.94), are indeed null hyper*planes* in the de Sitter space-time with metric tensor g_{ab} given via the line element (6.93) we note that with $X^\alpha = x^\alpha - w^\alpha(u)$ and $\rho = \dot{w}^\alpha X^\alpha$ it follows from (6.94) that

$$u_{,a} = \left(\frac{e^{-2\sqrt{2}kt}}{\sqrt{2}k\rho}, \frac{X^\alpha}{\rho} \right) ,$$
(6.108)

and hence

$$g^{ab} u_{,a} u_{,b} = 0 ,$$
(6.109)

as a consequence of (6.94). Making use of the formulas

$$X^\alpha{}_{,\beta} = \delta^\alpha_\beta - \frac{\dot{w}^\alpha X^\beta}{\rho} ,$$
(6.110)

$$\rho_{,\alpha} = \dot{w}^\alpha + (\ddot{w}^\beta X^\beta - \dot{w}^\beta \dot{w}^\beta) \frac{X^\alpha}{\rho} ,$$
(6.111)

and

$$\rho_{,0} = \frac{\partial \rho}{\partial t} = (\ddot{w}^\beta X^\beta - \dot{w}^\beta \dot{w}^\beta) u_{,0} ,$$
(6.112)

with $u_{,0}$ (and $u_{,\alpha}$) given in (6.108), we find that

$$u_{,\alpha;\beta} = -\frac{1}{\rho} \{ \dot{w}^\alpha u_{,\beta} + \dot{w}^\beta u_{,\alpha} + (\ddot{w}^\gamma X^\gamma - \dot{w}^\gamma \dot{w}^\gamma) u_{,\alpha} u_{,\beta} \} ,$$
(6.113)

$$u_{,\alpha;0} = -\sqrt{2}k u_{,\alpha} - \frac{1}{\rho} \{ \dot{w}^\alpha + (\ddot{w}^\gamma X^\gamma - \dot{w}^\gamma \dot{w}^\gamma) u_{,\alpha} \} u_{,0} ,$$
(6.114)

and

$$u_{,0;0} = -2\sqrt{2}k u_{,0} - \frac{1}{\rho} (\ddot{w}^\gamma X^\gamma - \dot{w}^\gamma \dot{w}^\gamma)(u_{,0})^2 .$$
(6.115)

Hence we can write

$$u_{a;b} = \psi_a u_{,b} + \psi_b u_{,a} ,$$
(6.116)

with

$$\psi_a = \left(-\sqrt{2}k - \frac{1}{2\rho} (\ddot{w}^\gamma X^\gamma - \dot{w}^\gamma \dot{w}^\gamma) u_{,0}, \ -\frac{\dot{w}^\alpha}{\rho} - \frac{1}{2\rho} (\ddot{w}^\gamma X^\gamma - \dot{w}^\gamma \dot{w}^\gamma) u_{,\alpha}, \right) .$$
(6.117)

It follows immediately from (6.116) that $u_{,a}$ is geodesic and shear-free, on account of the Robinson–Trautman [3] test for these properties, and in addition $u_{,a}$ is expansion-free since

$$g^{ab} \psi_a u_{,b} = 0 , \qquad (6.118)$$

which can be readily verified directly. Thus we have confirmed that in the de Sitter space-time with line element (6.93) the hypersurfaces $u = $ constant, with $u(t, x, y, z)$ given implicitly by (6.94) are *null hyperplanes*.

Following (6.107) we can write

$$X^1 = \frac{1}{\sqrt{2}k} e^{-\sqrt{2}kt} \sin\xi \cos\eta ,$$

$$X^2 = \frac{1}{\sqrt{2}k} e^{-\sqrt{2}kt} \sin\xi \sin\eta , \qquad (6.119)$$

$$X^3 = \frac{1}{\sqrt{2}k} e^{-\sqrt{2}kt} \cos\xi ,$$

and then

$$\rho = \frac{1}{\sqrt{2}k} e^{-\sqrt{2}kt} A \quad \text{with } A = \dot{w}^1 \sin\xi \cos\eta + \dot{w}^2 \sin\xi \sin\eta + \dot{w}^3 \cos\xi .$$
$$(6.120)$$

We shall make use of these in the form

$$X^\alpha = \frac{\rho}{A} (\sin\xi \cos\eta, \sin\xi \sin\eta, \cos\xi) \quad \text{with } \frac{\rho}{A} = \frac{1}{\sqrt{2}k} e^{-\sqrt{2}kt} . \qquad (6.121)$$

From this we find that

$$dx^\alpha dx^\alpha = \left(d\left(\frac{\rho}{A}\right) \right)^2 + \left(\frac{\rho}{A}\right)^2 (d\xi^2 + \sin^2\xi \, d\eta^2)$$
$$+ 2 \, d\rho \, du + \left(\dot{w}^\alpha \dot{w}^\alpha - \frac{2\rho}{A} \frac{\partial A}{\partial u} \right) du^2 . \qquad (6.122)$$

Substituting into the line element (6.93), after first noting that

$$d\left(\frac{\rho}{A}\right) = -\sqrt{2}k \left(\frac{\rho}{A}\right) dt \quad \text{and} \quad e^{\sqrt{2}kt} = \frac{1}{\sqrt{2}k} \left(\frac{A}{\rho}\right) , \qquad (6.123)$$

we arrive at

$$-ds^2 = \frac{1}{2\,k^2}\,(d\xi^2 + \sin^2\xi\,d\eta^2) + \frac{1}{k^2}\left(\frac{A}{\rho}\right)^2 d\rho\,du$$

$$+ \frac{1}{2\,k^2}\left(\frac{A}{\rho}\right)^2\left(\dot{w}^\alpha\,\dot{w}^\alpha - \frac{2\,\rho}{A}\,\dot{A}\right)du^2\,, \tag{6.124}$$

with the dot as always denoting differentiation with respect to u. Next make the coordinate transformation

$$\rho = -\frac{1}{2\,k^2\,r}\,, \tag{6.125}$$

and (6.124) becomes

$$-ds^2 = \frac{1}{2\,k^2}\,(d\xi^2 + \sin^2\xi\,d\eta^2) + 2\,A^2\,du\left\{dr + \left(k^2\,\dot{w}^\alpha\,\dot{w}^\alpha\,r^2 + A^{-1}\,\dot{A}\,r\right)du\right\}\,. \tag{6.126}$$

Finally put

$$k\,\zeta = e^{i\eta}\,\tan\frac{\xi}{2}\,, \tag{6.127}$$

which results in

$$A = (1 + k^2\,\zeta\,\bar{\zeta})^{-1}\left\{k\,(\dot{w}^1 - i\,\dot{w}^2)\,\zeta + k\,(\dot{w}^1 + i\,\dot{w}^2)\,\bar{\zeta} + \dot{w}^3\,(1 - k^2\,\zeta\,\bar{\zeta})\right\}$$

$$= p^{-1}\,q\,, \tag{6.128}$$

with

$$p = 1 + k^2\,\zeta\,\bar{\zeta}\,, \tag{6.129}$$
$$q = \bar{\beta}\,\zeta + \beta\,\bar{\zeta} + \alpha\,(1 - k^2\,\zeta\bar{\zeta})\,, \tag{6.130}$$

and

$$\beta = k\,(\dot{w}^1 + i\,\dot{w}^2)\,,\quad \alpha = \dot{w}^3\,. \tag{6.131}$$

Now (6.126) takes the Ozsvàth–Robinson–Rózga [4] form

$$-ds^2 = 2\,p^{-2}\,d\zeta\,d\bar{\zeta} + 2\,p^{-2}\,q^2\,du\left\{dr + \left(\frac{1}{2}\kappa\,r^2 + q^{-1}\dot{q}\,r\right)du\right\}\,, \tag{6.132}$$

with

$$\kappa = 2\,\beta\,\bar{\beta} + \frac{\Lambda}{3}\,\alpha^2 = 2\,k^2\,\dot{w}^\alpha\,\dot{w}^\alpha > 0\,. \tag{6.133}$$

6.5   Collision of Gravitational Waves and $\Lambda \neq 0$

Finding the space-time structure after the collision of gravitational and/or electro-magnetic waves is a difficult problem in general relativity due to the non-linearity of the field equations. The problem is simplified by specialising to impulsive and/or shock waves which are plane and homogeneous and then exact solutions can be found, with the Khan–Penrose [12] and Bell–Szekeres [13] solutions among the most famous. Up to recently no solution where a cosmological constant appears after the collision of two homogeneous, plane, impulsive gravitational waves has yet been found. We describe here a solution in which a cosmological constant occurs after the collision and in which the cosmological constant is necessarily constructed from the amplitudes of the incoming gravitational waves [14]. A remarkable property of the space-time following the head-on collision of the gravitational waves in this case is that it is a space-time of constant non-zero curvature. In other words it is a de Sitter or anti-de Sitter space-time and is thus curvature singularity-free, in contrast to the Khan–Penrose model. The solution derived here is not an extension of the Khan–Penrose solution since it has the property that if the cosmological constant vanishes then at least one of the incoming waves vanishes. The post collision model presented here can be explained in terms of a redistribution of the energy in the incoming waves and this is described in some detail. However it is an open question to discover a mechanism which triggers the transition from a vacuum to a region in which a cosmological constant must be non-zero. The products of the collision, in addition to a cosmological constant, include impulsive gravitational waves (as in the Khan–Penrose collision) and light-like shells of matter. When reasonable physical restrictions are invoked the post collision region of space-time is anti-de Sitter space-time.

Each incoming gravitational wave prior to the head-on collision is a homoge-neous, plane impulsive wave propagating in a vacuum. Such a wave is described in general relativity by a space-time with line element

$$ds^2 = -(1 + k\,u_+)^2 dx^2 - (1 - k\,u_+)^2 dy^2 + 2\,du\,dv\,, \tag{6.134}$$

where k is a constant (introduced for convenience) and $u_+ = u\,\vartheta(u)$ where $\vartheta(u) = 1$ for $u > 0$ and $\vartheta(u) = 0$ for $u < 0$ is the Heaviside step function. The metric given via this line element satisfies Einstein's vacuum field equations everywhere (in particular on $u = 0$). The only non-vanishing Newman–Penrose component of the Riemann curvature tensor on the tetrad given via the 1-forms $\vartheta^1 = (1 + k\,u_+)dx$, $\vartheta^2 = (1 - k\,u_+)dy$, $\vartheta^3 = dv$, $\vartheta^4 = du$ is

$$\Psi_4 = -k\,\delta(u)\,. \tag{6.135}$$

Thus the curvature tensor is type N (the radiative type) in the Petrov classification with the vector field $\partial/\partial v$ the degenerate principal null direction and therefore the propagation direction of the history of the wave (the null hypersurface $u = 0$) in space-time. The wave profile is the delta function, singular on $u = 0$, and thus the wave is an impulsive wave. There are two families of intersecting null hypersurfaces $u = $ constant and $v = $ constant in the space-time with line element (6.134). A homogeneous, plane impulsive gravitational wave propagating in a vacuum in the opposite direction to that with history $u = 0$ has history $v = 0$ and this is described by a space-time with line element

$$ds^2 = -(1 + l\,v_+)^2 dx^2 - (1 - l\,v_+)^2 dy^2 + 2\,du\,dv\,, \qquad (6.136)$$

where l is a convenient constant and $v_+ = v\,\vartheta(v)$. The Ricci tensor vanishes everywhere when calculated with the metric tensor given by this line element. The only non-vanishing Newman–Penrose component of the Riemann curvature tensor on the tetrad given via the 1-forms $\vartheta^1 = (1 + l\,v_+)dx$, $\vartheta^2 = (1 - l\,v_+)dy$, $\vartheta^3 = dv$, $\vartheta^4 = du$ is

$$\Psi_0 = -l\,\delta(v)\,, \qquad (6.137)$$

indicating a Petrov type N curvature tensor with degenerate principal null direction $\partial/\partial u$.

From the space-time point of view we visualise the collision problem as follows: we envisage a pre-collision vacuum region of space-time $v < 0$ with line element (6.134) and a pre-collision vacuum region of space-time $u < 0$ with line element (6.136) (with both line elements coinciding when $v < 0$ *and* $u < 0$). The waves collide at $u = v = 0$ and the post collision region of the space-time corresponds to $u > 0$ *and* $v > 0$. In this region the line element has the form [12, 15, 16]

$$ds^2 = -e^{-U}(e^V dx^2 + e^{-V} dy^2) + 2\,e^{-M} du\,dv\,, \qquad (6.138)$$

where U, V, M are each functions of u, v. These functions must satisfy the following conditions on the null hypersurface boundaries of the region $u > 0$, $v > 0$:

$$v = 0\,,\ u \geq 0\ \Rightarrow\ e^{-U} = 1 - k^2 u^2\,,\ e^V = \frac{1 + k\,u}{1 - k\,u}\,,\ M = 0\,, \qquad (6.139)$$

and

$$u = 0\,,\ v \geq 0\ \Rightarrow\ e^{-U} = 1 - l^2 v^2\,,\ e^V = \frac{1 + l\,v}{1 - l\,v}\,,\ M = 0\,. \qquad (6.140)$$

Einstein's field equations with a cosmological constant Λ in the region $u > 0$, $v > 0$ calculated with the metric tensor given by the line element (6.138) read:

$$U_{uv} = U_u\, U_v - \Lambda\, e^{-M}\ , \tag{6.141}$$

$$2\, V_{uv} = U_u\, V_v + U_v\, V_u\ , \tag{6.142}$$

$$2\, U_{uu} = U_u^2 + V_u^2 - 2\, M_u\, U_u\ , \tag{6.143}$$

$$2\, U_{vv} = U_v^2 + V_v^2 - 2\, M_v\, U_v\ , \tag{6.144}$$

$$2\, M_{uv} = V_u\, V_v - U_u\, U_v\ , \tag{6.145}$$

where the subscripts denote partial derivatives. To implement our strategy below for solving (6.141)–(6.165) subject to the boundary conditions (6.139) and (6.140) we will need to know V_v at $v = 0$, which we denote by $(V_v)_{v=0}$, and V_u at $u = 0$, which we denote by $(V_u)_{u=0}$. We already have from (6.139) and (6.140):

$$(V_u)_{v=0} = \frac{2\,k}{1 - k^2 u^2} \quad \text{and} \quad (V_v)_{u=0} = \frac{2\,l}{1 - l^2 v^2}\ , \tag{6.146}$$

and also

$$(U_u)_{v=0} = \frac{2\,k^2 u}{1 - k^2 u^2} \quad \text{and} \quad (U_v)_{u=0} = \frac{2\,l^2 v}{1 - l^2 v^2}\ . \tag{6.147}$$

In order to compute $(V_v)_{v=0}$ and $(V_u)_{u=0}$ we must first calculate $(U_v)_{v=0}$ and $(U_u)_{u=0}$. We obtain these latter quantities by evaluating (6.141) at $u = 0$ and at $v = 0$ and solving the resulting first order ordinary differential equations. The constants of integration which arise are determined from the fact that U_v and U_u both vanish when $u = 0$ *and* $v = 0$, which follows from (6.147). We then find that

$$(U_v)_{v=0} = -\frac{\Lambda\, u\, (1 - \frac{1}{3}k^2 u^2)}{1 - k^2 u^2} \quad \text{and} \quad (U_u)_{u=0} = -\frac{\Lambda\, v\, (1 - \frac{1}{3}l^2 v^2)}{1 - l^2 v^2}\ . \tag{6.148}$$

Now evaluating (6.142) at $v = 0$ and at $u = 0$ provides us with a pair of first order ordinary differential equations for $(V_v)_{v=0}$ and $(V_u)_{u=0}$. These equations are straightforward to solve and the resulting constants of integration are determined from the fact that $V_u = 2\,k$ and $V_v = 2\,l$ when $u = 0$ *and* $v = 0$, which follows from (6.146). The final results are:

$$(V_v)_{v=0} = \left(2\,l + \frac{\Lambda}{3\,k}\right)(1 - k^2 u^2)^{-1/2} - \frac{\Lambda}{3\,k}\left(\frac{1 + k^2 u^2}{1 - k^2 u^2}\right)\ , \tag{6.149}$$

$$(V_u)_{u=0} = \left(2\,k + \frac{\Lambda}{3\,l}\right)(1 - l^2 v^2)^{-1/2} - \frac{\Lambda}{3\,l}\left(\frac{1 + l^2 v^2}{1 - l^2 v^2}\right)\ . \tag{6.150}$$

Dividing (6.142) successively by V_u and by V_v and then differentiating the resulting equations and combining them we obtain

$$2 \frac{\partial^2}{\partial u \partial v} \log \frac{V_u}{V_v} = \left(U_u \frac{V_v}{V_u} \right)_u - \left(U_v \frac{V_u}{V_v} \right)_v . \tag{6.151}$$

This suggests that we examine the possibility of a separation of variables:

$$\frac{V_u}{V_v} = \frac{A(u)}{B(v)} , \tag{6.152}$$

for some functions $A(u)$ and $B(v)$. The resulting mathematical simplification is that (6.152) becomes a first order wave equation for V (see below) and that (6.151) becomes a second order wave equation for U. From a physical point of view we have shown [17] that if, as is the case in general, two systems of backscattered gravitational waves exist in the post collision region (one with propagation direction $\partial/\partial u$ in space-time and one with propagation direction $\partial/\partial v$) then (6.152) implies that there exists a frame of reference in which the energy densities of the two systems of waves are equal. Using (6.146), (6.149) and (6.150) determines the right hand side of (6.152) and the result is

$$\frac{V_u}{V_v} = \frac{k \left[\left(1 + \frac{\Lambda}{6kl} \right) \sqrt{1 - l^2 v^2} - \frac{\Lambda}{6kl}(1 + l^2 v^2) \right]}{l \left[\left(1 + \frac{\Lambda}{6kl} \right) \sqrt{1 - k^2 u^2} - \frac{\Lambda}{6kl}(1 + k^2 u^2) \right]} . \tag{6.153}$$

Hence this equation can be written as a first order wave equation

$$V_{\bar{u}} = V_{\bar{v}} , \tag{6.154}$$

with $\bar{u}(u)$ and $\bar{v}(v)$ given by the differential equations

$$\frac{d\bar{u}}{du} = k \left[\left(1 + \frac{\Lambda}{6kl} \right) \sqrt{1 - k^2 u^2} - \frac{\Lambda}{6kl}(1 + k^2 u^2) \right]^{-1} , \tag{6.155}$$

$$\frac{d\bar{v}}{dv} = l \left[\left(1 + \frac{\Lambda}{6kl} \right) \sqrt{1 - l^2 v^2} - \frac{\Lambda}{6kl}(1 + l^2 v^2) \right]^{-1} . \tag{6.156}$$

These two equations are interesting in general. However there are clearly two stand-out special cases: $\Lambda = 0$ and $\Lambda = -6kl$. The case $\Lambda = 0$ corresponds to the Khan–Penrose [12] space-time which is discussed in detail from the current point of view in [17].

With $\Lambda = -6kl$ we can solve (6.155) and (6.156), requiring $\bar{u} = 0$ when $u = 0$ and $\bar{v} = 0$ when $v = 0$, with

$$\bar{u} = \tan^{-1} k u , \quad \bar{v} = \tan^{-1} l v . \tag{6.157}$$

By (6.154) we have $V = V(\bar{u} + \bar{v})$ and the boundary condition (2.84) written in terms of $\bar{u}, \bar{v}$ reads: when $\bar{v} = 0$, $V = \log\left(\frac{1+\tan\bar{u}}{1-\tan\bar{u}}\right)$. Hence

$$V(\bar{u} + \bar{v}) = \log\left(\frac{1 + \tan(\bar{u} + \bar{v})}{1 - \tan(\bar{u} + \bar{v})}\right), \qquad (6.158)$$

and restoring the coordinates u, v we have

$$V(u, v) = \log\left(\frac{1 - k\,l\,u\,v + k\,u + l\,v}{1 - k\,l\,u\,v - k\,u - l\,v}\right), \qquad (6.159)$$

for $u \geq 0, v \geq 0$ provided $\Lambda = -6\,k\,l$. Next writing (6.142) in terms of the variables $\bar{u}, \bar{v}$ and using (6.154) and (6.158) we have

$$U_{\bar{u}} + U_{\bar{v}} = \frac{8\,\tan(\bar{u} + \bar{v})}{1 - \tan^2(\bar{u} + \bar{v})}, \qquad (6.160)$$

which is easily integrated to yield

$$e^{-U} = C(\bar{u} - \bar{v})\left(\frac{1 - \tan^2(\bar{u} + \bar{v})}{1 + \tan^2(\bar{u} + \bar{v})}\right), \qquad (6.161)$$

where $C(\bar{u} - \bar{v})$ is a function of integration. When $\bar{v} = 0$ the boundary condition (6.139) requires $e^{-U} = 1 - \tan^2\bar{u}$ and so $C(\bar{u}) = 1 + \tan^2\bar{u}$. Hence restoring the coordinates u, v we have $U(u, v)$ given by

$$e^{-U} = \frac{(1 - k\,l\,u\,v)^2 - (k\,u + l\,v)^2}{(1 + k\,l\,u\,v)^2}, \qquad (6.162)$$

for $u \geq 0, v \geq 0$. In the light of (6.161) we see that U is a linear combination of a function of $\bar{u} - \bar{v}$ and a function of $\bar{u} + \bar{v}$ and thus satisfies the second order wave equation $U_{\bar{u}\bar{u}} = U_{\bar{v}\bar{v}}$. This wave equation is the equation that (6.151) reduces to when (6.154) holds and the barred coordinates are used.

With $V(u, v)$ and $U(u, v)$ given by (6.159) and (6.162) we use the field equation (6.141) with $\Lambda = -6\,k\,l$ to calculate $M(u, v)$. The result is

$$M(u, v) = 2\,\log(1 + k\,l\,uv), \qquad (6.163)$$

and this clearly satisfies the boundary conditions (6.139) and (6.140). Now with V, U and M determined a lengthy calculation verifies that the remaining field equations (6.143)–(6.165) are automatically satisfied. Thus the line element (6.138) of the post collision region reads

$$ds^2 = \frac{-(1 - k\,l\,u\,v + k\,u + l\,v)^2 dx^2 - (1 - k\,l\,u\,v - k\,u - l\,v)^2 dy^2 + 2\,du\,dv}{(1 + k\,l\,u\,v)^2}.$$

$$(6.164)$$

If in the metric tensor components here we replace u, v by $u_+ = u\,\vartheta\,(u)$, $v_+ = v\,\vartheta\,(v)$ we obtain in a single line element the expressions (6.134) and (6.136) for the pre-collision regions and (6.164) for the post collision region. In particular this will enable us to calculate the physical properties of the boundaries $v = 0$, $u \geq 0$ and $u = 0$, $v \geq 0$ of the post collision region.

6.6 Post Collision Physical Properties

The Ricci tensor components of the space-time are given by

$$R_{ab} = 6\,k\,l\,\vartheta\,(u)\,\vartheta\,(v)\,g_{ab} + \frac{2\,k\,l\,u_+ (k^2 u_+^2 - 3)}{1 - k^2 u_+^2}\delta(v)\,\delta_a^3\delta_b^3$$

$$+ \frac{2\,k\,l\,v_+ (l^2 v_+^2 - 3)}{1 - l^2 v_+^2}\delta(u)\,\delta_a^4\delta_b^4\,. \tag{6.165}$$

This confirms that the space-time region $u > 0, v > 0$ is a solution of the field equations with a cosmological constant, $R_{ab} = -\Lambda\,g_{ab}$, with $\Lambda = -6\,k\,l$ and that there are light-like shells with the boundaries $v = 0, u \geq 0$ and $u = 0, v \geq 0$ as histories, corresponding to the delta function terms in (6.165). The light-like shells have no isotropic surface pressure [18] and the surface energy densities are $\mu_{(1)}$ and $\mu_{(2)}$ given by

$$8\,\pi\,\mu_{(1)} = \frac{\Lambda\,u}{3}\left(\frac{k^2 u^2 - 3}{1 - k^2 u^2}\right) \quad \text{on } v = 0,\ u \geq 0, \tag{6.166}$$

and

$$8\,\pi\,\mu_{(2)} = \frac{\Lambda\,v}{3}\left(\frac{l^2 v^2 - 3}{1 - l^2 v^2}\right) \quad \text{on } u = 0,\ v \geq 0. \tag{6.167}$$

The light-like shells must have positive surface energy densities. The only way to realise this on $v = 0, u \geq 0$ (respectively on $u = 0, v \geq 0$) is to have $kl > 0$ and $k^2 u^2 < 1$ (respectively $kl > 0$ and $l^2 v^2 < 1$). Thus the cosmological constant $\Lambda = -6\,k\,l$ must be negative. These restrictions on the coordinates are less restrictive than the condition $k^2 u^2 + l^2 v^2 < 1$ for $u \geq 0$ and $v \geq 0$ required in the Khan–Penrose post collision space-time on account of the presence of the curvature singularity. These restrictions on the coordinates also avoid infinite surface energy densities in the shells which are arguably as serious as a curvature singularity. Light-like shells did not appear in the Khan–Penrose model and their presence here is due to the non-zero cosmological constant.

The Newman–Penrose components of the Weyl conformal curvature tensor are given by

$$\Psi_0 = -\frac{l(1 + k^2 u_+^2)}{1 - k^2 u_+^2}\delta(v) \,, \quad \Psi_4 = -\frac{k(1 + l^2 v_+^2)}{1 - l^2 v_+^2}\delta(u) \,, \quad \Psi_1 = \Psi_2 = \Psi_3 = 0 \,.$$

$$(6.168)$$

Thus the boundaries $v = 0, 0 \le k^2 u^2 < 1$ and $u = 0, 0 \le l^2 v^2 < 1$ are the histories of impulsive gravitational waves corresponding to the delta function terms here. The post collision region $u > 0, v > 0$ is conformally flat and is a space-time of constant curvature with Riemann curvature tensor components given by

$$R_{abcd} = -2\,k\,l(g_{ac}\,g_{bd} - g_{ad}\,g_{bc}) \,. \tag{6.169}$$

Hence this region of space-time does not possess a curvature singularity, in striking contrast to the post collision region of the Khan–Penrose space-time.

For this model collision the energy in the incoming impulsive gravitational waves is re-distributed after the collision into two light-like shells of matter and two impulsive gravitational waves moving away from each other followed by a space-time of constant curvature (i.e. de Sitter or anti-de Sitter space-time). When the surface energy densities of the post collision light-like shells of matter are required to be positive the space-time of constant curvature must be anti-de Sitter space-time.

References

1. R. Penrose, *General Relativity*, papers in honour of J.L. Synge (Clarendon Press, Oxford, 1972), p. 112
2. P.A. Hogan, Int. J. Mod. Phys. D **27**, 1850045 (2018)
3. I. Robinson, A. Trautman, J. Math. Phys. **24**, 1425 (1983)
4. I. Ozsváth, I. Robinson, K. Rózga, J. Math. Phys. **26**, 1755 (1985)
5. H.V. Tran, Ph.D. Thesis, University of Texas at Dallas (1988)
6. I. Robinson, University of Texas at Dallas Internal Report (1983)
7. A. Trautman, *Recent Developments in General Relativity* (Pergamon Press Inc., New York, 1962), p. 459
8. R.P. Kerr, A. Schild, *Atti del convegno sulla relatività generale: problemi dell'energia e onde gravitationali* (Barbèra, Firenze, 1965), p. 222
9. R.P. Kerr, A. Schild, Proc. Symp. Appl. Math. **17**, 199 (1965)
10. I. Ozsváth, J. Math. Phys. **6**, 590 (1965)
11. I. Ozsváth, *Gravitation and Geometry*, ed. by A. Trautman, W. Rindler (Bibliopolis, Naples, 1987), p. 309
12. K.A. Khan, R. Penrose, Nature **229**, 185 (1971)
13. P. Bell, P. Szekeres, Gen. Rel. Grav. **5**, 275 (1974)
14. C. Barrabès, P.A. Hogan, Phys. Rev. D **92**, 044032 (2015)
15. P. Szekeres, Nature **228**, 1183 (1970)
16. P. Szekeres, J. Math. Phys. **13**, 286 (1972)
17. C. Barrabès, P.A. Hogan, *Singular Null Hypersurfaces in General Relativity* (World Scientific, Singapore, 2003)
18. C. Barrabès, W. Israel, Phys. Rev. D **43**, 1129 (1991)

Small Magnetic Black Hole 7

Abstract

A magnetic black hole in Einstein–Maxwell theory is described by the potential 1-form of a magnetic pole and a line element which coincides with the Reissner–Nordström line element dependent upon two parameters representing the mass of the magnetic pole and the magnetic monopole moment. When such an object interacts with external fields it behaves differently to an electric black hole. This can be exhibited most easily if one assumes that the magnetic black hole has small mass and small magnetic monopole moment (and is therefore a small magnetic black hole). In this case the equations of motion can be derived approximately with sufficient accuracy to include electromagnetic radiation reaction, external 4-force and tail terms.

7.1   Magnetic Poles

Notwithstanding the absence so far of an observation in nature of a magnetic monopole, there exist stimulating theoretical studies associated with such objects beginning with the seminal work of Dirac [1] and followed, most notably, by the Wu–Yang monopole [2] and the 't Hooft–Polyakov monopole [3, 4]. Magnetic monopole-like phenomena also occur in condensed matter physics (see, for example, [5], ch.17). The technique for studying the motion of a small magnetic black hole involves using the vacuum Einstein–Maxwell field equations together with the assumptions that the small magnetic black hole is isolated and the wave fronts produced by the motion of the black hole are smoothly deformed 2-spheres in the vicinity of the black hole. This technique, and its historical background, are described in detail in [6] (see also [7] and references therein). The application described below can be found in [8].

© The Author(s), under exclusive license to Springer Nature Switzerland AG 2021   153
P. A. Hogan, D. Puetzfeld, *Frontiers in General Relativity*, Lecture Notes
in Physics 984, https://doi.org/10.1007/978-3-030-69370-1_7

We begin with the Minkowskian line element in rectangular Cartesian coordinates and time $X^i = (T, X, Y, Z)$:

$$ds^2 = dT^2 - dX^2 - dY^2 - dZ^2 = \eta_{ij}\, dX^i\, dX^j \,. \tag{7.1}$$

For a magnetic field the Maxwell 2-form takes the form

$$F = \frac{1}{2} F_{ij}\, dX^i \wedge dX^j = B_1\, dZ \wedge dY + B_2\, dX \wedge dZ + B_3\, dY \wedge dX \,, \tag{7.2}$$

with

$$\mathbf{B} = (B_1, B_2, B_3) \,, \tag{7.3}$$

the magnetic 3-vector. For a magnetic pole located at $X = Y = Z = 0$ with monopole moment $g =$ constant the magnetic 3-vector is given by

$$\mathbf{B} = \frac{g}{r^3}\, (X, Y, Z) \quad \text{with} \quad r = \sqrt{X^2 + Y^2 + Z^2} \,. \tag{7.4}$$

Introducing spherical polars r, θ, ϕ in the usual way,

$$X = r\, \sin\theta\, \cos\phi \,, \quad Y = r\, \sin\theta\, \sin\phi \,, \quad Z = r\, \cos\theta \,, \tag{7.5}$$

(7.1), and (7.2) with (7.4), take the form

$$ds^2 = -r^2(d\theta^2 + \sin^2\theta\, d\phi^2) - dr^2 + dT^2 \,, \tag{7.6}$$

and

$$F = -g\, \sin\theta\, d\theta \wedge d\phi = d(g\, \cos\theta\, d\phi) \,, \tag{7.7}$$

respectively. The final equality in (7.7) is an exterior derivative. Introducing stereographic coordinates x, y and retarded time u via

$$x + i\, y = 2\, e^{i\phi} \cot\frac{\theta}{2} \,, \quad u = T - r \,, \tag{7.8}$$

we have

$$ds^2 = -r^2 p_0^{-2}(dx^2 + dy^2) + 2\, du\, dr + du^2 \quad \text{with} \quad p_0 = 1 + \frac{1}{4}(x^2 + y^2) \,, \tag{7.9}$$

and

$$F = g\, p_0^{-2} dx \wedge dy = dA \,, \tag{7.10}$$

with

$$A = g \left(\frac{\partial}{\partial x}(\log p_0)\, dy - \frac{\partial}{\partial y}(\log p_0)\, dx \right) , \tag{7.11}$$

determined up to a gauge term. In checking (7.10) we note that

$$p_0^2 \left(\frac{\partial^2}{\partial x^2} + \frac{\partial^2}{\partial y^2} \right) \log p_0 = 1 . \tag{7.12}$$

The world line in Minkowskian space-time of the magnetic pole above is the time-like geodesic $r = 0$. If the magnetic pole is accelerated in an external field its world line will have parametric equations $X^i = w^i(u)$ with $v^i = dw^i/du$ and $v_i v^i = +1$ (say), and thus u is proper-time along the world line, and $a^i = dv^i/du \neq 0$. In this case

$$ds^2 = -r^2 P_0^{-2}(dx^2 + dy^2) + 2\, du\, dr + (1 - 2\, h_0\, r)\, du^2 , \tag{7.13}$$

with

$$P_0 = \left\{ 1 + \frac{1}{4}(x^2 + y^2) \right\} v^0(u) + x\, v^1(u) + y\, v^2(u) + \left\{ 1 - \frac{1}{4}(x^2 + y^2) \right\} v^3(u) , \tag{7.14}$$

$$h_0 = a_i\, k^i = \frac{\partial}{\partial u}(\log P_0) , \tag{7.15}$$

$$k^i = \left(\left\{ 1 + \frac{1}{4}(x^2 + y^2) \right\} P_0^{-1}, -x\, P_0^{-1}, -y\, P_0^{-1}, -\left\{ 1 - \frac{1}{4}(x^2 + y^2) \right\} P_0^{-1} \right) , \tag{7.16}$$

and the potential 1-form (up to a gauge transformation) for the accelerating magnetic pole is

$$A = g \left(\frac{\partial}{\partial x}(\log P_0)\, dy - \frac{\partial}{\partial y}(\log P_0)\, dx \right) . \tag{7.17}$$

For future reference we note that P_0 in (7.14) satisfies

$$\Delta \log P_0 = P_0^2 \left(\frac{\partial^2}{\partial x^2} + \frac{\partial^2}{\partial y^2} \right) \log P_0 = 1 . \tag{7.18}$$

In terms of the 1-forms

$$\vartheta^{(1)} = r\, P_0^{-1} dx , \ \vartheta^{(2)} = r\, P_0^{-1} dy , \ \vartheta^{(3)} = dr + \left(\frac{1}{2} - h_0\, r \right) du , \ \vartheta^{(4)} = du , \tag{7.19}$$

we have

$$ds^2 = -(\vartheta^{(1)})^2 - (\vartheta^{(2)})^2 + 2\,\vartheta^{(3)}\,\vartheta^{(4)} = g_{(a)(b)}\,\vartheta^{(a)}\,\vartheta^{(b)}\,, \tag{7.20}$$

and

$$F = dA = \frac{g}{r^2}\,\vartheta^{(1)} \wedge \vartheta^{(2)} - \frac{g}{r}\,P_0\,\frac{\partial h_0}{\partial x}\,\vartheta^{(2)} \wedge \vartheta^{(4)} + \frac{g}{r}\,P_0\,\frac{\partial h_0}{\partial y}\,\vartheta^{(1)} \wedge \vartheta^{(4)}$$

$$= \frac{1}{2}F_{(a)(b)}\,\vartheta^{(a)} \wedge \vartheta^{(b)}\,. \tag{7.21}$$

The 1-forms $\vartheta^{(a)}$ define a half null tetrad. The tetrad indices are those inside round brackets. The components of the metric tensor on the tetrad, $g_{(a)(b)}$, are given by (7.20). With our sign conventions the 2-form dual to (7.21) is

$$^*F = -\frac{g}{r^2}\,\vartheta^{(3)} \wedge \vartheta^{(4)} - \frac{g}{r}\,P_0\,\frac{\partial h_0}{\partial x}\,\vartheta^{(1)} \wedge \vartheta^{(4)} - \frac{g}{r}\,P_0\,\frac{\partial h_0}{\partial y}\,\vartheta^{(2)} \wedge \vartheta^{(4)}$$

$$= -\frac{g}{r^2}\,dr \wedge du - g\,\frac{\partial h_0}{\partial x}\,dx \wedge du - g\,\frac{\partial h_0}{\partial y}\,dy \wedge du$$

$$= d\left(g\,(r^{-1} - h_0)\,du\right)\,, \tag{7.22}$$

with the final equality followed by an exterior derivative. Clearly Maxwell's equations ($d^*F = 0$) are satisfied. Also the dual field (7.22) coincides with the Liénard–Wiechert electromagnetic field of an accelerated charge g.

The solution of the vacuum Einstein–Maxwell field equations for a static magnetic monopole of constant mass m and monopole moment g is easily found to coincide with the Reissner–Nordström solution:

$$ds^2 = -r^2 p_0^{-2}(dx^2 + dy^2) + 2\,du\,dr + \left(1 - \frac{2\,m}{r} + \frac{g^2}{r^2}\right)du^2\,, \tag{7.23}$$

with the potential 1-form coinciding with (7.11) above and p_0 given in (7.9). This follows because the electromagnetic energy-momentum tensor $E_{ab} = (^*F_{ac}\,^*F_b{}^c + F_{ac}\,F_b{}^c)/2$ is invariant under the interchange of the Maxwell tensor F_{ab} and its dual $^*F_{ab}$ and so the electromagnetic energy-momentum tensor for the magnetic monopole field (of monopole moment g) coincides with the electromagnetic energy-momentum tensor for the field of a point charge g. Since the Reissner–Nordström solution with the Coulomb field as source is *an electric black hole*, we will refer to (7.23) with (7.11) as *a magnetic black hole*. The Maxwell field on the tetrad given via the 1-forms (7.19), with $P_0 = p_0$, has the following non-vanishing component:

$$F_{(1)(2)} = \frac{g}{r^2}\,, \tag{7.24}$$

on account of (7.10), while the gravitational field (the Weyl tensor) has only one non-vanishing Newman–Penrose component:

$$\Psi_2 = -\frac{1}{2} R_{(3)(4)(3)(4)} + E_{(3)(4)} = -\frac{m}{r^3} + \frac{g^2}{r^4} ,$$
(7.25)

where $R_{(a)(b)(c)(d)}$ are the tetrad components of the Riemann tensor and $E_{(a)(b)}$ are the tetrad components of the electromagnetic energy-momentum tensor.

7.2 Background Space-Time/External Fields

The external gravitational and electromagnetic fields in which we place the magnetic black hole are modelled by a potential 1-form and a space-time manifold which are solutions of the vacuum Einstein–Maxwell field equations. The space-time manifold contains a time-like world line ($r = 0$) on which the Maxwell field and the gravitational field (described by the Weyl conformal curvature tensor of the space-time) are non-singular. The space-time has the Fermi property of being Minkowskian in the neighbourhood of the world line $r = 0$ in the sense that the metric tensor is the Minkowskian metric tensor if we neglect terms of order r^2. The details are motivated and described in [9] and so we will briefly outline the results now. A convenient form of the line element for our purposes reads (see [10])

$$ds^2 = -(\vartheta^{(1)})^2 - (\vartheta^{(2)})^2 + 2\,\vartheta^{(3)}\,\vartheta^{(4)} = g_{(a)(b)}\,\vartheta^{(a)}\,\vartheta^{(b)} ,$$
(7.26)

with

$$\vartheta^{(1)} = r\,p^{-1}(e^\alpha \cosh\beta\,dx + e^{-\alpha}\sinh\beta\,dy + a\,du) ,$$
(7.27)

$$\vartheta^{(2)} = r\,p^{-1}(e^\alpha \sinh\beta\,dx + e^{-\alpha}\cosh\beta\,dy + b\,du) ,$$
(7.28)

$$\vartheta^{(3)} = dr + \frac{c}{2}\,du ,$$
(7.29)

$$\vartheta^{(4)} = du .$$
(7.30)

As above these 1-forms define a half null tetrad (two space-like vectors defined via $\vartheta^{(1)}$, $\vartheta^{(2)}$ and two null vectors defined via $\vartheta^{(3)}$, $\vartheta^{(4)}$). Tetrad indices, which include the index on $\vartheta^{(a)}$, will be lowered and raised using $g_{(a)(b)}$ and $g^{(a)(b)}$ respectively, with the latter defined by $g^{(a)(b)}\,g_{(b)(c)} = \delta_c^a$ and $g_{(a)(b)} = g_{(b)(a)}$ read off from (7.26) above. This line element is completely general, containing six functions $p, \alpha, \beta, a, b, c$ of the four coordinates x, y, r, u. The hypersurfaces $u = $ constant are null and are generated by the null geodesic integral curves of the vector field

$\partial/\partial r$ and r is an affine parameter along them. These null geodesics have complex shear

$$\sigma = \frac{\partial \alpha}{\partial r} \cosh 2\beta + i \frac{\partial \beta}{\partial r} , \qquad (7.31)$$

and real expansion

$$\rho = \frac{\partial}{\partial r} \log(r \, p^{-1}) . \qquad (7.32)$$

The special case of Minkowskian space-time is given by the line element (7.13)–(7.16) and so to implement the Fermi property mentioned above we take the expansions of the six functions in the neighbourhood of $r = 0$ to be:

$$p = P_0(1 + q_2 \, r^2 + q_3 \, r^3 + \dots) , \qquad (7.33)$$

$$\alpha = \alpha_2 \, r^2 + \alpha_3 \, r^3 + \dots , \qquad (7.34)$$

$$\beta = \beta_2 \, r^2 + \beta_3 \, r^3 + \dots , \qquad (7.35)$$

$$a = a_1 \, r + a_2 \, r^2 + \dots , \qquad (7.36)$$

$$b = b_1 \, r + b_2 \, r^2 + \dots , \qquad (7.37)$$

$$c = 1 - 2 \, h_0 \, r + c_2 \, r^2 + \dots , \qquad (7.38)$$

where P_0 and h_0 are given by (7.14) and (7.15) respectively and the remaining coefficients of the powers of r are functions of x, y, u. We note that now (7.31) and (7.32) satisfy

$$\sigma = O(r) \quad \text{and} \quad \rho = \frac{1}{r} + O(r) , \qquad (7.39)$$

indicating that near $r = 0$ the null hypersurfaces $u = \text{constant}$ are future null cones with vertices on $r = 0$. The potential 1-form of the background electromagnetic field can be written (modulo a gauge transformation)

$$A = L \, dx + M \, dy + K \, du , \qquad (7.40)$$

with L, M, K functions of x, y, r, u. In order to have the corresponding Maxwell field non-singular on $r = 0$, since this is intended to be the *external* electromagnetic field, we choose the following expansions of L, M, K in the neighbourhood of $r = 0$:

$$L = r^2 L_2 + r^3 L_3 + \dots \quad , \tag{7.41}$$

$$M = r^2 M_2 + r^3 M_3 + \dots \quad , \tag{7.42}$$

$$K = r K_1 + r^2 K_2 + \dots \quad . \tag{7.43}$$

The coefficients of the powers of r here are functions of x, y, u.

The relationship between the rectangular Cartesian coordinates and time X^i and the coordinates x, y, r, u, in the neighbourhood of the world line $r = 0$ is given by

$$X^i = w^i(u) + r k^i + O(r^2), \tag{7.44}$$

with k^i given by (7.16). On $r = 0$ the basis 1-forms (7.27)–(7.30) take the following forms when written in terms of the coordinates X^i (see [9], appendix A):

$$\vartheta^{(1)} = -P_0 \frac{\partial k_i}{\partial x} dX^i = -\vartheta_{(1)}, \tag{7.45}$$

$$\vartheta^{(2)} = -P_0 \frac{\partial k_i}{\partial y} dX^i = -\vartheta_{(2)}, \tag{7.46}$$

$$\vartheta^{(3)} = (v_i - \frac{1}{2}k_i) dX^i = \vartheta_{(4)}, \tag{7.47}$$

$$\vartheta^{(4)} = k_i dX^i = \vartheta_{(3)}. \tag{7.48}$$

The electromagnetic field calculated with the potential 1-form given by (7.40)–(7.43) has the form

$$F = dA = \frac{1}{2} F_{(a)(b)} \vartheta^{(a)} \wedge \vartheta^{(b)}, \tag{7.49}$$

with $F_{(a)(b)} = -F_{(b)(a)}$ and some of these tetrad components are given by

$$F_{(1)(3)} = -2 P_0 L_2 + O(r), \quad F_{(2)(3)} = -2 P_0 M_2 + O(r), \quad F_{(3)(4)} = K_1 + O(r). \tag{7.50}$$

Specialising these to the world line $r = 0$ and using (7.45)–(7.48) yields the leading terms in (7.41)–(7.43), namely,

$$L_2 = \frac{1}{2} F_{ij}(u) k^i \frac{\partial k^j}{\partial x}, \quad M_2 = \frac{1}{2} F_{ij}(u) k^i \frac{\partial k^j}{\partial y}, \quad K_1 = F_{ij}(u) k^i v^j, \tag{7.51}$$

where $F_{ij}(u) = -F_{ji}(u)$ are the components in coordinates X^i of the external Maxwell field calculated on $r = 0$. The leading, in powers of r, Maxwell equations $d^*F = 0$ require (7.51) to satisfy

$$K_1 = P_0^2 \left(\frac{\partial L_2}{\partial x} + \frac{\partial M_2}{\partial y} \right) , \tag{7.52}$$

$$\Delta K_1 + 2 P_0^2 \left(\frac{\partial L_2}{\partial x} + \frac{\partial M_2}{\partial y} \right) = 0 , \tag{7.53}$$

with

$$\Delta = P_0^2 \left(\frac{\partial^2}{\partial x^2} + \frac{\partial^2}{\partial y^2} \right) , \tag{7.54}$$

and

$$\frac{\partial K_1}{\partial x} + 2 L_2 - \frac{\partial}{\partial y} \left\{ P_0^2 \left(\frac{\partial M_2}{\partial x} - \frac{\partial L_2}{\partial y} \right) \right\} = 0 , \tag{7.55}$$

$$\frac{\partial K_1}{\partial y} + 2 M_2 + \frac{\partial}{\partial x} \left\{ P_0^2 \left(\frac{\partial M_2}{\partial x} - \frac{\partial L_2}{\partial y} \right) \right\} = 0 . \tag{7.56}$$

Clearly (7.53) is a consequence of (7.55) and (7.56). The dependence of L_2, M_2, K_1 in (7.51) on u is arbitrary (since the world line $r = 0$ is an arbitrary time-like world line), but their dependence on x, y is explicitly known because the dependence of k^i and P_0 on x, y is explicit in (7.14) and (7.16). Hence one can check that L_2, M_2, K_1 in (7.51) satisfy (7.52), (7.53), (7.55) and (7.56). For carrying this out we note the following useful formulas [9]:

$$\frac{\partial^2 k^i}{\partial x^2} = P_0^{-2}(v^i - k^i) - \frac{\partial}{\partial x}(\log P_0) \frac{\partial k^i}{\partial x} + \frac{\partial}{\partial y}(\log P_0) \frac{\partial k^i}{\partial y} , \tag{7.57}$$

$$\frac{\partial^2 k^i}{\partial y^2} = P_0^{-2}(v^i - k^i) + \frac{\partial}{\partial x}(\log P_0) \frac{\partial k^i}{\partial x} - \frac{\partial}{\partial y}(\log P_0) \frac{\partial k^i}{\partial y} , \tag{7.58}$$

$$\frac{\partial^2 k^i}{\partial x \partial y} = -\frac{\partial}{\partial y}(\log P_0) \frac{\partial k^i}{\partial x} - \frac{\partial}{\partial x}(\log P_0) \frac{\partial k^i}{\partial y} . \tag{7.59}$$

We also note (7.52) and (7.53) imply that K_1 satisfies

$$\Delta K_1 + 2 K_1 = 0 , \tag{7.60}$$

indicating that K_1 is an $l = 1$ spherical harmonic, since the operator Δ in (7.54) is the Laplacian on the unit sphere.

With the electromagnetic field above as source, Einstein's field equations for the background space-time are

$$R_{(a)(b)} = 2 E_{(a)(b)} , \qquad (7.61)$$

where $R_{(a)(b)}$ are the components of the Ricci tensor calculated on the half null tetrad defined via the 1-forms (7.27)–(7.30) and $E_{(a)(b)}$ are the tetrad components of the electromagnetic energy-momentum tensor, given in terms of the tetrad components of the Maxwell tensor $F_{(a)(b)}$ by

$$E_{(a)(b)} = F_{(c)(a)} F^{(c)}{}_{(b)} - \frac{1}{4} g_{(a)(b)} F_{(c)(d)} F^{(c)(d)} , \qquad (7.62)$$

with $g_{(a)(b)}$ given by (7.26). When the expansions (7.33)–(7.38) and (7.41)–(7.43) are used to calculate the Weyl conformal curvature tensor on $r = 0$ we arrive at the coefficients of the leading terms in (7.34)–(7.37) (the analogues of (7.51)):

$$\alpha_2 = \frac{1}{6} P_0^2 \, C_{ijkl}(u) \, k^i \, \frac{\partial k^j}{\partial x} \, k^k \, \frac{\partial k^l}{\partial x} , \qquad (7.63)$$

$$\beta_2 = \frac{1}{6} P_0^2 \, C_{ijkl}(u) \, k^i \, \frac{\partial k^j}{\partial x} \, k^k \, \frac{\partial k^l}{\partial y} , \qquad (7.64)$$

$$a_1 = \frac{2}{3} P_0^2 \left(C_{ijkl}(u) \, k^i \, v^j \, k^k \, \frac{\partial k^l}{\partial x} + F^p{}_i(u) \, F_{pj}(u) \, k^i \, \frac{\partial k^j}{\partial x} \right) , \qquad (7.65)$$

$$b_1 = \frac{2}{3} P_0^2 \left(C_{ijkl}(u) \, k^i \, v^j \, k^k \, \frac{\partial k^l}{\partial y} + F^p{}_i(u) \, F_{pj}(u) \, k^i \, \frac{\partial k^j}{\partial y} \right) , \qquad (7.66)$$

and

$$c_2 = -C_{ijkl}(u) \, k^i \, v^j \, k^k \, v^l - 2 \left\{ F^p{}_i(u) \, F_{pj}(u) - \frac{1}{4} \eta_{ij} \, F^{pq}(u) \, F_{pq}(u) \right\} k^i \, v^j$$

$$+ F^p{}_i(u) \, F_{pj}(u) \, k^i \, k^j . \qquad (7.67)$$

Here, as in (7.51), $F_{pj}(u) = -F_{jp}(u)$ (with $F^p{}_i(u) = \eta^{pq} F_{qi}(u)$) are the components of the Maxwell tensor in coordinates X^i calculated on $r = 0$ while $C_{ijkl}(u)$ are the components of the Weyl conformal curvature tensor in coordinates X^i calculated on $r = 0$. Hence C_{ijkl} satisfy $C_{ijkl} = -C_{jikl} = -C_{ijlk} = C_{klij}$, $C_{ijkl} + C_{iljk} + C_{iklj} = 0$ and $\eta^{il} C_{ijkl} = 0$.

The field equation $R_{(3)(3)} = 2 E_{(3)(3)}$ yields

$$q_2 = \frac{2}{3} P_0^2 (L_2^2 + M_2^2) = -\frac{1}{6} F^p{}_i(u) \, F_{pj}(u) \, k^i \, k^j . \qquad (7.68)$$

The final equality here follows from (7.51) and the useful expression (see [9], appendix A)

$$\eta^{ij} = -P_0^2 \left(\frac{\partial k^i}{\partial x} \frac{\partial k^j}{\partial x} + \frac{\partial k^i}{\partial y} \frac{\partial k^j}{\partial y} \right) + k^i v^j + k^j v^i - k^i k^j . \tag{7.69}$$

The remaining field equations (7.61) to be satisfied by (7.63)–(7.66) are given in [9] where they are confirmed to be satisfied using (7.57)–(7.59). For example using (7.57), (7.58) and (7.69) we calculate from the expression (7.67) for c_2 that c_2 satisfies the differential equation

$$\Delta c_2 + 6 c_2 = -4 \left(F^p{}_i(u) F_{pj}(u) - \frac{1}{4} \eta_{ij} F^{pq}(u) F_{pq}(u) \right) v^i v^j , \tag{7.70}$$

and this corresponds to the field equation $R_{(4)(4)} = 2 E_{(4)(4)}$ evaluated on $r = 0$.

7.3   Magnetic Black Hole Perturbation of Background

The magnetic black hole of small mass m and small monopole moment g is introduced as a perturbation of the background space-time which is singular on $r = 0$. This is carried out in such a way as to ensure that in the limits $m \to 0, g \to 0$ and $r \to 0$ in such a way the m/r and g/r remain finite the perturbed Weyl tensor is predominantly that of the magnetic black hole (7.25) and the perturbed Maxwell field is predominantly (7.24). To simplify matters further we found it useful in [9] to require the perturbed potential 1-form to coincide with the Liénard–Wiechert 1-form for small charge and small values of r. In the present case the Liénard–Wiechert 1-form is replaced by the 1-form (7.17) and this is predominantly the perturbed 1-form for small g and r if the perturbed 1-form is given by

$$A = \hat{L} \, dx + \hat{M} \, dy + \hat{K} \, du , \tag{7.71}$$

with

$$\hat{L} = -g \frac{\partial}{\partial y} (\log P_0) + \hat{L}_2 \, r^2 + \dots , \tag{7.72}$$

$$\hat{M} = g \frac{\partial}{\partial x} (\log P_0) + \hat{M}_2 \, r^2 + \dots , \tag{7.73}$$

$$\hat{K} = \hat{K}_1 \, r + \dots . \tag{7.74}$$

The coefficients of the powers of r are, as always, functions of x, y, u and satisfy

$$\hat{L}_2 = L_2 + l_2 + O_2 , \quad \hat{M}_2 = M_2 + m_2 + O_2 \quad \text{and} \quad \hat{K}_1 = K_1 + O_1 . \tag{7.75}$$

with L_2, M_2, K_1 given by (7.51) and $l_2 = O_1, m_2 = O_1$. We have labelled the O_1-terms in $\hat{L}_2$ and $\hat{M}_2$ explicitly as l_2, m_2 respectively because they will be required in the sequel. In contrast we shall not require a knowledge of the O_1-term in $\hat{K}_1$.

The perturbed line element is given by (7.26)–(7.30) with $p, \alpha, \beta, a, b, c$ replaced be their perturbed counterparts $\hat{p}, \hat{\alpha}, \hat{\beta}, \hat{a}, \hat{b}, \hat{c}$ having the forms

$$\hat{p} = \hat{P}_0(1 + \hat{q}_2 r^2 + \dots) , \tag{7.76}$$

with $\hat{q}_2 = q_2 + O_1$, q_2 given by (7.68) and

$$\hat{P}_0 = P_0(1 + Q_1 + Q_2 + O_3) \text{ with } Q_1 = O_1 \text{ and } Q_2 = O_2 . \tag{7.77}$$

Also

$$\hat{\alpha} = \hat{\alpha}_2 r^2 + \dots , \quad \hat{\beta} = \hat{\beta}_2 r^2 + \dots \text{ with } \hat{\alpha}_2 = \alpha_2 + O_1 \text{ and } \hat{\beta}_2 = \beta_2 + O_1 , \tag{7.78}$$

with α_2, β_2 given by (7.63) and (7.64),

$$\hat{a} = \frac{\hat{a}_{-1}}{r} + \hat{a}_0 + \hat{a}_1 r + \dots , \tag{7.79}$$

with

$$\hat{a}_{-1} = O_1 , \quad \hat{a}_0 = O_1 , \quad \hat{a}_1 = a_1 + O_1 , \tag{7.80}$$

$$\hat{b} = \frac{\hat{b}_{-1}}{r} + \hat{b}_0 + \hat{b}_1 r + \dots , \tag{7.81}$$

with

$$\hat{b}_{-1} = O_1 , \quad \hat{b}_0 = O_1 , \quad \hat{b}_1 = b_1 + O_1 , \tag{7.82}$$

with a_1, b_1 here given by (7.65) and (7.66) and finally

$$\hat{c} = \frac{g^2}{r^2} - \frac{2(m + 2\hat{f}_{-1})}{r} + \hat{c}_0 + \hat{c}_1 r + \hat{c}_2 r^2 + \dots , \tag{7.83}$$

with

$$\hat{f}_{-1} = O_2 , \quad \hat{c}_0 = 1 + O_1 , \quad \hat{c}_1 = -2 h_0 + O_1 , \quad \hat{c}_2 = c_2 + O_1 . \tag{7.84}$$

Here h_0 and c_2 are given by (7.15) and (7.70) respectively.

With these expansions the null hypersurfaces $u = $ constant in the perturbed space-time have the potential to play the role of the histories of the wave-fronts of the radiation produced by the motion of the magnetic black hole. These null hypersurfaces are approximately future null cones near $r = 0$. Hence the wave fronts can be approximately 2-spheres near the magnetic black hole. Neglecting $O(r^4)$-terms the line elements induced on these null hypersurfaces are given by

$$dl^2 = -r^2 \hat{P}_0^{-2}(dx^2 + dy^2) \text{ with } \hat{P}_0 = P_0(1 + Q_1 + Q_2 + O_3) . \quad (7.85)$$

We shall require the perturbations of these 2-spheres, described here by the functions $Q_1 = O_1$ and $Q_2 = O_2$, to be smooth, non-singular functions of x, y for $-\infty < x, y < +\infty$. We note here, for future reference, that if these functions are $l = 0$ or $l = 1$ spherical harmonics (a spherical harmonic Q of order l is a smooth solution of $\Delta Q + l(l+1) Q = 0$) then the perturbations described by them are trivial. This is because against a background of three dimensional Euclidean space Q_1 or Q_2 when $l = 0$ merely perturb the radius of the 2-sphere while Q_1 or Q_2 when $l = 1$ infinitesimally displace the origin of the sphere. Hence we will discard these cases as trivial in the sequel. It will follow from the Einstein–Maxwell field equations that *necessary conditions for the 2-surfaces with line elements (7.85) to be smooth, nontrivial deformations of 2-spheres will be the equations of motion of the magnetic black hole.*

7.4   Equations of Motion in First Approximation

Because we have chosen the leading terms in the expansions of L and M in (7.72) and (7.73) to coincide with those of the accelerating magnetic pole (7.17), Maxwell's vacuum field equations will be satisfied by the perturbed 2-form $F = dA$ to the accuracy we require. Hence we can concentrate on satisfying Einstein's field equations (7.61) by the perturbed metric tensor and Maxwell field to sufficient accuracy to enable us to derive the equations of motion of the magnetic black hole in first approximation. To this end we first calculate, to sufficient accuracy to determine the equations of motion in first approximation (i.e. with an O_2-error), the tetrad components of the perturbed Maxwell tensor (see Appendix E):

$$F_{(1)(2)} = \frac{1}{r^2}(g + O_2) + P_0^2 \left(\frac{\partial M_2}{\partial x} - \frac{\partial L_2}{\partial y} \right) + O_1 + O(r) , \quad (7.86)$$

$$F_{(1)(3)} = -2 P_0 L_2 + O_1 + O(r) , \quad (7.87)$$

$$F_{(2)(3)} = -2 P_0 M_2 + O_1 + O(r) , \quad (7.88)$$

$$F_{(1)(4)} = \frac{1}{r^2} \times O_2 + \frac{1}{r} \times O_1 + P_0 \left(\frac{\partial K_1}{\partial x} + L_2 \right) + O_1 + O(r) , \quad (7.89)$$

$$F_{(2)(4)} = \frac{1}{r^2} \times O_2 + \frac{1}{r} \times O_1 + P_0 \left(\frac{\partial K_1}{\partial y} + M_2 \right) + O_1 + O(r) \,,$$

$$(7.90)$$

$$F_{(3)(4)} = K_1 + O_1 + O(r) \,. \tag{7.91}$$

Substituting for L_2, M_2, K_1 into these components from (7.51) we can rewrite them in the form

$$F_{(1)(2)} = \frac{1}{r^2}(g + O_2) + P_0^2 \, F_{ij}(u) \, \frac{\partial k^i}{\partial x} \frac{\partial k^j}{\partial y} + O_1 + O(r) \,, \tag{7.92}$$

$$F_{(1)(3)} = -P_0 \, F_{ij}(u) \, k^i \, \frac{\partial k^i}{\partial x} + O_1 + O(r) \,, \tag{7.93}$$

$$F_{(2)(3)} = -P_0 \, F_{ij}(u) \, k^i \, \frac{\partial k^j}{\partial y} + O_1 + O(r) \,, \tag{7.94}$$

$$F_{(1)(4)} = \frac{1}{r^2} \times O_2 + \frac{1}{r} \times O_1 + P_0 \, F_{ij}(u) \left(\frac{1}{2} k^i - v^i \right) \frac{\partial k^j}{\partial x} + O_1 + O(r) \,,$$

$$(7.95)$$

$$F_{(2)(4)} = \frac{1}{r^2} \times O_2 + \frac{1}{r} \times O_1 + P_0 \, F_{ij}(u) \left(\frac{1}{2} k^i - v^i \right) \frac{\partial k^j}{\partial y} + O_1 + O(r) \,,$$

$$(7.96)$$

$$F_{(3)(4)} = F_{ij}(u) \, k^i \, v^j + O_1 + O(r) \,. \tag{7.97}$$

It is useful at this stage to rewrite these again in terms of the components, in coordinates X^i, of the dual of the tensor F_{ij} given by

$$^*F_{ij} = \frac{1}{2} \epsilon_{ijkl} \, F^{kl} \,, \tag{7.98}$$

where ϵ_{ijkl} is the Levi–Civita permutation symbol in four dimensions with the convention that $\epsilon_{0123} = -1$, and $F^{kl} = \eta^{ki} \eta^{lj} F_{ij}$. To facilitate this we note the following useful formulas:

$$\epsilon_{pqkl} \, k^p \, \frac{\partial k^q}{\partial y} = k_k \frac{\partial k_l}{\partial x} - k_l \frac{\partial k_k}{\partial x} \,, \tag{7.99}$$

$$\epsilon_{pqkl} \, k^p \, v^q = P_0^2 \left(-\frac{\partial k_k}{\partial x} \frac{\partial k_l}{\partial y} + \frac{\partial k_l}{\partial x} \frac{\partial k_k}{\partial y} \right) \,, \tag{7.100}$$

$$\epsilon_{pqkl} \, \frac{\partial k^p}{\partial x} \frac{\partial k^q}{\partial y} = P_0^{-2} (k_k \, v_l - k_l \, v_k) \,, \tag{7.101}$$

$$\epsilon_{pqkl} k^p \frac{\partial k^q}{\partial x} = -k_k \frac{\partial k_l}{\partial y} + \overset{\cdot}{k_l} \frac{\partial k_k}{\partial y} , \tag{7.102}$$

$$\epsilon_{pqkl} \frac{\partial k^p}{\partial x} v^q = -(v_k - k_k) \frac{\partial k_l}{\partial y} + (v_l - k_l) \frac{\partial k_k}{\partial y} , \tag{7.103}$$

$$\epsilon_{pqkl} \frac{\partial k^p}{\partial y} v^q = -\frac{\partial k_k}{\partial x} (v_l - k_l) + \frac{\partial k_l}{\partial x} (v_k - k_k) . \tag{7.104}$$

These can be checked by direct substitution of k^i from (7.16) or otherwise. Using these we can rewrite (7.92)–(7.97) as

$$F_{(1)(2)} = \frac{1}{r^2} (g + O_2) - {}^*F_{ij}(u) k^i v^j + O_1 + O(r) , \tag{7.105}$$

$$F_{(1)(3)} = -P_0 {}^*F_{ij}(u) k^i \frac{\partial k^i}{\partial y} + O_1 + O(r) , \tag{7.106}$$

$$F_{(2)(3)} = P_0 {}^*F_{ij}(u) k^i \frac{\partial k^j}{\partial x} + O_1 + O(r) , \tag{7.107}$$

$$F_{(1)(4)} = \frac{1}{r^2} \times O_2 + \frac{1}{r} \times O_1 - \frac{1}{2} P_0 {}^*F_{ij}(u) k^i \frac{\partial k^j}{\partial y}$$
$$- P_0 {}^*F_{ij} \frac{\partial k^i}{\partial y} v^j + O_1 + O(r) \tag{7.108}$$

$$F_{(2)(4)} = \frac{1}{r^2} \times O_2 + \frac{1}{r} \times O_1 + \frac{1}{2} P_0 {}^*F_{ij}(u) k^i \frac{\partial k^j}{\partial x}$$
$$+ P_0 {}^*F_{ij} \frac{\partial k^i}{\partial x} v^j + O_1 + O(r) , \tag{7.109}$$

$$F_{(3)(4)} = P_0^{2*}F_{ij}(u) \frac{\partial k^i}{\partial x} \frac{\partial k^j}{\partial y} + O_1 + O(r) . \tag{7.110}$$

Neglecting O_2-terms to simplify the presentation, since in this section we are calculating the equations of motion with an O_2-error (i.e. calculating the equations of motion in first approximation), the electromagnetic energy-momentum tensor has the following tetrad components:

$$E_{(1)(1)} = E_{(2)(2)} = \frac{g}{r^2} {}^*F_{ij}(u) k^i v^j + O\left(\frac{1}{r}\right) , \quad E_{(1)(2)} = O\left(\frac{1}{r}\right) , \tag{7.111}$$

$$E_{(1)(3)} = \frac{g}{r^2} P_0 {}^*F_{ij}(u) k^i \frac{\partial k^j}{\partial x} + O\left(\frac{1}{r}\right) , \tag{7.112}$$

$$E_{(2)(3)} = \frac{g}{r^2} P_0 {}^* F_{ij}(u) k^i \frac{\partial k^j}{\partial y} + O\left(\frac{1}{r}\right), \tag{7.113}$$

$$E_{(3)(3)} = O(r^0), \tag{7.114}$$

$$E_{(1)(4)} = -\frac{g}{r^2} P_0 {}^* F_{ij}(u)(v^i - \frac{1}{2}k^i)\frac{\partial k^j}{\partial x} + O\left(\frac{1}{r}\right), \tag{7.115}$$

$$E_{(2)(4)} = -\frac{g}{r^2} P_0 {}^* F_{ij}(u)(v^i - \frac{1}{2}k^i)\frac{\partial k^j}{\partial y} + O\left(\frac{1}{r}\right), \tag{7.116}$$

$$E_{(3)(4)} = \frac{g}{r^2} {}^* F_{ij}(u) k^i v^j + O\left(\frac{1}{r}\right), \quad E_{44} = O\left(\frac{1}{r}\right). \tag{7.117}$$

Now we turn to the perturbed version of the field equations (7.61). We first note that the leading term in $R_{(3)(3)} = 2 E_{(3)(3)}$ is $O(r^0)$ and this gives us (7.68). Next we find that

$$R_{(1)(3)} = -\frac{1}{r^2} P_0^{-1} \hat{a}_{-1} + O\left(\frac{1}{r}\right) \quad \text{and} \quad R_{(2)(3)} = -\frac{1}{r^2} P_0^{-1} \hat{b}_{-1} + O\left(\frac{1}{r}\right), \tag{7.118}$$

and so the field equations $R_{(1)(3)} = 2 E_{(1)(3)}$ and $R_{(2)(3)} = 2 E_{(2)(3)}$ yield

$$\hat{a}_{-1} = -2 g P_0^2 {}^* F_{ij}(u) k^i \frac{\partial k^j}{\partial x} \quad \text{and} \quad \hat{b}_{-1} = -2 g P_0^2 {}^* F_{ij}(u) k^i \frac{\partial k^j}{\partial y}. \tag{7.119}$$

An important property (see below) of $\hat{a}_{-1}$ and $\hat{b}_{-1}$ here, obtained with the use of the useful formulas (7.57)–(7.59), is that they satisfy the Cauchy–Riemann equations:

$$\frac{\partial \hat{a}_{-1}}{\partial x} - \frac{\partial \hat{b}_{-1}}{\partial y} = 0 = \frac{\partial \hat{a}_{-1}}{\partial y} + \frac{\partial \hat{b}_{-1}}{\partial x}. \tag{7.120}$$

Using (7.111) we have

$$R_{(1)(1)} + R_{(2)(2)} = 2(E_{(1)(1)} + E_{(2)(2)}) = \frac{4 g}{r^2} {}^* F_{ij} k^i v^j + O\left(\frac{1}{r}\right). \tag{7.121}$$

The perturbed Ricci tensor terms here are given by

$$R_{(1)(1)} + R_{(2)(2)} = \frac{1}{r^2} \left\{ 2 \hat{c}_0 - 2 - 2(\Delta Q_1 + 2 Q_1) \right.$$

$$\left. + 3 P_0^2 \left(\frac{\partial}{\partial x}(P_0^{-2}\hat{a}_{-1}) + \frac{\partial}{\partial y}(P_0^{-2}\hat{b}_{-1}) \right) \right\} + O\left(\frac{1}{r}\right), \tag{7.122}$$

with the operator Δ given by (7.54). With $\hat{a}_{-1}, \hat{b}_{-1}$ given by (7.119) and making use of (7.57) and (7.58) we have

$$
P_0^2 \left(\frac{\partial}{\partial x}(P_0^{-2}\hat{a}_{-1}) + \frac{\partial}{\partial y}(P_0^{-2}\hat{b}_{-1}) \right) = -2\, g \,{}^* F_{ij}\, k^i \,\Delta k^j = -4\, g \,{}^* F_{ij}\, k^i\, v^j \ .
$$

(7.123)

Hence we conclude from (7.121)–(7.123) that

$$
\hat{c}_0 = 1 + \Delta Q_1 + 2\, Q_1 + 8\, g \,{}^* F_{ij}\, k^i\, v^j \ ,
$$

(7.124)

neglecting O_2-terms. We also have from (7.111) that

$$
R_{(1)(1)} - R_{(2)(2)} = 2\,(E_{(1)(1)} - E_{(2)(2)}) = O\left(\frac{1}{r}\right) \ , \quad R_{(1)(2)} = 2\, E_{(1)(2)} = O\left(\frac{1}{r}\right) \ .
$$

(7.125)

But

$$
R_{(1)(1)} - R_{(2)(2)} = \frac{1}{r^2} \left\{ \frac{\partial \hat{a}_{-1}}{\partial x} - \frac{\partial \hat{b}_{-1}}{\partial y} \right\} + O\left(\frac{1}{r}\right) \ ,
$$

(7.126)

and

$$
R_{(1)(2)} = \frac{1}{2\, r^2} \left\{ \frac{\partial \hat{b}_{-1}}{\partial x} + \frac{\partial \hat{a}_{-1}}{\partial y} \right\} + O\left(\frac{1}{r}\right) \ ,
$$

(7.127)

and thus (7.125) are satisfied on account of (7.120). Using the first of (7.117) we have the field equation

$$
R_{(3)(4)} = \frac{2\, g}{r^2} \,{}^* F_{ij}\, k^i\, v^j + O\left(\frac{1}{r}\right) \ .
$$

(7.128)

But

$$
\begin{aligned}
R_{(3)(4)} &= -\frac{P_0^2}{2\, r^2} \left(\frac{\partial}{\partial x}(P_0^{-2}\hat{a}_{-1}) + \frac{\partial}{\partial y}(P_0^{-2}\hat{b}_{-1}) \right) + O\left(\frac{1}{r}\right) \\
&= \frac{2\, g}{r^2} \,{}^* F_{ij}\, k^i\, v^j + O\left(\frac{1}{r}\right) \ ,
\end{aligned}
$$

(7.129)

by (7.123) and so (7.128) is automatically satisfied with an $O(r^{-1})$-error. Next using (7.115) and (7.116) we have the two field equations

$$R_{(1)(4)} = -2\frac{g}{r^2}P_0^* F_{ij}(u)(v^i - \frac{1}{2}k^i)\frac{\partial k^j}{\partial x} + O\left(\frac{1}{r}\right),\qquad (7.130)$$

$$R_{(2)(4)} = -2\frac{g}{r^2}P_0^* F_{ij}(u)(v^i - \frac{1}{2}k^i)\frac{\partial k^j}{\partial y} + O\left(\frac{1}{r}\right).\qquad (7.131)$$

Calculation of the tetrad components of the Ricci tensor here yields

$$R_{(1)(4)} = \frac{P_0}{2r^2}\left\{P_0^{-2}\hat{a}_{-1} + \frac{\partial}{\partial y}\left[P_0^2\left(\frac{\partial}{\partial y}(P_0^{-2}\hat{a}_{-1}) - \frac{\partial}{\partial x}(P_0^{-2}\hat{b}_{-1})\right)\right]\right\}$$

$$+ O\left(\frac{1}{r}\right),\qquad (7.132)$$

$$R_{(2)(4)} = \frac{P_0}{2r^2}\left\{P_0^{-2}\hat{b}_{-1} - \frac{\partial}{\partial x}\left[P_0^2\left(\frac{\partial}{\partial y}(P_0^{-2}\hat{a}_{-1}) - \frac{\partial}{\partial x}(P_0^{-2}\hat{b}_{-1})\right)\right]\right\}$$

$$+ O\left(\frac{1}{r}\right).\qquad (7.133)$$

With $\hat{a}_{-1}, \hat{b}_{-1}$ given by (7.119) we have

$$\frac{\partial}{\partial y}(P_0^{-2}\hat{a}_{-1}) - \frac{\partial}{\partial x}(P_0^{-2}\hat{b}_{-1}) = 4g^* F_{ij}\frac{\partial k^i}{\partial x}\frac{\partial k^j}{\partial y}.\qquad (7.134)$$

Then making use of the formulas (7.57)–(7.59) we find that

$$\frac{\partial}{\partial y}\left[P_0^2\left(\frac{\partial}{\partial y}(P_0^{-2}\hat{a}_{-1}) - \frac{\partial}{\partial x}(P_0^{-2}\hat{b}_{-1})\right)\right] = 4g^* F_{ij}(k^i - v^i)\frac{\partial k^j}{\partial x},$$

$$(7.135)$$

$$\frac{\partial}{\partial x}\left[P_0^2\left(\frac{\partial}{\partial y}(P_0^{-2}\hat{a}_{-1}) - \frac{\partial}{\partial x}(P_0^{-2}\hat{b}_{-1})\right)\right] = -4g^* F_{ij}(k^i - v^i)\frac{\partial k^j}{\partial y}.$$

$$(7.136)$$

When these, along with (7.119), are substituted into (7.130) and (7.131) we see that (7.130) and (7.131) are automatically satisfied. Finally we consider the second of (7.117) and the field equation

$$R_{(4)(4)} = 2E_{(4)(4)} = O\left(\frac{1}{r}\right).\qquad (7.137)$$

Calculating the tetrad component $R_{(4)(4)}$ of the Ricci tensor we arrive at

$$R_{(4)(4)} = -\frac{1}{r^2}\left\{\frac{1}{2}\Delta\hat{c}_0 + 6\,m\,h_0 - \frac{1}{2}P_0^2\left(\frac{\partial}{\partial x}(P_0^{-2}\hat{a}_{-1}) + \frac{\partial}{\partial y}(P_0^{-2}\hat{b}_{-1})\right)\right\}$$

$$+ O\left(\frac{1}{r}\right)$$

$$= -\frac{1}{r^2}\left\{\frac{1}{2}\Delta\hat{c}_0 + 6\,m\,h_0 + 2\,g^{*}F_{ij}\,k^i\,v^j\right\} + O\left(\frac{1}{r}\right) , \qquad (7.138)$$

and thus (7.137) is satisfied, *neglecting O_2-terms*, provided

$$\Delta\hat{c}_0 = -12\,m\,h_0 - 4\,g^{*}F_{ij}\,k^i\,v^j . \qquad (7.139)$$

We note from (7.57) and (7.58) that $\Delta k^i + 2\,k^i = 2\,v^i$ and thus $h_0 = a_i\,k^i$ and $^{*}F_{ij}\,k^i\,v^j$ are $l = 1$ spherical harmonics since both a_i and $^{*}F_{ij}\,v^j$ are orthogonal to v^i.

We have now completed the calculation of the perturbed Einstein field equations and we are left with the two Eqs. (7.124) and (7.139) for further consideration. Substituting (7.124) into (7.139) results in

$$\Delta(\Delta Q_1 + 2\,Q_1) = -12\,m\,h_0 + 12\,g^{*}F_{ij}\,k^i\,v^j . \qquad (7.140)$$

Since $h_0 = a_i\,k^i$ and $^{*}F_{ij}\,k^i\,v^j$ are both $l = 1$ spherical harmonics we easily integrate (7.140) to read

$$\Delta Q_1 + 2\,Q_1 = 6\,m\,a_i\,k^i - 6\,g^{*}F_{ij}\,k^i\,v^j + A(u) , \qquad (7.141)$$

where $A(u) = O_1$ is an arbitrary $l = 0$ spherical harmonic. For Q_1 to be a bounded function of x, y for $-\infty < x, y < +\infty$ we must have

$$(m\,a_i - g^{*}F_{ij}\,v^j)k^i = 0 , \qquad (7.142)$$

for any k^i for which $k^i\,k_i = 0$ and $k^i\,v_i = 1$. Since $m\,a_i - g^{*}F_{ij}\,v^j$ is orthogonal to v^i, and thus space-like, it follows that we must have (restoring the O_2-error designation to emphasise that we are working in the linear approximation)

$$m\,a_i = g^{*}F_{ij}\,v^j + O_2 , \qquad (7.143)$$

which constitute *the equations of motion of the small magnetic black hole in first approximation*. It now follows from (7.141) with (7.143) that Q_1 can only be a linear combination of an $l = 0$ and an $l = 1$ spherical harmonic. This corresponds to a trivial perturbation of the 2-spheres (7.85) and so without loss of generality we take $A(u) = 0$ in (7.141) and $Q_1 = 0$.

7.5   Equations of Motion in Second Approximation

To obtain the equations of motion of the magnetic black hole in second approximation we require a knowledge of the perturbed space-time and the perturbed Maxwell
field which is more accurate than we needed for the equations of motion in first
approximation above. When the expansions (7.72)–(7.84) are substituted into the
tetrad components of the Maxwell tensor given in Appendix E the resulting tetrad
opponents of the perturbed Maxwell tensor read

$$F_{(1)(2)} = \frac{1}{r^2} \{g + O_3\} + P_0^2 \left(\frac{\partial M_2}{\partial x} - \frac{\partial L_2}{\partial y} \right)$$

$$+ P_0^2 \left(\frac{\partial m_2}{\partial x} - \frac{\partial l_2}{\partial y} \right) + 2\, g\, q_2 + O_2 + O(r)\,, \tag{7.144}$$

$$F_{(1)(3)} = -2\, P_0 \left(L_2 + l_2 + O_2 \right) + O(r)\,, \tag{7.145}$$

$$F_{(2)(3)} = -2\, P_0 \left(M_2 + m_2 + O_2 \right) + O(r)\,, \tag{7.146}$$

$$F_{(1)(4)} = \frac{1}{r^2} (g^2\, P_0\, L_2 - g\, P_0^{-1}\, \hat{b}_{-1} + O_3) + \frac{1}{r}\, P_0 \left(g\, \frac{\partial h_0}{\partial y} - 2m\, L_2 + O_2 \right)$$

$$+ P_0 \left(\frac{\partial K_1}{\partial x} + L_2 \right) + O_1 + O(r)\,, \tag{7.147}$$

$$F_{(2)(4)} = \frac{1}{r^2} (g^2\, P_0\, M_2 + g\, P_0^{-1}\, \hat{a}_{-1} + O_3) + \frac{1}{r}\, P_0 \left(-g\, \frac{\partial h_0}{\partial x} - 2m\, M_2 + O_2 \right)$$

$$+ P_0 \left(\frac{\partial K_1}{\partial y} + M_2 \right) + O_1 + O(r)\,, \tag{7.148}$$

$$F_{(3)(4)} = K_1 + O_1 + O(r)\,. \tag{7.149}$$

From these the perturbed components $E_{(1)(3)}$ and $E_{(2)(3)}$ of the electromagnetic
energy-momentum tensor are given by

$$E_{(1)(3)} = F_{(1)(2)}\, F_{(2)(3)} + F_{(1)(3)}\, F_{(3)(4)} = -\frac{2\, g\, P_0}{r^2}\, (M_2 + m_2 + O_2) + O\left(\frac{1}{r} \right)\,, \tag{7.150}$$

and

$$E_{(2)(3)} = -F_{(1)(2)}\, F_{(1)(3)} + F_{(2)(3)}\, F_{(3)(4)} = \frac{2\, g\, P_0}{r^2}\, (L_2 + l_2 + O_2) + O\left(\frac{1}{r} \right)\,. \tag{7.151}$$

Calculating $R_{(1)(3)}$ and $R_{(2)(3)}$ using (7.76)–(7.84) we find that

$$R_{(1)(3)} = -\frac{1}{r^2}\, P_0^{-1}\hat{a}_{-1} + O\left(\frac{1}{r}\right), \tag{7.152}$$

$$R_{(2)(3)} = -\frac{1}{r^2}\, P_0^{-1}\hat{b}_{-1} + O\left(\frac{1}{r}\right), \tag{7.153}$$

Hence Einstein's field equations

$$R_{(1)(3)} - 2\, E_{(1)(3)} = \frac{1}{r^2} \times O_3 + O\left(\frac{1}{r}\right), \tag{7.154}$$

and

$$R_{(2)(3)} - 2\, E_{(2)(3)} = \frac{1}{r^2} \times O_3 + O\left(\frac{1}{r}\right), \tag{7.155}$$

yield

$$\hat{a}_{-1} = 4\,g\,P_0^2\,M_2 + 4\,g\,P_0^2\,m_2 + O_3 = 2\,g\,P_0^2\,F_{ij}\,k^i\,\frac{\partial k^j}{\partial y} + 4\,g\,P_0^2\,m_2$$

$$= -2\,g\,P_0^2\,{}^*F_{ij}\,k^i\,\frac{\partial k^j}{\partial x} + 4\,g\,P_0^2\,m_2 + O_3\,, \tag{7.156}$$

and

$$\hat{b}_{-1} = -4\,g\,P_0^2\,M_2 - 4\,g\,P_0^2\,l_2 + O_3 = -2\,g\,P_0^2\,F_{ij}\,k^i\,\frac{\partial k^j}{\partial y} - 4\,g\,P_0^2\,m_2$$

$$= -2\,g\,P_0^2\,{}^*F_{ij}\,k^i\,\frac{\partial k^j}{\partial y} - 4\,g\,P_0^2\,l_2 + O_3\,, \tag{7.157}$$

For convenience we note, using (7.51), (7.68), (7.99), (7.102) and

$$F^p{}_i\,F_{pj} = {}^*F^p{}_i\,{}^*F_{pj} - \frac{1}{2}\eta_{ij}\,{}^*F^{pq}\,{}^*F_{pq}\,, \tag{7.158}$$

that

$$L_2 = \frac{1}{2}{}^*F_{ij}\,k^i\,\frac{\partial k^j}{\partial y}\,, \quad M_2 = -\frac{1}{2}{}^*F_{ij}\,k^i\,\frac{\partial k^j}{\partial x} \quad \text{and} \quad q_2 = -\frac{1}{6}{}^*F^p{}_i\,{}^*F_{pj}\,k^i\,k^j\,. \tag{7.159}$$

We can now write the tetrad components $F_{(a)(b)}$ of the perturbed Maxwell field (7.144)–(7.149) more explicitly as

$$F_{(1)(2)} = \frac{1}{r^2}(g + O_3) - {}^*F_{ij}\,k^i\,v^j + P_0^2\left(\frac{\partial m_2}{\partial x} - \frac{\partial l_2}{\partial y}\right)$$

$$-\frac{1}{3}g\,{}^*F^p{}_i\,{}^*F_{pj}\,k^i\,k^j + O_2 + O(r)\,, \tag{7.160}$$

$$F_{(1)(3)} = -P_0\,{}^*F_{ij}\,k^i\,\frac{\partial k^j}{\partial y} - 2\,P_0\,l_2 + O_2 + O(r)\,, \tag{7.161}$$

$$F_{(2)(3)} = P_0\,{}^*F_{ij}\,k^i\,\frac{\partial k^j}{\partial x} - 2\,P_0\,m_2 + O_2 + O(r)\,, \tag{7.162}$$

$$F_{(1)(4)} = \frac{1}{r^2}\left(\frac{5}{2}\,g^2\,P_0\,{}^*F_{ij}\,k^i\,\frac{\partial k^j}{\partial y} + O_3\right)$$

$$+\frac{1}{r}\left(g\,P_0\,a_i\,\frac{\partial k^i}{\partial y} - m\,P_0\,{}^*F_{ij}\,k^i\,\frac{\partial k^j}{\partial y} + O_2\right)$$

$$-\frac{1}{2}\,P_0\,{}^*F_{ij}\,k^i\,\frac{\partial k^j}{\partial y} - P_0\,{}^*F_{ij}\,\frac{\partial k^i}{\partial y}\,v^j + O_1 + O(r)\,, \tag{7.163}$$

$$F_{(2)(4)} = \frac{1}{r^2}\left(-\frac{5}{2}\,g^2\,P_0\,{}^*F_{ij}\,k^i\,\frac{\partial k^j}{\partial x} + O_3\right)$$

$$+\frac{1}{r}\left(-g\,P_0\,a_i\,\frac{\partial k^i}{\partial x} + m\,P_0\,{}^*F_{ij}\,k^i\,\frac{\partial k^j}{\partial x} + O_2\right)$$

$$+\frac{1}{2}\,P_0\,{}^*F_{ij}\,k^i\,\frac{\partial k^j}{\partial x} + P_0\,{}^*F_{ij}\,\frac{\partial k^i}{\partial x}\,v^j + O_1 + O(r)\,, \tag{7.164}$$

and

$$F_{(3)(4)} = K_1 + O_1 + O(r)\,. \tag{7.165}$$

Turning to the field equations $R_{(A)(B)} = 2\,E_{(A)(B)}$ with $A, B = 1, 2$ we first find, using (7.76)–(7.84), that the tetrad components of the Ricci tensor here are given by

$$R_{(1)(2)} = \frac{1}{r^2}\left\{\frac{1}{2}\left(\frac{\partial\hat{a}_{-1}}{\partial y} + \frac{\partial\hat{b}_{-1}}{\partial x}\right) + \frac{1}{2}P_0^{-2}\hat{a}_{-1}\,\hat{b}_{-1} + 2\,g^2\,\beta_2\right\}$$

$$+\frac{1}{r}\left\{\frac{\partial\hat{a}_0}{\partial y} + \frac{\partial\hat{b}_0}{\partial x} - 8\,m\,\beta_2 + O_2\right\} + O(r^0)\,, \tag{7.166}$$

$$R_{(1)(1)} - R_{(2)(2)} = \frac{1}{r^2} \left\{ \frac{\partial \hat{a}_{-1}}{\partial x} - \frac{\partial \hat{b}_{-1}}{\partial y} + \frac{1}{2} P_0^{-2}(\hat{a}_{-1}^2 - \hat{b}_{-1}^2) + 4 g^2 \alpha_2 \right\}$$

$$+ \frac{1}{r} \left\{ 2 \left(\frac{\partial \hat{a}_0}{\partial x} - \frac{\partial \hat{b}_0}{\partial y} \right) - 16 m \, \alpha_2 + O_2 \right\} + O(r^0) ,$$

$$(7.167)$$

and

$$R_{(A)(A)} = -\frac{2 g^2}{r^4} + \frac{2}{r^2} \left\{ -1 + \hat{c}_0 - (\Delta Q_2 + 2 Q_2) + \frac{1}{4} P_0^{-2}(\hat{a}_{-1}^2 + \hat{b}_{-1}^2) \right.$$

$$+ \frac{3}{2} P_0^2 \left(\frac{\partial}{\partial x}(P_0^{-2}\hat{a}_{-1}) + \frac{\partial}{\partial y}(P_0^{-2}\hat{b}_{-1}) \right) \right\} + \frac{4}{r} \left\{ \hat{c}_1 + 2 h_0 \right.$$

$$+ P_0^2 \left(\frac{\partial}{\partial x}(P_0^{-2}\hat{a}_0) + \frac{\partial}{\partial y}(P_0^{-2}\hat{b}_0) \right) + O_2 \right\} + O(r^0) . \qquad (7.168)$$

We have consistently neglected O_3-terms throughout. Using these we examine the field equations

$$R_{(1)(1)} - R_{(2)(2)} + 2 i \, R_{(1)(2)} = 2 \left(E_{(1)(1)} - E_{(2)(2)} + 2 i \, E_{(1)(2)} \right) . \qquad (7.169)$$

The left hand side is given by (7.166) and (7.167). This can be written in a convenient form using the complex variable $\zeta = x + i \, y$, and its complex conjugate denoted by a bar, as

$$R_{(1)(1)} - R_{(2)(2)} + 2 i \, R_{(1)(2)} = \frac{1}{r^2} \left\{ 2 \frac{\partial}{\partial \bar{\zeta}}(\hat{a}_{-1} + i \, \hat{b}_{-1}) + \frac{1}{2} P_0^{-2}(\hat{a}_{-1} + i \, \hat{b}_{-1})^2 \right.$$

$$+ 4 g^2(\alpha_2 + i \, \beta_2) \right\} + \frac{1}{r} \left\{ 4 \frac{\partial}{\partial \bar{\zeta}}(\hat{a}_0 + i \hat{b}_0) - 16 m \, (\alpha_2 + i \, \beta_2) + O_2 \right\}$$

$$+ O(r^0) . \qquad (7.170)$$

From (7.156) and (7.157) we have

$$\hat{a}_{-1} + i \, \hat{b}_{-1} = -4 g \, P_0^{2*} F_{ij} \, k^i \frac{\partial k^j}{\partial \bar{\zeta}} + 4 g \, P_0^2(m_2 - i \, l_2) = -4 g \, P_0^{2*} F_{ij} \, k^i \frac{\partial k^j}{\partial \bar{\zeta}} + O_2 ,$$

$$(7.171)$$

and thus

$$\frac{1}{2} P_0^{-2}(\hat{a}_{-1} + i \, \hat{b}_{-1})^2 = 8 g^2 P_0^2 \left({}^* F_{ij} \, k^i \frac{\partial k^j}{\partial \bar{\zeta}} \right)^2 + O_3 . \qquad (7.172)$$

In terms of the complex variable ζ and its complex conjugate we can rewrite the useful formulas (7.57)–(7.59) compactly as

$$\frac{\partial}{\partial \zeta}\left(P_0^2 \frac{\partial k^i}{\partial \zeta}\right) = 0 \quad \text{and} \quad \frac{\partial^2 k^i}{\partial \zeta \partial \bar{\zeta}} = \frac{1}{2} P_0^{-2} (v^i - k^i),$$

(7.173)

and (7.69) as

$$\eta^{ij} = -2 P_0^2 \left(\frac{\partial k^i}{\partial \zeta} \frac{\partial k^j}{\partial \bar{\zeta}} + \frac{\partial k^i}{\partial \bar{\zeta}} \frac{\partial k^j}{\partial \zeta}\right) + k^i v^j + k^j v^i - k^i k^j.$$

(7.174)

Using these we can write

$$\frac{1}{2} P_0^{-2}(\hat{a}_{-1} + i\,\hat{b}_{-1})^2 = -8\,g^2\,\frac{\partial}{\partial \zeta}\left\{P_0^2\,{}^*F^p{}_i\,{}^*F_{pj}\,k^i\,\frac{\partial k^j}{\partial \bar{\zeta}}\right.$$

$$\left. +2\,P_0^2\,{}^*F_{ij}\,k^i\,\frac{\partial k^j}{\partial \bar{\zeta}}\,{}^*F_{pq}\,k^p\,v^q\right\},$$

(7.175)

and

$$\frac{\partial}{\partial \zeta}(\hat{a}_{-1} + i\,\hat{b}_{-1}) = 4\,g\,\frac{\partial}{\partial \zeta}\{P_0^2\,(m_2 - i\,l_2)\}.$$

(7.176)

In addition with α_2 and β_2 given by (7.63) and (7.64) we have

$$\alpha_2 + i\,\beta_2 = -\frac{\partial}{\partial \bar{\zeta}}\left\{P_0^4\,\frac{\partial}{\partial \zeta}(P_0^{-2}(\alpha_2 + i\,\beta_2))\right\} = -\frac{\partial}{\partial \bar{\zeta}}\left\{\frac{2}{3}\,P_0^2\,C_{ijkl}\,k^i\,v^j\,k^k\,\frac{\partial k^l}{\partial \bar{\zeta}}\right\},$$

(7.177)

which incidentally is a consequence of Einstein's field equations satisfied by the background space-time (see [9], Eq. (2.41)). Hence we can write (7.170) as

$$R_{(1)(1)} - R_{(2)(2)} + 2\,i\,R_{(1)(2)} = \frac{1}{r^2}\left\{\frac{\partial}{\partial \bar{\zeta}}\left[2\,(\hat{a}_{-1} + i\,\hat{b}_{-1})\right.\right.$$

$$-8\,g^2\,P_0^2\,{}^*F^p{}_i\,{}^*F_{pj}\,k^i\,\frac{\partial k^j}{\partial \bar{\zeta}} - 16\,g^2\,P_0^2\,{}^*F_{ij}\,k^i\,\frac{\partial k^j}{\partial \bar{\zeta}}$$

$$\left.\left.-\frac{8}{3}\,g^2\,P_0^2\,C_{ijkl}\,k^i\,v^j\,k^k\,\frac{\partial k^l}{\partial \bar{\zeta}}\right]\right\}$$

$$+\frac{1}{r}\left\{\frac{\partial}{\partial \bar{\zeta}}\left[4\,(\hat{a}_0 + i\,\hat{b}_0) + \frac{32}{3}\,m\,P_0^2\,C_{ijk}\,k^i\,v^j\,k^k\,\frac{\partial k^l}{\partial \bar{\zeta}} + O_2\right]\right\}$$

$$+O(r^0).$$

(7.178)

Next we require $E_{(1)(1)} - E_{(2)(2)} + 2\,i\,E_{(1)(2)}$. To aid the calculation of this complex variable we first note from (7.161)–(7.164) that we can write

$$F_{(1)(3)} = F^0_{(1)(3)} + O(r)\,,\ F_{(2)(3)} = F^0_{(2)(3)} + O(r)\,, \tag{7.179}$$

and

$$F_{(1)(4)} = \frac{F^2_{(1)(4)}}{r^2} + \frac{F^1_{(1)(4)}}{r} + F^0_{(1)(4)} + O(r)\,, \tag{7.180}$$

$$F_{(2)(4)} = \frac{F^2_{(2)(4)}}{r^2} + \frac{F^1_{(2)(4)}}{r} + F^0_{(1)(4)} + O(r)\,, \tag{7.181}$$

with

$$F^0_{(1)(3)} + i\,F^0_{(2)(3)} = 2\,i\,P_0\,{}^*F_{ij}\,k^i\,\frac{\partial k^j}{\partial \bar{\zeta}} - 2\,i\,P_0(m_2 - i\,l_2) + O_2\,, \tag{7.182}$$

$$F^2_{(1)(4)} + i\,F^2_{(2)(4)} = -5\,i\,g^2\,P_0\,{}^*F_{ij}\,k^i\,\frac{\partial k^j}{\partial \bar{\zeta}} + O_3\,, \tag{7.183}$$

$$F^1_{(1)(4)} + i\,F^1_{(2)(4)} = -2\,i\,g\,P_0\,a_i\,\frac{\partial k^i}{\partial \bar{\zeta}} + 2\,m\,i\,P_0\,{}^*F_{ij}\,k^i\,\frac{\partial k^j}{\partial \bar{\zeta}} + O_2\,, \tag{7.184}$$

$$F^0_{14} + i\,F^0_{(2)(4)} = i\,P_0\,{}^*F_{ij}\,k^i\,\frac{\partial k^j}{\partial \bar{\zeta}} + 2\,i\,P_0\,{}^*F_{ij}\,\frac{\partial k^i}{\partial \bar{\zeta}}\,v^j + O_1\,. \tag{7.185}$$

Now

$$E_{(1)(1)} - E_{(2)(2)} + 2\,i\,E_{(1)(2)} = \frac{T_2}{r^2} + \frac{T_1}{r} + O(r^0)\,, \tag{7.186}$$

with (neglecting O_3-terms in order to simplify the presentation at this stage)

$$\begin{aligned}
T_2 &= 2\,(F^0_{(1)(3)} + i\,F^0_{(2)(3)})(F^2_{(1)(4)} + i\,F^2_{(2)(4)}) \\
&= 20\,g^2\left(P_0\,{}^*F_{ij}\,k^i\,\frac{\partial k^j}{\partial \bar{\zeta}}\right)^2 \\
&= \frac{\partial}{\partial \bar{\zeta}}\left\{-20\,g^2\,P_0^2\,{}^*F^p{}_i\,{}^*F_{pj}\,k^i\,\frac{\partial k^j}{\partial \bar{\zeta}} - 40\,g^2\,P_0^2\,{}^*F_{ij}\,k^i\,\frac{\partial k^j}{\partial \bar{\zeta}}\,{}^*F_{pq}\,k^p\,v^q\right\}\,.
\end{aligned} \tag{7.187}$$

We also find that

$$
T_1 = 2 \, (F^0_{(1)(3)} + i \, F^0_{(2)(3)})(F^1_{(1)(4)} + i \, F^1_{(2)(4)})
$$

$$
= 8 \, g \, P_0^2 \, a_i \, \frac{\partial k^i}{\partial \zeta} \, {}^* F_{pq} \, k^p \, \frac{\partial k^q}{\partial \zeta} - 8 \, m \left(P_0 \, {}^* F_{ij} \, k^i \, \frac{\partial k^j}{\partial \bar{\zeta}} \right)^2 + O_2
$$

$$
= 8 \, \frac{g^2}{m} \, P_0^2 \, {}^* F_{ij} \, \frac{\partial k^i}{\partial \bar{\zeta}} \, v^j \, {}^* F_{pq} \, k^p \, \frac{\partial k^q}{\partial \bar{\zeta}} - 8 \, m \left(P_0 \, {}^* F_{ij} \, k^i \, \frac{\partial k^j}{\partial \bar{\zeta}} \right)^2 + O_2 \, ,
$$

$$
\tag{7.188}
$$

where we have made use of the equations of motion in first approximation (7.143).
We can rewrite (7.188) as

$$
T_1 = \frac{\partial}{\partial \bar{\zeta}} \left\{ \left(\frac{8 \, g^2}{m} + 16 \, m \right) P_0^2 \, {}^* F_{ij} \, k^i \, \frac{\partial k^j}{\partial \bar{\zeta}} \, {}^* F_{pq} \, k^p \, v^q \right.
$$

$$
\left. + 8 \, m \, P_0^2 \, {}^* F^p{}_i \, {}^* F_{pj} \, k^i \, \frac{\partial k^j}{\partial \bar{\zeta}} \right\} + O_2 \, .
\tag{7.189}
$$

Now with $R_{(A)(B)} = 2 \, E_{(A)(B)}$ we obtain from (7.178), (7.186), (7.187) and (7.189):

$$
m_2 - i \, l_2 = \frac{1}{3} \, g \, C_{ijkl} \, k^i \, v^j \, k^k \, \frac{\partial k^l}{\partial \bar{\zeta}} - 4 \, g \, {}^* F^p{}_i \, {}^* F_{pj} \, k^i \, \frac{\partial k^j}{\partial \bar{\zeta}}
$$

$$
- 8 \, g \, {}^* F_{ij} \, k^i \, \frac{\partial k^j}{\partial \bar{\zeta}} \, {}^* F_{pq} \, k^p \, v^q + P_0^{-2}(U(u) - i \, V(u)) + O_2 \, ,
$$

$$
= O_1 \, ,
\tag{7.190}
$$

where $U(u) = O_1$ and $V(u) = O_1$ are functions of integration, and

$$
P_0^{-2}(\hat{a}_0 + i \, \hat{b}_0) = -\frac{8}{3} \, m \, C_{ijkl} \, k^i \, v^j \, k^k \, \frac{\partial k^l}{\partial \bar{\zeta}} + 4 \, m \, {}^* F^p{}_i \, {}^* F_{pj} \, k^i \, \frac{\partial k^j}{\partial \bar{\zeta}}
$$

$$
+ \left(\frac{4 \, g^2}{m} + 8 \, m \right) {}^* F_{ij} \, k^i \, \frac{\partial k^j}{\partial \bar{\zeta}} \, {}^* F_{pq} \, k^p \, v^q
$$

$$
+ P_0^{-2}(A_0(u) + i \, B_0(u)) + O_2 \, ,
$$

$$
= O_1 \, ,
\tag{7.191}
$$

where $A_0(u) = O_1$ and $B_0(u) = O_1$ are functions of integration. We have chosen functions of integration $A_0(u)$ and $B_0(u)$ so that $P_0^{-1} \hat{a}_0$ and $P_0^{-1} \hat{b}_0$ are guaranteed to be free of singularities in x, y for $-\infty < x, y < +\infty$. This ensures that the components of the metric tensor of the perturbed space-time are free of such singularities since $\hat{a}_0$, $\hat{b}_0$ appear in the metric tensor components in the form $P_0^{-1}\hat{a}_0$ and $P_0^{-1}\hat{b}_0$. The same applies to the choice of the functions of integration $U(u)$ and $V(u)$ in (7.190) which will ensure that

$$P_0^2 \left(\frac{\partial m_2}{\partial x} - \frac{\partial l_2}{\partial y} \right) ,$$

which appears in the metric tensor component involving $\hat{c}_0$ (see below) is free of singularities in x, y for $-\infty < x, y < +\infty$.

For later use we need to derive from (7.190) and (7.191) the expressions

$$P_0^2 \left(\frac{\partial m_2}{\partial x} - \frac{\partial l_2}{\partial y} \right) = P_0^2 \left\{ \frac{\partial}{\partial \zeta}(m_2 - i\, l_2) + \frac{\partial}{\partial \bar{\zeta}}(m_2 + i\, l_2) \right\} , \tag{7.192}$$

and

$$P_0^2 \left(\frac{\partial}{\partial x}(P_0^{-2}\hat{a}_0) + \frac{\partial}{\partial y}(P_0^{-2}\hat{b}_0) \right) = P_0^2 \left\{ \frac{\partial}{\partial \zeta}(P_0^{-2}(\hat{a}_0 + i\,\hat{b}_0)) \right.$$

$$\left. + \frac{\partial}{\partial \bar{\zeta}}(P_0^{-2}(\hat{a}_0 - i\,\hat{b}_0)) \right\} . \tag{7.193}$$

We can express the results neatly by making use of the following quantities:

$$S_{(1)} = C_{ijkl}\, k^i\, v^j\, k^k\, v^l , \tag{7.194}$$

$$S_{(2)} = (^*F_{ij}\, k^i\, v^j)^2 + \frac{1}{3}\,^*F^P{}_i\,^*F_{pj}\, v^i\, v^j , \tag{7.195}$$

$$S_{(3)} = {}^*F^P{}_i\,^*F_{pj}\, k^i\, k^j - 2\,^*F^P{}_i\,^*F_{pj}\, k^i\, v^j + \frac{1}{3}\,^*F^{ij}\,^*F_{ij}$$

$$+ \frac{2}{3}\,^*F^P{}_i\,^*F_{pj}\, v^i\, v^j , \tag{7.196}$$

$$S_{(4)} = h_0^2 + \frac{1}{3}\, a_i\, a^i , \tag{7.197}$$

$$L_{(1)} = {}^*F^P{}_i\,^*F_{pj}\, k^i\, v^j - {}^*F^P{}_i\,^*F_{pj}\, v^i\, v^j , \tag{7.198}$$

$$L_{(2)} = \dot{a}_i\, k^i + a_i\, a^i . \tag{7.199}$$

A dot, as in (7.199), indicates differentiation with respect to u. The significance of these quantities is that, using the useful formulas (7.57)–(7.59) the S's are $l = 2$ spherical harmonics (i.e. each S satisfies $\Delta S + 6\,S = 0$) while the L's are $l = 1$

spherical harmonics (i.e. each L satisfies $\Delta L + 2\, L = 0$). In addition we note that as a consequence of the equations of motion in first approximation (7.143) we have

$$m^2\, L_{(2)} = m\, g\, {}^*\dot{F}_{ij}\, k^i\, v^j - g^2\, L_{(1)} + O_3\,, \tag{7.200}$$

which is useful for deriving (7.218) below. Using the second of (7.173) and also using (7.174) we now find that

$$P_0^2 \left(\frac{\partial m_2}{\partial x} - \frac{\partial l_2}{\partial y} \right) = -12\, g\, S_{(2)} + 6\, g\, S_{(3)} + \frac{1}{2}\, g\, S_{(1)}$$

$$-2\, U(u)\, \frac{\partial}{\partial x}(\log P_0) + 2\, V(u)\, \frac{\partial}{\partial y}(\log P_0)\,, \tag{7.201}$$

and

$$P_0^2 \left(\frac{\partial}{\partial x}(P_0^{-2}\hat{a}_0) + \frac{\partial}{\partial y}(P_0^{-2}\hat{b}_0) \right) = -4\, m\, S_{(1)} + 12\, m\, S_{(2)} + 6\, m\, S_{(4)}$$

$$-6\, m\, S_{(3)} + 2\, g\, {}^*\dot{F}_{ij}\, k^i\, v^j - 2\, m\, L_{(2)}$$

$$-2\, A_0(u)\, \frac{\partial}{\partial x}(\log P_0) - 2\, B_0(u)\, \frac{\partial}{\partial y}(\log P_0)\,. \tag{7.202}$$

It is helpful for later to note that ${}^*\dot{F}_{ij}\, k^i\, v^j$, $\frac{\partial}{\partial x}(\log P_0)$ and $\frac{\partial}{\partial y}(\log P_0)$ are $l = 1$ spherical harmonics with the latter two following from differentiating (7.18) with respect to x and with respect to y respectively. In addition the derivation of (7.202) from (7.191) has involved the following simplifications using the equations of motion in first approximation (7.143) along with (7.197) and (7.199):

$$\frac{6\, g^2}{m}\, ({}^*F_{ij}\, k^i\, v^j)^2 = 6\, m\, h_0^2 + O_2 = 6\, m\, S_{(4)} - 2\, m\, a_i\, a^i + O_2\,, \tag{7.203}$$

and

$$\frac{2\, g^2}{m}\, {}^*F^p{}_i\, {}^*F_{pj}\, k^i\, v^j = -2\, m\, L_{(2)} + 2\, g\, {}^*\dot{F}_{ij}\, k^i\, v^j + 2\, m\, a_i\, a^i + O_2\,. \tag{7.204}$$

Next we consider the field equation $R_{(A)(A)} = 2\, E_{(A)(A)}$ with $R_{(A)(A)}$ given by (7.168) and $E_{(A)(A)} = -F_{(1)(2)}^2 - F_{(3)(4)}^2$. Using (7.160) the latter reads explicitly

$$E_{(A)(A)} = -\frac{1}{r^4}\, (g^2 + O_4) + \frac{1}{r^2} \left\{ 2\, g\, {}^*F_{ij}\, k^i\, v^j - 2\, g\, P_0^2 \left(\frac{\partial m_2}{\partial x} - \frac{\partial l_2}{\partial y} \right) \right.$$

$$\left. + \frac{2}{3}\, g^2\, {}^*F^p{}_i\, {}^*F_{pj}\, k^i\, k^j + O_3 \right\} + \frac{1}{r} \times O_1 + O(r^0)\,. \tag{7.205}$$

Hence in order to satisfy

$$R_{(A)(A)} - 2\,E_{(A)(A)} = \frac{1}{r^4} \times O_3 + \frac{1}{r^2} \times O_3 + \frac{1}{r} \times O_1 + O(r^0)\,, \qquad (7.206)$$

we must have $\hat{c}_1 = -2\,h_0 + O_1$ (which will be sufficient accuracy for our purposes) and

$$\hat{c}_0 = 1 + 8\,g\,{}^*F_{ij}\,k^i\,v^j - 8\,g\,P_0^2\left(\frac{\partial m_2}{\partial x} - \frac{\partial l_2}{\partial y}\right) + \frac{5}{3}\,g^2\,{}^*F^p{}_i\,{}^*F_{pj}\,k^i\,k^j$$

$$+\Delta Q_2 + 2\,Q_2$$

$$= 1 + 8\,g\,{}^*F_{ij}\,k^i\,v^j - \frac{5}{9}\,g^2\,{}^*F^{ij}\,{}^*F_{ij} + \frac{20}{9}\,g^2\,{}^*F^p{}_i\,{}^*F_{pj}\,v^i\,v^j$$

$$+\frac{10}{3}\,g^2\,L_{(1)} - \frac{139}{3}\,g^2\,S_{(3)} + 96\,g^2\,S_{(2)} - 4\,g^2\,S_{(1)} + 16\,g\,U(u)\,\frac{\partial}{\partial x}(\log P_0)$$

$$-16\,g\,V(u)\,\frac{\partial}{\partial y}(\log P_0) + \Delta Q_2 + 2\,Q_2 + O_3\,. \qquad (7.207)$$

We note for use below that now

$$-\frac{1}{2}\Delta\hat{c}_0 = 8\,g\,{}^*F_{ij}\,k^i\,v^j - \frac{1}{2}\Delta(\Delta Q_2 + 2\,Q_2) + \frac{10}{3}\,g^2\,L_{(1)}$$

$$-139\,g^2\,S_{(3)} + 288\,g^2\,S_{(2)} - 12\,g^2\,S_{(1)} + 16\,g\,U(u)\,\frac{\partial}{\partial x}(\log P_0)$$

$$-16\,g\,V(u)\,\frac{\partial}{\partial y}(\log P_0)\,. \qquad (7.208)$$

At this point in the derivation of the equations of motion in second approximation of the magnetic black hole we have obtained all of the required perturbations in the metric tensor and the potential 1-form appearing in (7.72)–(7.75) and (7.76)–(7.83) with the exception of the functions $\hat{f}_{-1} = O_2$ and $Q_2 = O_2$. The latter function determines the behaviour of the geometry of the wave fronts of the radiation generated by the motion of the magnetic black hole and thus is key to our derivation of the equations of motion. However we need $\hat{f}_{-1}$ in order to derive Q_2. Both of these functions are obtained from the field equation $R_{(4)(4)} = 2\,E_{(4)(4)}$. The leading term on both sides of this equation is $O(r^{-3})$, neglecting O_3-terms. We find that

$$R_{(4)(4)} = \frac{1}{r^3}\left\{2\,\Delta\hat{f}_{-1} + 4\,g^2\,h_0 - 2\,m\,P_0^2\left(\frac{\partial}{\partial x}(P_0^{-2}\hat{a}_{-1}) + \frac{\partial}{\partial y}(P_0^{-2}\hat{b}_{-1})\right)\right\}$$

$$+\frac{W}{r^2} + O\left(\frac{1}{r}\right)$$

$$= \frac{1}{r^3}\,(2\,\Delta\hat{f}_{-1} + 4\,g^2\,h_0 + 8\,m\,g\,{}^*F_{ij}\,k^i\,v^j) + \frac{W}{r^2} + O\left(\frac{1}{r}\right)\,, \qquad (7.209)$$

and

$$
W = 4\frac{\partial \hat{f}_{-1}}{\partial u} - 6\,m\,h_0 - \frac{1}{2}\Delta \hat{c}_0 + \frac{1}{2}\hat{c}_0\,P_0^2\left(\frac{\partial}{\partial x}(P_0^{-2}\hat{a}_{-1}) + \frac{\partial}{\partial y}(P_0^{-2}\hat{b}_{-1})\right)
$$
$$
+ \frac{5}{2}\,g^2\,P_0^2\left(\frac{\partial}{\partial x}(P_0^{-2}\,a_1) + \frac{\partial}{\partial y}(P_0^{-2}\,b_1)\right) - 3\,m\,P_0^2\left(\frac{\partial}{\partial x}(P_0^{-2}\hat{a}_0)\right.
$$
$$
\left. + \frac{\partial}{\partial y}(P_0^{-2}\hat{b}_0)\right) - 12\,h_0\,\hat{f}_{-1} + 8\,g^2({}^*F_{ij}\,k^i\,v^j)^2 - 6(g^2 + 2\,m^2)\,q_2\,,
$$

$$(7.210)$$

while, again neglecting O_3-terms,

$$
E_{(4)(4)} = \frac{1}{r^2}\left\{g^2\,S_{(4)} + 7\,g^2\,S_{(2)} + \left(m^2 - \frac{5}{2}\,g^2\right)S_{(3)}\right.
$$
$$
+ 2(m^2 + g^2)\,L_{(1)} + \frac{2}{3}\,g^2\,a_i\,a^i + \left(\frac{5}{6}g^2 - \frac{1}{3}m^2\right){}^*F^{ij}\,{}^*F_{ij}
$$
$$
\left. + \frac{4}{3}(g^2 + m^2)\,{}^*F^p{}_i\,{}^*F_{pj}\,v^i\,v^j\right\} + \frac{1}{r}\times O_1 + O(r^0)\,.\qquad(7.211)
$$

All of the preliminary work for substitution into W in (7.210) has been done at this stage with the exception of the evaluation of the term involving a_1 and b_1 which are given by (7.65) and (7.66). We find that

$$
P_0^2\left(\frac{\partial}{\partial x}(P_0^{-2}\,a_1) + \frac{\partial}{\partial y}(P_0^{-2}\,b_1)\right) = 2\,S_{(1)} - 2\,S_{(3)} - \frac{4}{3}\,L_{(1)}\,.\qquad(7.212)
$$

We satisfy $R_{(4)(4)} = 2\,E_{(4)(4)}$ (writing all orders of magnitude explicitly now) approximately in the sense that

$$
R_{(4)(4)} - 2\,E_{(4)(4)} = \frac{1}{r^4}\times O_4 + \frac{1}{r^3}\times O_3 + \frac{1}{r^2}\times O_3 + \frac{1}{r}\times O_1 + O_1 + O(r)\,,\qquad(7.213)
$$

by first requiring $\hat{f}_{-1}$ to satisfy the differential equation

$$
\Delta \hat{f}_{-1} = -2\,g^2\,a_i\,k^i - 4\,m\,g\,{}^*F_{ij}\,k^i\,v^j\,,\qquad(7.214)
$$

where we have written $h_0 = a_i\,k^i$ as in (7.14). Both terms on the right hand side here are $l = 1$ spherical harmonics and so we can immediately solve this equation for an $\hat{f}_{-1}$ which is well behaved for $-\infty < x, y < +\infty$ by taking

$$
\hat{f}_{-1} = g^2\,a_i\,k^i + 2\,m\,g\,{}^*F_{ij}\,k^i\,v^j + G(u)\,,\qquad(7.215)
$$

where $G(u) = O_2$ is a function of integration and an $l = 0$ spherical harmonic. The remaining differential equation for Q_2 is thus found to be

$$-\frac{1}{2}\Delta(\Delta Q_2 + 2\,Q_2) = \Sigma_{(0)} + \Sigma_{(1)} + \Sigma_{(2)} \, , \qquad (7.216)$$

where

$$\Sigma_{(0)} = -4\,\dot{G} - 4\,g^2 \left({}^*F^p{}_i\,{}^*F_{pj} - \frac{1}{2}\eta_{ij}\,{}^*F^{pq}\,{}^*F_{pq} \right) v^i\,v^j$$

$$= -4\,\dot{G} - 4\,g^2 F^p{}_i\,F_{pj}\,v^i\,v^j \, , \qquad (7.217)$$

is an $l = 0$ spherical harmonic,

$$\Sigma_{(1)} = -4\,g^2 L_{(2)} - 8\,m\,g\,{}^*\dot{F}_{ij}\,k^i\,v^j + 16\,g^2 L_{(1)} + 6\,m\,h_0 - 6\,g\,{}^*F_{ij}\,k^i\,v^j$$

$$+12\,G\,h_0 - 6\,\hat{U}(u)\,\frac{\partial}{\partial x}(\log P_0) - 6\,\hat{V}(u)\,\frac{\partial}{\partial y}(\log P_0) \, , \qquad (7.218)$$

is an $l = 1$ spherical harmonic (we have written $2\,g\,U(u) + m\,A_0(u) = \hat{U}(u) = O_2$ and $-2\,g\,V(u) + m\,B_0(u) = \hat{V}(u) = O_2$ for convenience), and

$$\Sigma_{(2)} = (6\,g^2 - 12\,m^2)S_{(1)} + (36\,m^2 - 242\,g^2)S_{(2)} + (50\,m^2 + 2\,g^2)S_{(4)}$$

$$+(126\,g^2 - 18\,m^2)S_{(3)} \, , \qquad (7.219)$$

is an $l = 2$ spherical harmonic.

We can solve (7.216) for $\Delta Q_2 + 2\,Q_2$ ensuring that this expression is well behaved for $-\infty < x, y < +\infty$ by taking the $l = 0$ spherical harmonic $\Sigma_{(0)} = 0$. We thus have $G(u)$ given by

$$\dot{G} = -g^2 F^p{}_i\,F_{pj}\,v^i\,v^j \, , \qquad (7.220)$$

and $\Delta Q_2 + 2\,Q_2$ given by

$$\Delta Q_2 + 2\,Q_2 = \frac{1}{3}\,\Sigma_{(2)} + \Sigma_{(1)} \, . \qquad (7.221)$$

To have Q_2 free of singularities for $-\infty < x, y < +\infty$ the $l = 1$ term on the right hand side of this equation must vanish (i.e. $\Sigma_{(1)} = 0$, remembering that we are neglecting O_3-terms). In this case

$$Q_2 = -\frac{1}{12}\,\Sigma_{(2)} \, . \qquad (7.222)$$

We note that the addition of $l = 0$ and/or $l = 1$ terms to Q_2 are trivial, as explained following (7.85) above. The equation $\Sigma_{(1)} = 0$ will give us the equations of motion of the small magnetic black hole. To see this we first define

$$p^i = h^i_j k^j \quad \text{with} \quad h^i_j = \delta^i_j - v^i v_j .\tag{7.223}$$

Here p^i is the unit, space-like projection of the null vector k^i orthogonal to v^i. Since k^i is arbitrary it follows that p^i points in any direction orthogonal to v^i. For substitution into (7.218) we can write the following quantities in terms of $p^i = k^i - v^i$:

$$L_{(2)} = h^j_i \, \dot{a}_j \, p^i \, , \quad {}^*\dot{F}_{ij} k^i v^j = {}^*\dot{F}_{ij} \, p^i v^j \, , \quad L_{(1)} = {}^*F^{p}{}_k \, {}^*F_{pj} h^k_i v^j p^i \, ,$$

$$h_0 = a_i \, p^i \, , \quad {}^*F_{ij} k^i v^j = {}^*F_{ij} \, p^i v^j \, , \quad \frac{\partial}{\partial x}(\log P_0) = c_i \, p^i \, , \quad \frac{\partial}{\partial y}(\log P_0) = d_i \, p^i \, ,$$

$$\tag{7.224}$$

with

$$c_i = \left(\frac{1}{2}(v^3 - v^4), 0, -\frac{1}{2}v^1, \frac{1}{2}v^1 \right) \quad \text{and} \quad d_i = \left(0, \frac{1}{2}(v^3 - v^4), -\frac{1}{2}v^2, \frac{1}{2}v^2 \right).$$

$$\tag{7.225}$$

Now (7.218) becomes

$$\Sigma_{(1)} = 6 \left\{ m \, a_i - g \, {}^*F_{ij} \, v^j - \frac{2}{3} g^2 \, h^j_i \dot{a}_j + \frac{8}{3} g^2 \, {}^*F^{p}{}_k \, {}^*F_{pj} \, h^k_i \, v^j \right.$$

$$\left. - \frac{4}{3} m \, g \, {}^*\dot{F}_{ij} \, v^j + 2 \, G \, a_i - \hat{U}(u) \, c_i - \hat{V}(u) \, d_i \right\} p^i .\tag{7.226}$$

This is the scalar product of p^i, which is *any* unit space-like vector orthogonal to v^i, with a vector orthogonal to v^i and it is required to vanish for all p^i. Hence we must have

$$m \, a_i = g \, {}^*F_{ij} \, v^j + \frac{2}{3} g^2 \, h^j_i \dot{a}_j - \frac{8}{3} g^2 \, {}^*F^{p}{}_k \, {}^*F_{pj} \, h^k_i \, v^j$$

$$+ \frac{4}{3} m \, g \, {}^*\dot{F}_{ij} \, v^j - 2 \, G \, a_i + \hat{U}(u) \, c_i + \hat{V}(u) \, d_i .\tag{7.227}$$

We note that $G(u)$ is given by (7.220) and thus $G(u)$ can be written as an integral with respect to u for which we would naturally choose as range of integration $(-\infty, u]$ which results in $-2 \, G \, a_i$ in (7.227) contributing to a "tail term". We can write $\hat{U} \, c_i + \hat{V} \, d_i = \Omega_{ij}(u) \, v^j$ with $\Omega_{ij} = -\Omega_{ji}$ vanishing except for $\Omega_{13} = -\Omega_{14} = \hat{U}/2$ and $\Omega_{23} = -\Omega_{24} = \hat{V}/2$. Define $\omega_{ij}(u) = -\omega_{ji}(u)$ by $\dot{\omega}_{ij} = \Omega_{ij}$.

Now make the 1-parameter family of infinitesimal Lorentz transformations on the tangent vector, $v^i \rightarrow \bar{v}^i = v^i - (4/3)\, g * F_{ij}\, v^j - (1/m)\, \omega^i{}_j\, v^j$. When this is done and we then drop the bars we arrive at the final form of the equations of motion of the small magnetic black hole in second approximation (re-introducing the designation of the O_3-error which we have been leaving understood for convenience):

$$m\, a_i = g * F_{ij}\, v^j + \frac{2}{3} g^2\, h_i^j \dot{a}_j - \frac{8}{3} g^2 * F^p{}_k * F_{pj}\, h_i^k\, v^j + \mathcal{T}_i + O_3\,, \qquad (7.228)$$

with

$$\mathcal{T}_i = \frac{g}{m}\{-\omega_{ij} * F^j{}_k + \omega_{kj} * F^j{}_i - 2\,G * F_{ik}\}\, v^k = O_2\,. \qquad (7.229)$$

Since ω_{ij} can be expressed as an integral with respect to u, whose range we naturally take to be $(-\infty, u]$, we see that (7.229) is a "tail term" (depending on the past history of the source) in the equations of motion (7.228). The first term on the right hand side of (7.228) is the external 4-force in first approximation, the second term is the radiation reaction due to the electromagnetic radiation emitted by the accelerating magnetic black hole and the third term on the right hand side of (7.228) is a second order correction to the external 4-force. All of these terms have analogues in the case of a small charged black hole moving in an external field [9]. In particular it is interesting to note that the second order contribution to the external 4-force in (7.228) is

$$-\frac{8}{3} g^2 * F^p{}_k * F_{pj}\, h_i^k\, v^j = -\frac{8}{3} g^2\, F^p{}_k\, F_{pj}\, h_i^k\, v^j\,, \qquad (7.230)$$

on account of (7.158). The charged analogue of this term is [9]

$$+\frac{4}{3} e^2\, F^p{}_k\, F_{pj}\, h_i^k\, v^j\,, \qquad (7.231)$$

where e is the electric charge of the small black hole.

7.6   Review of Approximations

In the Appendix E the exact Maxwell field equations are given by (E.7)–(E.10) and for our model these are satisfied approximately as indicated in (E.23)–(E.26). The exact Einstein field equations are $R_{(a)(b)} - 2\,E_{(a)(b)} = 0$ and again for our model (i.e. to derive the equations of motion in second approximation of a small magnetic black hole) these are satisfied approximately as follows:

$$R_{(3)(3)} - 2\,E_{(3)(3)} = O_1 + O(r)\,, \qquad (7.232)$$

$$R_{(A)(3)} - 2\,E_{(A)(3)} = \frac{1}{r^2} \times O_3 + \frac{1}{r} \times O_1 + O_1 + O(r)\,, \qquad (7.233)$$

$$R_{(1)(1)} - R_{(2)(2)} + 2i\,R_{(1)(2)} = 2\left(E_{(1)(1)} - E_{(2)(2)} + 2i\,E_{(1)(2)}\right)$$

$$+\frac{1}{r^2} \times O_3 + \frac{1}{r} \times O_2 + O_1 + O(r)\,, \qquad (7.234)$$

$$R_{(A)(A)} - 2E_{(A)(A)} = \frac{1}{r^4} \times O_3 + \frac{1}{r^2} \times O_3 + \frac{1}{r} \times O_1 + O_1 + O(r)\,,$$
$$(7.235)$$

$$R_{(3)(4)} - 2E_{(3)(4)} = \frac{1}{r^4} \times O_4 + \frac{1}{r^3} \times O_3 + \frac{1}{r^2} \times O_2 + \frac{1}{r} \times O_1$$
$$+O_1 + O(r)\,, \qquad (7.236)$$

$$R_{(A)(4)} - 2E_{(A)(4)} = \frac{1}{r^4} \times O_3 + \frac{1}{r^3} \times O_2 + \frac{1}{r^2} \times O_2 + \frac{1}{r} \times O_1$$
$$+O_1 + O(r)\,, \qquad (7.237)$$

$$R_{(4)(4)} - 2E_{(4)(4)} = \frac{1}{r^4} \times O_4 + \frac{1}{r^3} \times O_3 + \frac{1}{r^2} \times O_3 + \frac{1}{r} \times O_1$$
$$+O_1 + O(r)\,, \qquad (7.238)$$

where capital letters take values 1, 2 and the summation convention, as always, applies in (7.235). By way of comparison, for deriving the equations of motion in first approximation of the small magnetic black hole in Sect. 7.4, these approximations are relaxed in the sense that the coefficients of r^{-2} in (7.233), (7.234), (7.235) and (7.238) are only required to be O_2 and the coefficient of r^{-3} in (7.238) can also be relaxed to O_2. It is in this sense that the accuracy with which the field equations must be satisfied is determined by the accuracy with which the equations of motion are required.

We have seen explicitly that it is relatively straightforward, but of course increasingly more complicated, to proceed from requiring the equations of motion in first approximation to requiring the equations of motion in second approximation. This is because the expansions in powers of r of the functions appearing in the perturbed potential 1-form and the perturbed metric tensor components have the property that the coefficients of the powers of r merely have to be determined more accurately in order to pass from one degree of approximation to the next. If however the equations of motion are required with still greater accuracy then the expansions of the functions in powers of r may have to be modified with the introduction of additional powers of r in each case. This represents a highly non-trivial future challenge. As a final comment we note that no infinities requiring "renormalisation" arise in our approach and nor has there been any slow-motion assumption made, as has already been pointed out in [6, 9] and [7].

References

1. P.A.M. Dirac, Proc. R. Soc. A **133**, 60 (1931)
2. T.T. Wu, C.N. Yang, *Properties of Matter under Unusual Conditions*, ed. by H. Mark, S. Fernbach (Interscience, New York, 1968), p. 349
3. G. 't Hooft, Nucl. Phys. **79**, 276 (1974)
4. A.M. Polyakov, JETP Lett. **20**, 194 (1974)
5. G.E. Volovik, *The Universe in a Helium Droplet*. International Series of Monographs on Physics No. 117 (Oxford University Press, Oxford, 2003)
6. H. Asada, T. Futamase, P.A. Hogan, *Equations of Motion in General Relativity*. International Series of Monographs on Physics No. 148 (Oxford University Press, Oxford, 2010)
7. P.A. Hogan, *Equations of Motion in Relativistic Gravity*, vol. 179, ed. by D. Puetzfeld et al. Fundamental Theories of Physics, (Springer, Berlin, 2015), p. 265
8. C.G. Böhmer, P.A. Hogan, Preprint (2020)
9. T. Futamase, P.A. Hogan, Y. Itoh, Phys. Rev. D **78**, 104014 (2008)
10. P.A. Hogan, A. Trautman, *Gravitation and Geometry*, ed. by A. Trautman, W. Rindler (Bibliopolis, Naples, 1987), p. 215

Run-Away Reissner–Nordström Particle

8

Abstract

In the absence of external fields a Reissner–Nordström particle of mass m and charge e for which m and e^2 are small of first order performs run-away motion. Since in particular $m^2 < e^2$ there is no event horizon associated with this object and so we refer to it as a Reissner–Nordström particle. A Reissner–Nordström particle of mass m and magnetic monopole moment g for which m and g^2 are small of first order behaves in precisely the same way.

8.1 Robinson–Trautman Solutions of the Einstein–Maxwell Equations

It follows from Sect. 8.2 of the previous chapter that the relevant form of line element in the absence of external fields is given by (2.1)–(2.5) with the functions α, β, a, b vanishing. The result is a Robinson–Trautman [1] line element

$$ds^2 = -r^2 p^{-2}(dx^2 + dy^2) + 2\,du\,dr + c\,du^2$$
$$= -(\vartheta^{(1)})^2 - (\vartheta^{(2)})^2 + 2\,\vartheta^{(3)}\,\vartheta^{(4)} = g_{(a)(b)}\,\vartheta^{(a)}\,\vartheta^{(b)}\,, \qquad (8.1)$$

with the basis 1-forms taken to be

$$\vartheta^{(1)} = r\,p^{-1}dx\,, \ \ \vartheta^{(2)} = r\,p^{-1}dy\,, \ \vartheta^{(3)} = dr + \frac{1}{2}c\,du\,, \ \ \vartheta^{(4)} = du\,, \qquad (8.2)$$

and the Robinson–Trautman potential 1-form

$$A = e(u)\left(\frac{1}{r} + w\right)du\,. \qquad (8.3)$$

Here $p = p(x, y, u)$, $c = c(x, y, r, u)$ and $w = w(x, y, u)$. The corresponding Maxwell 2-form is the exterior derivative of (8.3), namely,

$$F = dA = -\frac{e}{r^2} \vartheta^{(3)} \wedge \vartheta^{(4)} + \frac{e\,p}{r} \frac{\partial w}{\partial x} \vartheta^{(1)} \wedge \vartheta^{(4)} + \frac{e\,p}{r} \frac{\partial w}{\partial y} \vartheta^{(2)} \wedge \vartheta^{(4)}$$

$$= \frac{1}{2} F_{(a)(b)} \vartheta^{(a)} \wedge \vartheta^{(b)} , \tag{8.4}$$

and thus $F_{(a)(b)} = -F_{(b)(a)}$ are the components of the Maxwell tensor on the half null tetrad defined via the 1-forms (8.2). The Hodge dual of this 2-form is the 2-form

$$*F = \frac{e}{r^2} \vartheta^{(1)} \wedge \vartheta^{(2)} + \frac{e\,p}{r} \frac{\partial w}{\partial x} \vartheta^{(2)} \wedge \vartheta^{(4)} - \frac{e\,p}{r} \frac{\partial w}{\partial y} \vartheta^{(1)} \wedge \vartheta^{(4)} . \tag{8.5}$$

Maxwell's vacuum field equations $d*F = 0$ now yield [1]

$$\Delta w = -e^{-1}\dot{e} + 2H \quad \text{with} \quad \Delta = p^2 \left(\frac{\partial^2}{\partial x^2} + \frac{\partial^2}{\partial y^2} \right) \quad \text{and} \quad H = \frac{\partial}{\partial u}(\log p) = p^{-1}\dot{p} , \tag{8.6}$$

with the dot indicating partial differentiation with respect to u. As before the electromagnetic energy-momentum tensor has components $E_{(a)(b)} = E_{(b)(a)}$ on the half null tetrad given via the 1-forms (8.2):

$$E_{(a)(b)} = F_{(a)(c)} F_{(b)}{}^{(c)} - \frac{1}{4} g_{(a)(b)} F_{(d)(c)} F^{(d)(c)} , \tag{8.7}$$

with $F_{(a)(b)}$ given by (8.4). Also $g_{(a)(b)}$ are the components of the metric tensor on the half null tetrad as given by (8.1) and tetrad indices are raised and lowered by $g^{(a)(b)}$ and $g_{(a)(b)}$ respectively, with $g^{(a)(b)}$ the inverse of $g_{(a)(b)}$ defined via $g^{(a)(b)} g_{(b)(c)} = \delta^{(a)}_{(c)}$. With a subscript denoting partial differentiation, $E_{(a)(b)}$ is found to be

$$(E_{(a)(b)}) = \begin{pmatrix} -\frac{e^2}{2r^4} & 0 & 0 & \frac{e^2}{r^3}\,p\,w_x \\ 0 & -\frac{e^2}{2r^4} & 0 & \frac{e^2}{r^3}\,p\,w_y \\ 0 & 0 & 0 & -\frac{e^2}{2r^4} \\ \frac{e^2}{r^3}\,p\,w_x & \frac{e^2}{r^3}\,p\,w_y & -\frac{e^2}{2r^4} & -\frac{e^2}{r^2}\,p^2\,(w_x^2 + w_y^2) \end{pmatrix} . \tag{8.8}$$

Einstein's field equations

$$R_{(a)(b)} = 2\,E_{(a)(b)} , \tag{8.9}$$

with $R_{(a)(b)}$ the components of the Ricci tensor on the half null tetrad calculated with the metric given via the line element (8.1), result in [1]

$$c = K - 2 H r - \frac{2 M}{r} + \frac{e^2}{r^2} , \tag{8.10}$$

with

$$K = \Delta \log p , \quad M = m(u) - 2 e^2 w , \tag{8.11}$$

and

$$\frac{1}{4} \Delta K = \dot{M} - 3 H M + e^2 p^2 (w_x^2 + w_y^2) . \tag{8.12}$$

For the Reissner–Nordström spherically symmetric and static solution e, m are constants,

$$w = 0 , \quad p = p_0 = 1 + \frac{1}{4}(x^2 + y^2) , \quad H = 0 , \quad K = 1 , \tag{8.13}$$

and so the Maxwell 2-form reads

$$F = -\frac{e^2}{r^2} \vartheta^{(3)} \wedge \vartheta^{(4)} , \tag{8.14}$$

with e the charge on the source, and

$$ds^2 = -r^2 p_0^{-2}(dx^2 + dy^2) + 2 du \, dr + \left(1 - \frac{2 m}{r} + \frac{e^2}{r^2}\right) du^2 , \tag{8.15}$$

with m the mass of the source. Hence it is possible to view the space-time as a perturbation of Minkowskian space-time with

$$ds^2 = ds_0^2 + \gamma_{ij} \, dx^i \, dx^j \quad \text{with} \quad x^i = (x, y, r, u) , \tag{8.16}$$

where

$$ds_0^2 = -r^2 p_0^{-2}(dx^2 + dy^2) + 2 du \, dr + du^2 , \tag{8.17}$$

is the line element of Minkowskian space-time and $r = 0$ is a time-like geodesic in this space-time with u proper-time or arc length along it, and

$$\gamma_{ij} \, dx^i \, dx^j = \left(-\frac{2 m}{r} + \frac{e^2}{r^2}\right) du^2 , \tag{8.18}$$

may be a small perturbation of Minkowskian space-time which is singular on $r = 0$ in the "background" Minkowskian space-time.

8.2   Accelerating Reissner–Nordström Particle

We generalise the latter part of the introduction above by first writing the Minkowskian line element in a coordinate system based on an arbitrary non-geodesic, time-like world line $r = 0$ (say). If $X^i = (X, Y, Z, T)$ are rectangular Cartesians and time the Minkowskian line element reads

$$ds_0^2 = -(dX)^2 - (dY)^2 - (dZ)^2 + (dT)^2 = \eta_{ij}\, dX^i\, dX^j \,, \tag{8.19}$$

so that $\eta_{ij} = \mathrm{diag}(-1, -1, -1, +1)$. Let the parametric equations of $r = 0$ be $X^i = w^i(u)$. Define

$$v^i = \frac{dw^i}{du} \quad \text{with} \quad \eta_{ij}\, v^i\, v^j = +1 = v_j\, v^j \,. \tag{8.20}$$

Hence v^i are the components of the unit time-like tangent to $r = 0$ and u is proper-time or arc length along $r = 0$. Thus v^i is the 4-velocity of the particle with world line $r = 0$. Its 4-acceleration has components

$$a^i = \frac{dv^i}{du} \quad \Rightarrow \quad \eta_{ij}\, a^i\, v^j = 0 = a_j\, v^j \,, \tag{8.21}$$

with the implication following from the assumption above that v^i is a unit vector. The world line $r = 0$ is a time-like *geodesic* if $a^i = 0$. Now write the position 4-vector X^i of a point of Minkowskian space-time relative to the world line $r = 0$ in the form

$$X^i = w^i(u) + r\, k^i \quad \text{with} \quad k_i\, k^i = 0 \quad \text{and} \quad v_i\, k^i = +1 \,. \tag{8.22}$$

Thus k^i is a future directed null vector field normalised with the use of v^i here. We note that (8.22) defines r by

$$r = \eta_{ij}\, (X^i - w^i(u))\, (X^j - w^j(u)) \geq 0 \,, \tag{8.23}$$

with equality if and only if X^i lies on $r = 0$. Hence r is a good measure of distance from the world line $r = 0$. The null vector field k^i is constrained by two conditions in (8.22) and so has only two independent components. Hence it can be parametrized by two real parameters x, y and written for example in the form

$$- P_0\, k^i = \left(x, y, 1 - \frac{1}{4}(x^2 + y^2), 1 + \frac{1}{4}(x^2 + y^2) \right) \,, \tag{8.24}$$

with the real-valued function P_0 determined by the normalisation in (8.22) to be given by

$$P_0 = x\,v^1(u) + y\,v^2(u) + \left(1 - \frac{1}{4}(x^2 + y^2)\right) v^3(u) + \left(1 + \frac{1}{4}(x^2 + y^2)\right) v^4(u) .$$

$$(8.25)$$

Defining the operator

$$\underset{0}{\Delta} = P_0^2 \left(\frac{\partial^2}{\partial x^2} + \frac{\partial^2}{\partial y^2}\right) ,$$

$$(8.26)$$

we can easily verify that

$$\underset{0}{\Delta} \log P_0 = v_i\, v^i = +1 .$$

$$(8.27)$$

Also we can easily see that

$$\frac{\partial}{\partial u}(\log P_0) = \eta_{ij}\, a^i\, k^j = a_j\, k^j = h_0(x, y, u) \ \text{(say)} .$$

$$(8.28)$$

The formula (8.22) can now be viewed as a coordinate transformation relating the coordinates X^i to a new set of coordinates x, y, r, u. Conversely it can be viewed as giving x, y, r, u implicitly as scalar functions of X^i. The partial derivatives of these four scalars with respect to X^i (denoted by a comma) are useful and can be obtained by initially differentiating (8.22) with respect to X^j to arrive at

$$\delta^i_j = (v^i - r\,h_0\,k^i)\,u_{,j} + k^i\,r_{,j} + r\,\frac{\partial k^i}{\partial x}\,x_{,j} + r\,\frac{\partial k^i}{\partial y}\,y_{,j} .$$

$$(8.29)$$

Here we have used

$$\frac{\partial k^i}{\partial u} = -h_0\,k^i ,$$

$$(8.30)$$

which follows from (8.24) and the definition of h_0 given in (8.28). Multiplying (8.29) successively by k_i, v_i, $\partial k_i/\partial x$ and $\partial k_i/\partial y$ results in

$$u_{,j} = k_j , \quad r_{,j} = v_j - (1 - r\,h_0)\,k_j , \quad x_{,j} = -\frac{P_0^2}{r}\frac{\partial k_j}{\partial x} \ \text{and} \ y_{,j} = -\frac{P_0^2}{r}\frac{\partial k_j}{\partial y} .$$

$$(8.31)$$

Substituting these into (8.29) and raising the covariant index with the inverse of the Minkowskian metric tensor $\eta^{ij} = \text{diag}(-1, -1, -1, +1)$ (since $\eta^{ij} \eta_{jk} = \delta^i_k$) results in the useful formula

$$\eta^{ij} = -P_0^2 \left(\frac{\partial k^i}{\partial x} \frac{\partial k^j}{\partial x} + \frac{\partial k^i}{\partial y} \frac{\partial k^j}{\partial y} \right) + k^i v^j + k^j v^i - k^i k^j . \qquad (8.32)$$

In deriving (8.31) we have made use of the fact that in (8.24) with (8.25) the null vector field k^i is given explicitly in terms of x and y and with its dependence on u unspecified since the world line $r = 0$ is unspecified. The following formulas, which have been listed in (7.57)–(7.59) in a previous chapter already, for the second partial derivatives of k^i with respect to x, y are useful in the sequel:

$$\frac{\partial^2 k^i}{\partial x^2} = P_0^{-2} (v^i - k^i) - \frac{\partial}{\partial x}(\log P_0) \frac{\partial k^i}{\partial x} + \frac{\partial}{\partial y}(\log P_0) \frac{\partial k^i}{\partial y} , \qquad (8.33)$$

$$\frac{\partial^2 k^i}{\partial y^2} = P_0^{-2} (v^i - k^i) + \frac{\partial}{\partial x}(\log P_0) \frac{\partial k^i}{\partial x} - \frac{\partial}{\partial y}(\log P_0) \frac{\partial k^i}{\partial y} , \qquad (8.34)$$

$$\frac{\partial^2 k^i}{\partial x \partial y} = -\frac{\partial}{\partial y}(\log P_0) \frac{\partial k^i}{\partial x} - \frac{\partial}{\partial x}(\log P_0) \frac{\partial k^i}{\partial y} . \qquad (8.35)$$

We note from these that

$$\underset{0}{\Delta} k^i + 2 k^i = 2 v^i , \qquad (8.36)$$

and so from (8.28)

$$\underset{0}{\Delta} h_0 + 2 h_0 = 0 , \qquad (8.37)$$

indicating that h_0 is an $l = 1$ spherical harmonic (since $\underset{0}{\Delta}$ is the Laplacian operator on the unit 2-sphere).

Substituting (8.22) into the line element (8.19) we arrive at a generalisation of (8.17), namely,

$$ds_0^2 = -r^2 P_0^{-2}(dx^2 + dy^2) + 2\, du\, dr + (1 - 2 h_0 r)\, du^2 . \qquad (8.38)$$

Here $r = 0$ is an arbitrary time-like world line with u proper time or arc length along it. We now wish to generalise (8.15) to an accelerating source. We do this in the next section assuming that the charge e and the mass m are small in the sense that e^2 and m are small of first order, writing $e^2 = O_1$ and $m = O_1$. More strictly speaking if $a = (-a_i\, a^i)^{1/2}$ is the magnitude of the 4-acceleration (with dimension of inverse length in the units we are using) then the dimensionless quantities $a^2 e^2$ and $a\, m$ are small of first order. This will ensure that the 4-acceleration of the source is a zeroth order quantity. If an external field is present, as in the previous chapter, the Lorentz 4-force is linear in the charge or the magnetic monopole moment and so assuming $e = O_1$ in that case ensures that the 4-acceleration is a zeroth order quantity. With our assumptions of $e^2 = O_1$ and $m = O_1$ here the basic object, the source of the Reissner–Nordström field, is not a black hole because $e^2 > m^2$, and so we refer to it as a Reissner–Nordström particle.

8.3 Perturbations of a Reissner–Nordström Particle

In the absence of external fields we perturb the Reissner–Nordström line element and potential 1-form consistent with the Robinson–Trautman forms of these quantities (8.1) and (8.3). We do this by assuming the expansions:

$$p = P_0 (1 + Q_1 + Q_2 + O_3) \quad \text{with} \quad Q_1 = O_1 \text{ and } Q_2 = O_2 , \tag{8.39}$$

with P_0 given by (8.25), Q_1, Q_2 functions of x, y, u, and

$$w = w_0 + w_1 + O_2 \quad \text{with} \quad w_0 = O_0 , \ w_1 = O_1 , \tag{8.40}$$

with w_0, w_1 functions of x, y, u. This will be sufficient accuracy for us to determine the equations of motion (differential equations for the world line $r = 0$) in second approximation (with an O_3-error). Now

$$p^2 = P_0^2(1 + 2 Q_1 + 2 Q_2 + Q_1^2 + O_3) , \tag{8.41}$$

$$\Delta = (1 + 2 Q_1 + 2 Q_2 + Q_1^2 + O_3)\underset{0}{\Delta} , \tag{8.42}$$

and

$$\log p = \log P_0 + Q_1 + Q_2 - \frac{1}{2} Q_1^2 + O_3 . \tag{8.43}$$

Hence H and K in (8.6) and (8.11) take the approximate forms (remembering that a dot indicates partial differentiation with respect to u)

$$H = h_0 + \dot{Q}_1 + \dot{Q}_2 - Q_1 \dot{Q}_1 + O_3 , \tag{8.44}$$

and

$$K = 1 + K_1 + K_2 + O_3 , \tag{8.45}$$

with

$$K_1 = \underset{0}{\Delta} Q_1 + 2\,Q_1 \,, \tag{8.46}$$

$$K_2 = \underset{0}{\Delta} Q_2 + 2\,Q_2 - \frac{1}{2}\underset{0}{\Delta} Q_1^2 + 2\,Q_1 \underset{0}{\Delta} Q_1 + Q_1^2 \,, \tag{8.47}$$

and thus

$$\Delta K = \underset{0}{\Delta} K_1 + \underset{0}{\Delta} K_2 + 2\,Q_1 \underset{0}{\Delta} K_1 + O_3 \,. \tag{8.48}$$

The Maxwell equation (8.6) provides differential equations for w_0 and w_1, namely,

$$\underset{0}{\Delta} w_0 = -e^{-1}\,\dot{e} + 2\,h_0 \,, \tag{8.49}$$

and

$$\underset{0}{\Delta} w_1 = 2\,\dot{Q}_1 - 2\,Q_1 \underset{0}{\Delta} w_0 \,. \tag{8.50}$$

The first term on the right hand side of (8.49) is an $l = 0$ spherical harmonic while the second term on the right hand side of (8.49) is an $l = 1$ spherical harmonic (on account of (8.37)). To have a solution w_0 of (8.49) which is well behaved (non-singular) for $-\infty < x, y < +\infty$ we must have

$$\dot{e} = 0 \quad \Rightarrow \quad e = \text{constant} \,. \tag{8.51}$$

Hence

$$w_0 = -h_0 \,, \tag{8.52}$$

up to the addition of an arbitrary function of u equivalent to a gauge term in the potential 1-form (8.3) and so we shall neglect it. Neglecting O_1-terms the potential 1-form (8.3) reads

$$A = e\left(\frac{1}{r} - h_0\right)du \,. \tag{8.53}$$

Up to a gauge transformation this is the Liénard–Wiechert potential 1-form since it follows from (8.31) that

$$dr = v_j\,dX^j - (1 - r\,h_0)\,du \quad \Leftrightarrow \quad \left(\frac{1}{r} - h_0\right)du + \frac{1}{r}\,dr = \frac{v_j}{r}\,dX^j \,. \tag{8.54}$$

Now the differential equation (8.50) for w_1 becomes

$$\underset{0}{\Delta} w_1 = 2\,\dot{Q}_1 - 4\,h_0\,Q_1 \,. \tag{8.55}$$

Turning now to the field equation (8.12) we have

$$M = m + 2\,e^2\,h_0 - 2\,e^2\,w_1 + O_3 \,, \tag{8.56}$$

and

$$\dot{M} = \dot{m} + 2\,e^2\,\dot{h}_0 - 2\,e^2\,w_1 + O_3 \,. \tag{8.57}$$

With h_0 given by (8.28) and satisfying (8.30) we can write

$$\dot{h}_0 = \dot{a}_i\,k^i - h_0^2 \,. \tag{8.58}$$

It will be useful to introduce the projection of k^i orthogonal to v^i using the projection tensor $h^i{}_j = \delta^i{}_j - v^i\,v_j$. This projection will be denoted

$$p^i = h^i{}_j\,k^i = k^i - v^i \;\Rightarrow\; p_i\,p^i = -1 \text{ and } p_i\,v^i = 0 \,. \tag{8.59}$$

It follows from (8.36) that each component p^i of the projection is an $l = 1$ spherical harmonic:

$$\underset{0}{\Delta} p^i + 2\,p^i = 0 \,. \tag{8.60}$$

Since $\dot{a}_i\,v^i = -a_i\,a^i$, on account of the orthogonality of the 4-velocity and the 4-acceleration, we can write (8.57) as

$$\dot{M} = \dot{m} + 2\,e^2\,\dot{a}_i\,p^i - 2\,e^2\,a_i\,a^i - 2\,e^2\,h_0^2 - 2\,e^2\,\dot{w}_1 + O_3 \,. \tag{8.61}$$

With H given by (8.44), M by (8.56) and $\dot{M}$ by (8.61) we have

$$\dot{M} - 3\,H\,M = \dot{m} - 3\,m\,h_0 + 2\,e^2\,\dot{a}_i\,p^i - 2\,e^2\,a_i\,a^i - 8\,e^2\,h_0^2 - 2\,e^2\,\dot{w}_1$$
$$+6\,e^2\,h_0\,w_1 - 3\,m\,\dot{Q}_1 - 6\,e^2\,h_0\dot{Q}_1 + O_3 \,. \tag{8.62}$$

The equations of motion for the world line $r = 0$ in second approximation will have the general form

$$m\,a_i = \underset{1}{A_i} + \underset{2}{A_i} + O_3 \,, \tag{8.63}$$

with

$$A_i = O_1 \,,\, A_i = O_2 \text{ and } A_i v^i = 0 = A_i v^i \,. \tag{8.64}$$

Hence

$$m\,h_0 = m\,a_i\,k^i = m\,a_i\,p^i = A_i\,p^i + A_i\,p^i + O_3 \,. \tag{8.65}$$

Now (8.62) reads

$$\dot{M} - 3\,H\,M = \dot{m} - 3\,A_i\,p^i + 2\,e^2\,\dot{a}_i\,p^i - 2\,e^2\,a_i\,a^i - 8\,e^2\,h_0^2$$
$$- 3\,A_i\,p^i - 2\,e^2\,\dot{w}_1 + 6\,e^2\,h_0\,w_1 - 3\,m\,\dot{Q}_1 - 6\,e^2\,h_0\dot{Q}_1$$
$$+ O_3 \,. \tag{8.66}$$

We note that the first line on the right hand side here consists only of O_1-terms while the second line consists only of O_2-terms. With w given by (8.40), p^2 given by (8.41) and using (8.32) to show that

$$P_0^2\left[\left(\frac{\partial h_0}{\partial x}\right)^2 + \left(\frac{\partial h_0}{\partial y}\right)^2\right] = -a_i\,a^i - h_0^2 \,, \tag{8.67}$$

we find that

$$e^2\,p^2\,(w_x^2 + w_y^2) = -e^2\,a_i\,a^i - e^2\,h_0^2$$
$$- 2\,e^2\,P_0^2\left(\frac{\partial h_0}{\partial x}\frac{\partial w_1}{\partial x} + \frac{\partial h_0}{\partial x}\frac{\partial w_1}{\partial y}\right) - 2\,e^2\,Q_1\,(a_i\,a^i + h_0^2)$$
$$+ O_3 \,. \tag{8.68}$$

In similar fashion to (8.66) the first line on the right hand side in (8.68) consists only of O_1-terms while the terms on the second line are all O_2-terms. Now substituting (8.48), (8.66) and (8.68) into the field equation (8.12) results in the following equations:

$$\frac{1}{4}\,\Delta K_1 = \dot{m} - 3\,A_i\,p^i + 2\,e^2\,\dot{a}_i\,p^i - 3\,e^2\,a_i\,a^i - 9\,e^2\,h_0^2 \,, \tag{8.69}$$

with K_1 given by (8.46), and

$$\frac{1}{4} \underset{0}{\Delta} K_2 = -3 A_i \, p^i - 2 \, e^2 \, \dot{w}_1 + 6 \, e^2 \, h_0 \, w_1 - 3 \, m \, \dot{Q}_1 - 6 \, e^2 \, h_0 \, \dot{Q}_1$$

$$-2 \, e^2 \, P_0^2 \left\{ \frac{\partial h_0}{\partial x} \frac{\partial w_1}{\partial x} + \frac{\partial h_0}{\partial y} \frac{\partial w_1}{\partial y} \right\} - 2 \, e^2 \, Q_1 \, (a_i \, a^i + h_0^2)$$

$$-\frac{1}{2} \, Q_1 \underset{0}{\Delta} K_1 \,, \tag{8.70}$$

with K_2 given by (8.47).

8.4 Equations of Motion of a Reissner–Nordström Particle

We begin by solving (8.69) for $K_1(x, y, u)$. We seek a solution which is well behaved for $-\infty < x, y < +\infty$. To find this we rewrite (8.69) in the form

$$\frac{1}{4} \underset{0}{\Delta} K_1 = \dot{m} + (-3 A_i + 2 h_i^j \, \dot{a}_j) \, p^i - 9 \, e^2 \, V_1 \,. \tag{8.71}$$

The first term on the right hand side here is an $l = 0$ spherical harmonic. Hence to have a solution K_1 which is non-singular for $-\infty < x, y < +\infty$ we must have

$$\dot{m} = 0 \quad \Rightarrow \quad m = \text{constant} \,. \tag{8.72}$$

We have introduced the projection tensor $h^i_{\ j} = \delta^i_{\ j} - v^i \, v_j$ to emphasise that the second term on the right hand side here is the Minkowskian scalar product of p^i, which is an arbitrary unit space-like vector orthogonal to v^i, with a fixed space-like vector orthogonal to v^i. We note that on account of (8.60) p^i is an $l = 1$ spherical harmonic. Also

$$V_1 = h_0^2 + \frac{1}{3} \, a_i \, a^i \,, \tag{8.73}$$

satisfies

$$\underset{0}{\Delta} V_1 + 6 \, V_1 = 0 \,, \tag{8.74}$$

so that V_1 is an $l = 2$ spherical harmonic. With these observations we obtain the required solution of (8.71):

$$K_1 = \underset{0}{\Delta} Q_1 + 2 \, Q_1 = (6 A_i - 4 \, e^2 \, h_i^j \, \dot{a}_j) \, p^i + 6 \, e^2 \, V_1 + \underset{1}{C}(u) \,, \tag{8.75}$$

where $C(u)$ is an arbitrary O_1-function of u. The first equality here follows from
(8.46) and results in a differential equation for $Q_1(x, y, u)$. In the space-time with
line element (8.1) the hypersurfaces $u =$ constant are null and represent the histories
of the wave fronts of radiation produced by the behaviour of the source. The induced
line elements on $u =$ constant read, with the approximation (8.39),

$$dl^2 = r^2 \, P_0^{-2} (1 - 2\,Q_1 - 2\,Q_2 + 3\,Q_1^2 + O_3)(dx^2 + dy^2) \,. \tag{8.76}$$

These are smooth perturbations of 2-spheres provided Q_1, Q_2 are well behaved
functions of x, y, u. In particular to have a solution Q_1 of (8.75) which is non-
singular for $-\infty < x, y < +\infty$ the $l = 1$ term on the right hand side must vanish,
i.e.

$$(6\,A_i - 4\,e^2\,h_i^j\,\dot{a}_j)\,p^i = 0 \quad \text{for all} \quad p^i \quad \Rightarrow \quad A_i = \frac{2}{3}\,e^2\,h_i^j\,\dot{a}_j \,. \tag{8.77}$$

Substituting into (8.63) we have the equations of motion of the Reissner–Nordström
particle in first approximation:

$$m\,a_i = \frac{2}{3}\,e^2\,h_i^j\,\dot{a}_j + O_2 = \frac{2}{3}\,e^2\,\{\dot{a}_i + (a_j\,a^j)\,v_i\} + O_2 \,, \tag{8.78}$$

which is the well-known Lorentz–Dirac equation (compare the derivation here with
[2, 3] and [4] for example). Now (8.75) becomes

$$\underset{0}{\Delta} Q_1 + 2\,Q_1 = 6\,e^2\,V_1 + \underset{1}{C}(u) \,, \tag{8.79}$$

and this has the well behaved solution

$$Q_1 = -\frac{3}{2}\,e^2\,V_1 + \frac{1}{2}\,\underset{1}{C}(u) \,, \tag{8.80}$$

up to the addition of an $l = 1$ spherical harmonic. The second term on the right
hand side of (8.80) is an $l = 0$ spherical harmonic. In general perturbations Q of a
2-sphere which are $l = 0$ or $l = 1$ spherical harmonics are trivial in the sense that
the 2-sphere remains a 2-sphere under such perturbations. When viewed against a
background of three dimensional Euclidean space such perturbations are either an
infinitesimal change in the radius of the 2-sphere (for $l = 0$) or an infinitesimal
displacement of the centre of the 2-sphere (for $l = 1$). We will consistently neglect
such trivial cases and so the non-trivial perturbations of the wave fronts in first
approximation are described by

$$Q_1 = -\frac{3}{2}\,e^2\,V_1 \,, \tag{8.81}$$

Now with $\underset{1}{C}(u) = 0$ and (8.77) holding we have

$$K_1 = 6\,e^2\,V_1\,. \tag{8.82}$$

Turning now to the derivation of the equations of motion in second approximation we proceed as follows: (i) substitute Q_1 in (8.81) into (8.55) and solve for $w_1(x, y, u)$; (ii) substitute Q_1 and w_1 into (8.70) and solve for K_2; (iii) substitute K_2 and Q_1 into (8.47) and solve for Q_2. At each stage of this process we require functions which are well behaved for $-\infty < x, y < +\infty$ and this requirement (equivalent to requiring that the wave fronts be smooth deformations of 2-spheres) will yield the equations of motion as it did in the case of the first approximation. The calculations become quite extensive and in addition to the $l = 2$ spherical harmonic V_1 introduced in (8.73) the following $l = 2$ spherical harmonics (verified using the formulas (8.33)–(8.35)) are useful:

$$V_2 = h_0\,\dot{a}_i\,p^i + \frac{1}{3}\dot{a}_i\,a^i\,, \tag{8.83}$$

$$V_3 = h_0\,\ddot{a}_i\,p^i + \frac{1}{3}a_i\,\ddot{a}^i\,, \tag{8.84}$$

$$V_4 = (\dot{a}_i\,p^i)^2 + \frac{1}{3}\dot{a}_i\,\dot{a}^i - \frac{1}{3}(a_i\,a^i)^2\,. \tag{8.85}$$

Next we shall need the following $l = 3$ spherical harmonics:

$$S_1 = h_0^3 + \frac{3}{5}a_i\,a^i\,h_0\,, \tag{8.86}$$

$$S_2 = \dot{a}_i\,p^i\,h_0^2 + \frac{2}{5}\dot{a}_i\,a^i\,h_0 + \frac{1}{5}(a_j\,a^j)\,\dot{a}_i\,p^i\,. \tag{8.87}$$

Finally we shall require the $l = 4$ spherical harmonic:

$$U_1 = h_0^4 + \frac{6}{7}a_i\,a^i\left(h_0^2 + \frac{1}{3}a_j\,a^j\right) - \frac{1}{5}(a_i\,a^i)^2\,. \tag{8.88}$$

To implement the strategy described above we start with part (i) involving the Maxwell equation (8.55). We first note that with Q_1 given by (8.81) we have

$$\dot{Q}_1 = -3\,e^2\,V_2 + 3\,e^2\,S_1 + \frac{6}{5}e^2\,a_i\,a^i\,h_0\,, \tag{8.89}$$

and thus (8.55) reads

$$\underset{0}{\Delta}w_1 = -6\,e^2\,V_2 + 12\,e^2\,S_1 + \frac{4}{5}e^2\,a_i\,a^i\,h_0\,. \tag{8.90}$$

The well behaved solution of this is (remembering that h_0 is an $l = 1$ spherical harmonic)

$$w_1 = e^2 V_2 - e^2 S_1 - \frac{2}{5} e^2 a_i a^i h_0 , \qquad (8.91)$$

modulo the addition of an arbitrary function of u (an $l = 0$ spherical harmonic) which amounts to a gauge term in the potential 1-form (8.3) and so we shall neglect it. We now consider part (ii) of the strategy which involves Eq. (8.70) for K_2. We will calculate each of the terms on the right hand side of (8.70), starting with the second term, and then substitute the results into (8.70). With Q_1 and w_1 given by (8.81) and (8.91) we find, the first three terms of the seven terms taken in order on the right hand side of (8.70) after the first term, that

$$- 2 e^2 \dot{w}_1 = -6 e^4 U_1 + 10 e^4 S_2 - 2 e^4 V_3 - \frac{34}{7} e^4 (a_i a^i) V_1 - 2 e^4 V_4$$

$$+ 2 e^4 (\dot{a}_i a^i) h_0 + 2 e^4 (a_j a^j) \dot{a}_i p^i - \frac{8}{15} e^4 (a_i a^i)^2 , \qquad (8.92)$$

$$6 e^2 h_0 w_1 = -6 e^4 U_1 + 6 e^4 S_2 - \frac{6}{7} e^4 (a_i a^i) V_1 - \frac{6}{5} e^4 (a_j a^j) \dot{a}_i p^i$$

$$- \frac{2}{5} e^4 (\dot{a}_i a^i) h_0 + \frac{4}{5} e^4 (a_i a^i)^2 , \qquad (8.93)$$

$$-3 m \dot{Q}_1 = -9 m e^2 S_1 + 9 m e^2 V_2 - \frac{18}{5} m e^2 (a_i a^i) h_0 ,$$

$$= -6 e^4 S_2 + 6 e^4 V_4 - \frac{12}{5} e^4 (a_j a^j) \dot{a}_i p^i + O_3 , \qquad (8.94)$$

where in the latter we have made use of the equations of motion in first approximation (8.78) in the forms

$$m a_i a^i = \frac{2}{3} e^2 (\dot{a}_i a^i) + O_2 , \qquad (8.95)$$

and

$$m h_0 = \frac{2}{3} e^2 \dot{a}_i p^i + O_2 . \qquad (8.96)$$

Next

$$- 6 e^2 h_0 \dot{Q}_1 = -18 e^4 U_1 + 18 e^4 S_2 - \frac{18}{7} e^4 (a_i a^i) V_1 - \frac{18}{5} e^4 (a_j a^j) \dot{a}_i p^i$$

$$- \frac{6}{5} e^4 (\dot{a}_i a^i) h_0 + \frac{12}{5} e^4 (a_i a^i)^2 , \qquad (8.97)$$

$$-2e^2 P_0^2 \left(\frac{\partial h_0}{\partial x} \frac{\partial w_1}{\partial x} + \frac{\partial h_0}{\partial y} \frac{\partial w_1}{\partial y} \right) = -6 e^4 U_1 + 4 e^4 S_2 - \frac{20}{7} e^4 (a_i a^i) V_1$$

$$+ \frac{2}{5} e^4 (\dot{a}_i a^i) h_0 + \frac{6}{5} e^4 (a_j a^j) \dot{a}_i p^i$$

$$- \frac{8}{15} e^4 (a_i a^i)^2 , \tag{8.98}$$

$$-2 e^2 Q_1 (a_i a^i + h_0^2) = 3 e^4 U_1 + \frac{10}{7} e^4 (a_i a^i) V_1 + \frac{4}{15} e^4 (a_i a^i)^2 , \tag{8.99}$$

$$-\frac{1}{2} Q_1 \underset{0}{\Delta} K_1 = -27 e^4 U_1 + \frac{36}{7} e^4 (a_i a^i) V_1 - \frac{12}{5} e^4 (a_i a^i)^2. \tag{8.100}$$

When (8.92)–(8.94) and (8.97)–(8.100) are substituted into the right hand side of (8.70) the $l = 0$ terms (the final terms in (8.92), (8.93) and (8.97)–(8.100)) importantly cancel and we are left with

$$\underset{0}{\Delta} K_2 = \{ -12 \underset{2}{A}_i + \frac{16}{5} e^4 (\dot{a}_j a^j) a_i - 16 e^4 (a_j a^j) h_i^k \dot{a}_k \} p^i$$

$$+ 16 e^4 V_4 - 8 e^4 V_3 - \frac{128}{7} e^4 (a_i a^i) V_1$$

$$+ 128 e^4 S_2$$

$$- 240 e^4 U_1 . \tag{8.101}$$

As in the case of (8.71) above we have introduced the projection tensor $h_j^i = \delta_j^i - v^i v_j$ to emphasise that the first line on the right hand side here, which is an $l = 1$ spherical harmonic, is the Minkowskian scalar product of p^i, which is an arbitrary unit space-like vector orthogonal to v^i, with a fixed space-like vector orthogonal to v^i. The second line on the right hand side is an $l = 2$ spherical harmonic, the third line is an $l = 3$ spherical harmonic and the fourth line is an $l = 4$ spherical harmonic. Hence the well behaved solution for x, y in the range $(-\infty, +\infty)$ is given by

$$K_2 = \left\{ 6 \underset{2}{A}_i - \frac{8}{5} e^4 (\dot{a}_j a^j) a_i + 8 e^4 (a_j a^j) h_i^k \dot{a}_k \right\} p^i - \frac{8}{3} e^4 V_4$$

$$+ \frac{4}{3} e^4 V_3 + \frac{64}{21} e^4 (a_i a^i) V_1 - \frac{32}{3} e^4 S_2 + 12 e^4 U_1 + \underset{2}{C}(u) , \tag{8.102}$$

where $C_2(u) = O_2$ is an arbitrary function of u. From (8.47) with Q_1 given by (8.81) we have

$$
\begin{aligned}
K_2 &= \underset{0}{\Delta} Q_2 + 2\,Q_2 - \frac{1}{2}\underset{0}{\Delta} Q_1^2 + 2\,Q_1 \underset{0}{\Delta} Q_1 + Q_1^2 \\
&= \underset{0}{\Delta} Q_2 + 2\,Q_2 - \frac{9}{4}\,e^4\,U_1 + \frac{24}{7}\,e^4\,(a_i\,a^i)\,V_1 - \frac{11}{5}\,e^4\,(a_i\,a^i)^2\,.
\end{aligned}
\tag{8.103}
$$

From these last two equations we arrive at the differential equation for Q_2:

$$
\begin{aligned}
\underset{0}{\Delta} Q_2 + 2\,Q_2 &= \left\{ 6\,\underset{2}{A_i} - \frac{8}{5}\,e^4\,(\dot{a}_j\,a^j)\,a_i + 8\,e^4\,(a_j\,a^j)\,h_i^k\,\dot{a}_k \right\} p^i \\
&\quad + \frac{57}{4}\,e^4\,U_1 - \frac{32}{3}\,e^4\,S_2 + \frac{4}{3}\,e^4\,V_3 - \frac{8}{21}\,e^4\,V_1 \\
&\quad - \frac{8}{3}\,e^4\,V_4 + \underset{2}{C}(u) + \frac{11}{5}\,e^4\,(a_i\,a^i)^2\,.
\end{aligned}
\tag{8.104}
$$

The $l = 1$ spherical harmonic on the first line on the right hand side here must vanish for all space-like vectors p^i in order to have Q_2 a well behaved function for $-\infty < x, y < +\infty$. This means that

$$
\underset{2}{A_i} = \frac{4}{15}\,e^4\,(\dot{a}_j\,a^j)\,a_i - \frac{4}{3}\,e^4\,(a_j\,a^j)\,h_i^k\,\dot{a}_k\,.
\tag{8.105}
$$

This together with $\underset{1}{A_i}$ given by (8.77) substituted into (8.63) results in the equations of motion of the Reissner–Nordström particle in second approximation:

$$
m\,a_i = \frac{2}{3}\,e^2\,h_i^j\,\dot{a}_j + \frac{4}{15}\,e^4\,(\dot{a}_j\,a^j)\,a_i - \frac{4}{3}\,e^4\,(a_j\,a^j)\,h_i^k\,\dot{a}_k + O_3\,.
\tag{8.106}
$$

This can be simplified by noting from it that

$$
\frac{4}{15}\,e^4\,\dot{a}_j\,a^j = \frac{2}{5}\,e^2\left(\frac{2}{3}\,e^2\,h_j^k\,\dot{a}_k \right) a^j = \frac{2}{5}\,m\,e^2\,a_j\,a^j + O_3\,,
\tag{8.107}
$$

and so we can rewrite (8.106) in the final form

$$
m\,a_i = \frac{2}{3}\,e^2\left(1 - \frac{8}{5}\,e^2\,(a_k\,a^k) \right) h_i^j\,\dot{a}_j + O_3\,.
\tag{8.108}
$$

By taking

$$
\underset{2}{C}(u) = -\frac{11}{5}\,e^4\,(a_i\,a^i)^2\,,
\tag{8.109}
$$

we can ensure that no $l = 0$ term appears in Q_2 since, as pointed out following (8.80), such a term along with an $l = 1$ spherical harmonic appearing in Q_2 correspond to trivial perturbations of the wave fronts. Therefore the non-trivial, well behaved solution of (8.104) is

$$Q_2 = -\frac{19}{24} e^4 U_1 + \frac{16}{15} e^4 S_2 - \frac{1}{3} e^4 V_3 + \frac{2}{21} e^4 (a_i a^i) V_1 + \frac{2}{3} e^4 V_4 . \qquad (8.110)$$

8.5   Energy Radiation Rate and Run-Away Motion

We see from the expression (8.8) for the electromagnetic energy-momentum tensor that for large positive values of r (in the neighbourhood of future null infinity) we have approximately, in coordinates X^i,

$$E_{ij} = -\frac{e^2}{r^2} p^2 (w_x^2 + w_y^2) k_i k_j + O\left(\frac{1}{r^3}\right) , \qquad (8.111)$$

with $k_i = u_{,i}$ as in (8.31). Noting from (8.31) that $r_{,i} = -p_i + r h_0 k_i$, the rate, with respect to proper time u, at which the field 4-momentum P^i is crossing $r = $ constant in the *outward* radial direction, in the limit $r \to +\infty$, is given by Synge [5]

$$\frac{dP^i}{du} = \lim_{r \to +\infty} \left(\frac{r^2}{4\pi} \int E^{ij} r_{,j} d\omega\right) = \frac{1}{4\pi} \int e^2 p^2 (w_x^2 + w_y^2) k^i d\omega , \qquad (8.112)$$

where $d\omega$ is the element of solid angle and the factor $1/4\pi$ is necessary to compensate for having absorbed a factor of 4π into E_{ab} as indicated following (8.12) above. With $e^2 p^2 (w_x^2 + w_y^2)$ given by (8.68) and using (8.98) and (8.99) we obtain

$$e^2 p^2 (w_x^2 + w_y^2) = -e^2 a_i a^i - e^2 h_0^2 - \frac{4}{15} e^4 (a_i a^i)^2 + \frac{2}{5} e^4 (\dot{a}_i a^i) a_j p^j$$

$$+ \frac{6}{5} e^4 (a_i a^i) \dot{a}_j p^j - \frac{10}{7} e^4 (a_i a^i) V_1 + 4 e^4 S_2 - 3 e^4 U_1$$

$$+ O_3 . \qquad (8.113)$$

Making use of the formulas:

$$\int d\omega = 4\pi , \quad \int p^i p^j d\omega = -\frac{4\pi}{3} h^{ij} ,$$

$$\int p^i p^j p^k p^l d\omega = \frac{4\pi}{15} (h^{ij} h^{kl} + h^{ik} h^{jl} + h^{il} h^{jk}) , \qquad (8.114)$$

where $h^{ij} = \eta^{ij} - v^i v^j$ is the projection tensor we find that

$$\frac{1}{4\pi} \int e^2 p^2 (w_x^2 + w_y^2) \, d\omega = -\frac{2}{3} e^2 (a_k \, a^k) \left(1 + \frac{2}{5} e^2 a_j \, a^j\right), \tag{8.115}$$

and

$$\frac{1}{4\pi} \int e^2 p^2 (w_x^2 + w_y^2) \, p^i \, d\omega = -\frac{2}{15} e^4 (\dot{a}_j \, a^j) a^i - \frac{2}{5} e^4 (a_j \, a^j) h_k^i \, \dot{a}_k + O_3,$$

$$= -\frac{4}{5} m \, e^2 (a_j \, a^j) a^i + O_3, \tag{8.116}$$

with the second equality here a consequence of the equations of motion in first approximation (8.78). Hence (8.112) finally reads

$$\frac{dP^i}{du} = -\frac{2}{3} e^2 (a_k \, a^k) \left(1 + \frac{2}{5} e^2 a_j \, a^j\right) v^i - \frac{4}{5} m \, e^2 (a_j \, a^j) a^i + O_3. \tag{8.117}$$

This is a generalisation of the well-known invariant form (see, for example, [6]) of the classical Larmor formula in electromagnetic theory, namely,

$$\frac{dP^i}{du} = -\frac{2}{3} e^2 (a_k \, a^k) v^i. \tag{8.118}$$

We note that with our choice of metric signature $a_i \, a^i \leq 0$ since a^i is space-like.

It is interesting to exhibit explicitly the so called run-away motion implied by the equations of motion (8.108). To achieve this we shall extend to the case of (8.108) an argument due to Synge [6] applied to the Lorentz–Dirac equation (8.78). We look for a solution of (8.108) having the property that when $u = 0$,

$$v^i = \delta_4^i \ \text{and} \ a^i = (0, 0, B, 0), \tag{8.119}$$

with B a real constant. Putting

$$\mu = \frac{3 \, m}{2 \, e^2}, \tag{8.120}$$

we rewrite (8.108) in the form

$$\mu \, a^i = \dot{a}^i + (a_j \, a^j) \, v^i - \frac{8}{5} e^2 a_k \, a^k \, (\dot{a}^i + (a_j \, a^j) \, v^i) + O_2. \tag{8.121}$$

From this we derive the following two equations:

$$\mu \, a_i \, a^i = \frac{1}{2} \frac{d}{du} (a_i \, a^i) - \frac{4}{5} e^2 \, a_k \, a^k \, \frac{d}{du} (a_i \, a^i) \,, \tag{8.122}$$

$$\mu \, (a^i \, v^j - a^j \, v^i) = \frac{d}{du} (a^i \, v^j - a^j \, v^i) - \frac{8}{5} e^2 \, (a_k \, a^k) \, \frac{d}{du} (a^i \, v^j - a^j \, v^i) \,, \tag{8.123}$$

with an O_2-error understood in each case from now on. Solving the first of these with the initial condition (8.119) results in

$$a_i \, a^i = -B^2 \, e^{2 \mu u} + \frac{8}{5} e^2 \, B^4 \, e^{4 \mu u} - \frac{8}{5} e^2 \, B^4 \, e^{2 \mu u} \,. \tag{8.124}$$

The general solution of (8.123) reads

$$a^i \, v^j - a^j \, v^i = K^{ij} \, e^{\mu u} - \frac{4}{5} e^2 \, B^2 \, K^{ij} \, e^{3 \mu u} + \Lambda^{ij}_1 \, e^{\mu u} \,, \tag{8.125}$$

where $K^{ij} = -K^{ji} = O_0$ are constants and $\Lambda^{ij}_1 = -\Lambda^{ji}_1 = O_1$ are constants, all to be determined. Writing out (8.125) for $i, j = 1, 2, 3, 4$ and imposing the initial conditions (8.119) we find that

$$K^{ij} = 0 \quad \text{except} \quad K^{34} = -K^{43} = B \,, \tag{8.126}$$

and

$$\Lambda^{ij}_1 = 0 \quad \text{except} \quad \Lambda^{34}_1 = -\Lambda^{43}_1 = \frac{4}{5} e^2 \, B^3 = O_1 \,. \tag{8.127}$$

Now writing out (8.125) explicitly we have

$$a^1 \, v^2 - a^2 \, v^1 = 0 \,, \tag{8.128}$$

$$a^1 \, v^3 - a^3 \, v^1 = 0 \,, \tag{8.129}$$

$$a^1 \, v^4 - a^4 \, v^1 = 0 \,, \tag{8.130}$$

$$a^2 \, v^3 - a^3 \, v^2 = 0 \,, \tag{8.131}$$

$$a^2 \, v^4 - a^4 \, v^2 = 0 \,, \tag{8.132}$$

$$a^3 \, v^4 - a^4 \, v^3 = B \, e^{\mu u} - \frac{4}{5} e^2 \, B^3 \, e^{3 \mu u} + \frac{4}{5} e^2 \, B^3 \, e^{\mu u} \,. \tag{8.133}$$

Since $a^3(u) \neq 0$ and $v^4(u) \neq 0$ it is straightforward to show from (8.128)–(8.132) that

$$v^1(u) = 0 = v^2(u) , \tag{8.134}$$

for all values of u. Hence in addition to (8.133) we are left with (8.124) which now reads

$$- (a^3)^2 + (a^4)^2 = -B^2 e^{2\mu u} + \frac{8}{5} e^2 B^4 e^{4\mu u} - \frac{8}{5} e^2 B^4 e^{2\mu u} . \tag{8.135}$$

But this equation is a consequence of (8.133) (neglecting O_2-terms) since

$$-(v^3)^2 + (v^4)^2 = 1 \text{ and } - a^3 v^3 + a^4 v^4 = 0 \Rightarrow (a^3)^2 - (a^4)^2 = (a^3 v^4 - a^4 v^3)^2 . \tag{8.136}$$

To solve (8.133) it is useful to define

$$k_1 = v^4 + v^3 \text{ and } k_2 = v^4 - v^3 \Rightarrow k_1 k_2 = 1 , \tag{8.137}$$

then

$$\frac{d}{du}(\log k_1) = a^3 v^4 - a^4 v^3 = B e^{\mu u} - \frac{4}{5} e^2 B^3 e^{3\mu u} + \frac{4}{5} e^2 B^3 e^{\mu u} . \tag{8.138}$$

Integrating and using the initial condition that $k_1 = 1$ when $u = 0$ we arrive at

$$k_1 = v^4 + v^3 = \exp\left[\frac{B}{\mu} \left(e^{\mu u} - 1 \right) - \frac{4 e^2 B^3}{15 \mu} e^{3\mu u} + \frac{4 e^2 B^3}{5 \mu} e^{\mu u} - \frac{8 e^2 B^3}{15 \mu} \right] , \tag{8.139}$$

We can simplify this to read

$$v^4 + v^3 = \exp\left\{ \frac{B}{\mu}(e^{\mu u} - 1) \right\} \left[1 - \frac{4 e^2 B^3}{15 \mu}(e^{\mu u} - 1)^2(e^{\mu u} + 2) + O_2 \right] . \tag{8.140}$$

Since $v^4 - v^3 = k_2 = (k_1)^{-1}$ we have

$$v^4 - v^3 = \exp\left\{ -\frac{B}{\mu}(e^{\mu u} - 1) \right\} \left[1 + \frac{4 e^2 B^3}{15 \mu}(e^{\mu u} - 1)^2(e^{\mu u} + 2) + O_2 \right] . \tag{8.141}$$

From (8.140) and (8.141) we finally obtain

$$
v^3 = \sinh\left\{\frac{B}{\mu}(e^{\mu\,u} - 1)\right\} - \frac{4\,e^2\,B^3}{15\,\mu}(e^{\mu\,u} - 1)^2(e^{\mu\,u} + 2)\cosh\left\{\frac{B}{\mu}(e^{\mu\,u} - 1)\right\} + O_2\,,
$$

$$\tag{8.142}$$

$$
v^4 = \cosh\left\{\frac{B}{\mu}(e^{\mu\,u} - 1)\right\} - \frac{4\,e^2\,B^3}{15\,\mu}(e^{\mu\,u} - 1)^2(e^{\mu\,u} + 2)\sinh\left\{\frac{B}{\mu}(e^{\mu\,u} - 1)\right\} + O_2\,.
$$

$$\tag{8.143}$$

The exponential growth in u of both v^3 and v^4 is symptomatic of run-away behaviour in which the object accelerates to the speed of light in the limit $v^3/v^4 \to 1$.

8.6 Run-Away Magnetic Reissner–Nordström Particle

The Einstein–Maxwell field of a run-away *magnetic* Reissner–Nordström particle is described by a Robinson–Trautman line element (8.1) with a Maxwell 2-form which is the dual of the Maxwell 2-form (8.4). In this case the potential 1-form (8.3) is replaced by the potential 1-form

$$
A = \frac{\partial G}{\partial y}\,dx - \frac{\partial G}{\partial x}\,dy\,, \tag{8.144}
$$

with $G = G(x, y, u)$. On the half null tetrad defined via the basis 1-forms (8.2) the Maxwell 2-form is the exterior derivative of (8.144) given by

$$
F = dA = -\frac{p}{r}\,\dot{G}_y\,\vartheta^{(1)}\wedge\vartheta^{(4)} + \frac{p}{r}\,\dot{G}_x\,\vartheta^{(2)}\wedge\vartheta^{(4)} - \frac{1}{r^2}\,\Delta G\,\vartheta^{(1)}\wedge\vartheta^{(2)}\,, \tag{8.145}
$$

where the dot, as always, indicates partial differentiation with respect to u, partial derivatives with respect with x and y are indicated by the use of subscripts and the operator Δ is given in (8.6). The Hodge dual of this 2-form is the 2-form

$$
\begin{aligned}
{}^*F &= -\frac{p}{r}\,\dot{G}_x\,\vartheta^{(1)}\wedge\vartheta^{(4)} - \frac{p}{r}\,\dot{G}_y\,\vartheta^{(2)}\wedge\vartheta^{(4)} - \frac{1}{r^2}\,\Delta G\,\vartheta^{(3)}\wedge\vartheta^{(4)} \\
&= -\dot{G}_x\,dx\wedge du - \dot{G}_y\,dy\wedge du - \frac{1}{r^2}\,\Delta G\,dr\wedge du\,.
\end{aligned} \tag{8.146}
$$

Now Maxwell's vacuum field equations $d\,{}^*F = 0$ result in

$$
\Delta G = -g(u)\,, \tag{8.147}
$$

for some real-valued function of u. Consequently

$$F = -\frac{p}{r}\,\dot{G}_y\,\vartheta^{(1)} \wedge \vartheta^{(4)} + \frac{p}{r}\,\dot{G}_x\,\vartheta^{(2)} \wedge \vartheta^{(4)} + \frac{g}{r^2}\,\vartheta^{(1)} \wedge \vartheta^{(2)}\ . \tag{8.148}$$

The partial derivative of (8.147) with respect to u gives

$$\Delta\dot{G} = -\dot{g} + 2\,g\,H\ , \tag{8.149}$$

with H given by (8.6). If we put

$$\dot{G} = g\,w\ , \tag{8.150}$$

for some $w(x, y, u)$ then (8.149) becomes

$$\Delta w = -g^{-1}\dot{g} + 2\,H\ , \tag{8.151}$$

and (8.148) reads now

$$F = \frac{g}{r^2}\,\vartheta^{(1)} \wedge \vartheta^{(2)} + \frac{g\,p}{r}\,\frac{\partial w}{\partial x}\,\vartheta^{(2)} \wedge \vartheta^{(4)} - \frac{g\,p}{r}\,\frac{\partial w}{\partial y}\,\vartheta^{(1)} \wedge \vartheta^{(4)}\ . \tag{8.152}$$

Here (8.151) coincides with the Maxwell equation (8.6) and (8.152) coincides with the dual field (8.5), with now the electric charge e of the source replaced by the magnetic monopole moment g in each case. Since the electromagnetic energy momentum tensor is invariant under the interchange of the Maxwell field and its dual field, the remaining field equation (8.12), along with (8.10) and (8.11), is unchanged except for the replacement of e with g. Hence the equations of motion of a magnetic Reissner–Nordström particle in first or second approximation will be given again by (8.78) and (8.108) respectively but with e replaced by g.

References

1. I. Robinson, A. Trautman, Proc. R. Soc. A **265**, 463 (1962)
2. I. Robinson, P.A. Hogan, Found. Phys. **15**, 617 (1985)
3. E.T. Newman, R. Posadas, Phys. Rev. **187**, 1784 (1969)
4. P.A. Hogan, M. Imaeda, J. Phys. A **12**, 1061 (1979)
5. J.L. Synge, Annali di Matematica Pura ed Applicata (IV) **LXXXIV**, 33 (1970)
6. J.L. Synge, *Relativity: The Special Theory* (North-Holland Publishing Company, Amsterdam, 1965)

Congruences of World Lines

A

A.1 Properties of η_{abcd}

1. $\eta_{abcd} = \sqrt{-g}\,\epsilon_{abcd}$ are the components of a tensor:

First define

$$\Lambda_{a'b'c'd'} = \frac{\partial x^p}{\partial x^{a'}}\frac{\partial x^q}{\partial x^{b'}}\frac{\partial x^r}{\partial x^{c'}}\frac{\partial x^s}{\partial x^{d'}}\eta_{pqrs}\,. \tag{A.1}$$

This quantity is totally skew-symmetric. Its one independent component is

$$\Lambda_{0'1'2'3'} = \sqrt{-g}\,\frac{\partial x^p}{\partial x^{0'}}\frac{\partial x^q}{\partial x^{1'}}\frac{\partial x^r}{\partial x^{2'}}\frac{\partial x^s}{\partial x^{3'}}\epsilon_{pqrs} = \sqrt{-g}\,\det\left(\frac{\partial x^p}{\partial x^{a'}}\right)\,. \tag{A.2}$$

But

$$g_{a'b'} = \frac{\partial x^p}{\partial x^{a'}}\frac{\partial x^q}{\partial x^{b'}}g_{pq} \;\Rightarrow\; g' = \left(\det\frac{\partial x^p}{\partial x^{a'}}\right)^2 g \;\Rightarrow\; \det\frac{\partial x^p}{\partial x^{a'}} = \frac{\sqrt{-g'}}{\sqrt{-g}}\,. \tag{A.3}$$

Hence

$$\Lambda_{0'1'2'3'} = \sqrt{-g'}\,, \tag{A.4}$$

and thus

$$\Lambda_{a'b'c'd'} = \sqrt{-g'}\,\epsilon_{a'b'c'd'} = \eta_{a'b'c'd'}\,. \tag{A.5}$$

P. A. Hogan, D. Puetzfeld, *Frontiers in General Relativity*, Lecture Notes
in Physics 984, https://doi.org/10.1007/978-3-030-69370-1

2. How is η^{abcd} related to η_{abcd}?:

$$\eta^{abcd} = g^{ar} g^{bs} g^{cp} g^{dq} \eta_{rspq} , \tag{A.6}$$

is totally skew-symmetric. Hence it has only one independent component

$$\eta^{0123} = \sqrt{-g} g^{0r} g^{1s} g^{2p} g^{3q} \epsilon_{pqrs} = \sqrt{-g} \, g^{-1} = -\frac{1}{\sqrt{-g}} , \tag{A.7}$$

and thus

$$\eta^{abcd} = -\frac{1}{\sqrt{-g}} \epsilon_{abcd} . \tag{A.8}$$

3. η_{abcd} is covariantly constant:

Since η_{abcd} has the important property, written as a determinant, that

$$\eta^{apqr} \eta_{clmn} = - \begin{vmatrix} \delta^a_c & \delta^a_l & \delta^a_m & \delta^a_n \\ \delta^p_c & \delta^p_l & \delta^p_m & \delta^p_n \\ \delta^q_c & \delta^q_l & \delta^q_m & \delta^q_n \\ \delta^r_c & \delta^r_l & \delta^r_m & \delta^r_n \end{vmatrix} , \tag{A.9}$$

and since $\delta^a_{b;c} = 0$ we have

$$\eta_{abcd;f} \, \eta^{pqrs} + \eta_{abcd} \, \eta^{pqrs}{}_{;f} = 0 . \tag{A.10}$$

It also follows from (A.9) that $\eta_{abcd} \, \eta^{abcd} = -24$ and so multiplying (A.10) by η^{pqrs} yields

$$\eta_{abcd;f} = 0 . \tag{A.11}$$

Bateman Waves in the Linear Approximation   **B**

B.1   Covariant Treatment on General Space-Time

Let u^a with $g_{ab}\, u^a\, u^b = 1$ be a unit time-like vector field in a space-time with metric tensor components g_{ab} in coordinates x^a. The integral curves of u^a constitute a time-like congruence in the space-time, described in Chap. 1. A Maxwell field on the space-time is described by a real bivector with components $F_{ab} = -F_{ba}$ which satisfies Maxwell's source-free field equations:

$$F^{ab}{}_{;b} = 0 \quad \text{and} \quad {}^*F^{ab}{}_{;b} = 0 \, , \tag{B.1}$$

with ${}^*F_{ab} = \frac{1}{2}\eta_{abcd}\, F^{cd}$ and $\eta_{abcd} = \sqrt{-g}\, \epsilon_{abcd}$ as in Chap. 2. With respect to the congruence tangent to u^a we define the *electric part* of F_{ab} by the space-like covariant vector field

$$E_a = F_{ab}\, u^b \quad (\Rightarrow \ E_a\, u^a = 0) \, , \tag{B.2}$$

and the *magnetic part* of F_{ab} by the space-like covariant vector field

$$B_a = {}^*F_{ab}\, u^b \quad (\Rightarrow \ B_a\, u^a = 0) \, . \tag{B.3}$$

From (B.3) we have

$$B_a\, \eta^{apqr} = \frac{1}{2}\eta_{abcd}\, \eta^{apqr}\, F^{cd}\, u^b = u^p\, F^{rq} + u^q\, F^{pr} + u^r\, F^{qp} \, , \tag{B.4}$$

and thus we can write F_{ab} in terms of E_a, H_a as

$$F_{ab} = E_a\, u_b - E_b\, u_a + \eta_{abcd}\, u^c\, B^d \, . \tag{B.5}$$

P. A. Hogan, D. Puetzfeld, *Frontiers in General Relativity*, Lecture Notes
in Physics 984, https://doi.org/10.1007/978-3-030-69370-1

Substituting this into (B.3) gives

$$^*F_{ab} = B_a\, u_b - B_b\, u_a - \eta_{abcd}\, u^c\, E^d \ . \tag{B.6}$$

From (B.5) and (B.6) we can write the invariants

$$\frac{1}{2} F_{ab}\, F^{ab} = E_a\, E^a - B_a\, B^a \quad \text{and} \quad \frac{1}{2} F_{ab}\, ^*F^{ab} = 2\, E_a\, B^a \ . \tag{B.7}$$

Thus for pure electromagnetic radiation we must have

$$E_a\, E^a = B_a\, B^a \quad \text{and} \quad E_a\, B^a = 0 \ . \tag{B.8}$$

Now define the vectors

$$e^a = \frac{E^a}{(-E_b\, E^b)^{1/2}} \ , \quad b^a = \frac{B^a}{(-E_b\, E^b)^{1/2}} \ ,$$

$$p^a = \frac{\eta^{abcd}\, E_b\, u_c\, B_d}{E^f\, E_f} = -\eta^{abcd}\, e_b\, u_c\, b_d \ . \tag{B.9}$$

These are mutually orthogonal, unit space-like vectors (with $e_a\, e^a = b_a\, b^a = p_a\, p^a = -1$) and each is orthogonal to u^a. Checking that $p_a\, p^a = -1$ involves the use of the identity

$$\eta^{abcd}\, \eta_{arst} = - \begin{vmatrix} \delta_r^b & \delta_s^b & \delta_t^b \\ \delta_r^c & \delta_s^c & \delta_t^c \\ \delta_r^d & \delta_s^d & \delta_t^d \end{vmatrix} \ . \tag{B.10}$$

Making use of the more general identity

$$\eta^{apqr}\, \eta_{clmn} = - \begin{vmatrix} \delta_c^a & \delta_l^a & \delta_m^a & \delta_n^a \\ \delta_c^p & \delta_l^p & \delta_m^p & \delta_n^p \\ \delta_c^q & \delta_l^q & \delta_m^q & \delta_n^q \\ \delta_c^r & \delta_l^r & \delta_m^r & \delta_n^r \end{vmatrix} \ , \tag{B.11}$$

we find that

$$p^a\, p_c = -\delta_c^a - e^a\, e_c + u^a\, u_c - b^a\, b_c \ , \tag{B.12}$$

which we rewrite as

$$g_{ac} = u_a\, u_c - e_a\, e_c - b_a\, b_c - p_a\, p_c \ . \tag{B.13}$$

Thus the vectors u^a, e^a, b^a, p^a constitute an orthonormal tetrad. It is useful to note, using (B.13), that

$$\eta_{abcd}\, u^c\, B^d = \delta_a^l\, \delta_b^m\, \eta_{lmcd}\, u^c\, B^d$$
$$= (p_a\, e_b - p_b\, e_a)\, p^l\, \eta_{lmcd}\, e^m\, u^c\, B^d$$
$$= (p_a\, e_b - p_b\, e_a)\, (-E^f\, E_f)^{1/2}\,, \tag{B.14}$$

and thus

$$\eta_{abcd}\, u^c\, B^d = p_a\, E_b - p_b\, E_a\,. \tag{B.15}$$

Similarly

$$\eta_{abcd}\, u^c\, E^d = -(p_a\, B_b - p_b\, B_a)\,. \tag{B.16}$$

When (B.15) and (B.16) are substituted into (B.5) and (B.6) we find that

$$F_{ab} = E_a\, k_b - E_b\, k_a \quad\text{and}\quad {}^*F_{ab} = B_a\, k_b - B_b\, k_a\,, \tag{B.17}$$

where

$$k_a = u_a - p_a \quad\Rightarrow\quad k_a\, k^a = 0\,. \tag{B.18}$$

The null vector field k^a, which satisfies

$$F_{ab}\, k^b = 0 = {}^*F_{ab}\, k^b\,, \tag{B.19}$$

is the degenerate principal null direction associated with the bivector F_{ab} and is thus the propagation direction in space-time of the electromagnetic waves.

Substituting (B.17) into Maxwell's equations (B.1), contracting each in turn with e_a and b_a and then adding and subtracting the results, we arrive at

$$k_{a;b}\, e^a\, e^b = k_{a;b}\, b^a\, b^b\,, \tag{B.20}$$

and

$$(E_f\, E^f)_{,b}\, k^b + 2\,(E_f\, E^f)\{k^b{}_{;b} + k_{a;b}\, e^a\, e^b\} = 0\,, \tag{B.21}$$

from the first of (B.1) and

$$k_{a;b}\, e^a\, b^b + k_{a;b}\, b^a\, e^b = 0\,, \tag{B.22}$$

and

$$k_{a;b}\, e^a\, b^b - k_{a;b}\, b^a\, e^b = e_a\, b^a{}_{;b}\, k^b - b_a\, e^a{}_{;b}\, k^b \,, \tag{B.23}$$

from the second of (B.1). We shall require (B.21) in a slightly different form below. This is obtained by first noting from (B.20) and using (B.13) that

$$\begin{aligned}
k_{a;b}\, e^a\, e^b &= \frac{1}{2} k_{a;b}(e^a\, e^b + b^a\, b^b) \\
&= \frac{1}{2} k_{a;b}(-g^{ab} + u^a\, u^b - p^a\, p^b) \\
&= -\frac{1}{2}\, k^a{}_{;a} + \frac{1}{2}(k_{a;b}\, u^a\, u^b - k_{a;b}\, p^a\, p^b)\,. \tag{B.24}
\end{aligned}$$

With k^a given by (B.18) we have

$$k_{a;b}\, u^a\, u^b - k_{a;b}\, p^a\, p^b = -p_{a;b}\, u^a\, u^b - u_{a;b}\, p^a\, p^b = u_{a;b}\, p^a\, k^b = -u_{a;b}\, k^a\, k^b \,, \tag{B.25}$$

and so (B.24) becomes

$$k_{a;b}\, e^a\, e^b = -\frac{1}{2}\, k^a{}_{;a} - \frac{1}{2}\, u_{a;b}\, k^a\, k^b \,. \tag{B.26}$$

Using this we can write (B.21) as

$$(E_f\, E^f)_{,b}\, k^b + (E_f\, E^f)\, k^b{}_{;b} = u_{b;c}\, k^b\, k^c\, (E_f\, E^f)\,. \tag{B.27}$$

We can make use of this immediately by noting that since the electromagnetic energy tensor

$$E^{ab} = F^a{}_c\, F^{bc} - \frac{1}{4}\, g^{ab}\, F_{dc}\, F^{dc} \,, \tag{B.28}$$

satisfies

$$E^{ab}{}_{;b} = 0 \,, \tag{B.29}$$

as a consequence of Maxwell's equations (B.1), we have in the present case of (B.17),

$$E^{ab} = (E_f\, E^f)\, k^a\, k^b \,, \tag{B.30}$$

and thus (B.29) yields

$$E_c \, E^c \, k^a{}_{;b} \, k^b + \left\{ (E_c \, E^c)_{,b} \, k^b + E_c \, E^c \, k^b{}_{;b} \right\} k^a = 0 \,. \tag{B.31}$$

Using (B.27) this simplifies to

$$k^a{}_{;b} \, k^b = -(u_{b;c} \, k^b \, k^c) \, k^a \,, \tag{B.32}$$

and we have recovered Mariot's [1] result, namely, the propagation direction of electromagnetic radiation in space-time is *geodesic*.

Expanding $k_{a;b} + k_{b;a}$ on the orthonormal basis u^a, e^a, b^a, p^a and using (B.18), (B.20), (B.22) and (B.32) we find that

$$k_{a;b} + k_{b;a} = \lambda \, g_{ab} + \xi_a \, k_b + \xi_b \, k_a \,, \tag{B.33}$$

with

$$\lambda = k^a{}_{;a} + u_{a;b} \, k^a \, k^b \,, \tag{B.34}$$

and

$$\xi_a = (A + B) \, u_a - k_{b;c} \, (e^b \, p^c + e^c \, p^b) \, e_a - k_{b;c} \, (b^b \, p^c + b^c \, p^b) \, b_a + (A - B) \, p_a \,, \tag{B.35}$$

with

$$A = -\frac{1}{2} \, k^a{}_{;a} - u_{a;b} \, k^b \, k^c \,, \tag{B.36}$$

$$B = \frac{1}{2} \, k_{a;b} \, (u^a \, p^b + u^b \, p^a) \,. \tag{B.37}$$

As a result of establishing (B.33) we have recovered, using the Robinson–Trautman [2] test, Robinson's [3] result that the propagation direction of electromagnetic radiation in space-time is not only geodesic but is *shear-free*.

B.2   Ricci Identities and Bianchi Identities

We are concerned here only with vacuum space-times for which $R_{ab} = 0$, where R_{ab} are the components of the Ricci tensor. In this case the Weyl tensor coincides with the Riemann tensor and thus $C_{abcd} = R_{abcd}$ in Chap. 2. We are also interested in identifying a time-like congruence in the space-time consisting of the integral curves of a unit time-like vector field u^a (with $u_a \, u^a = 1$) as in Chap. 1, but one which is geodesic ($\dot{u}^a = u^a{}_{;b} \, u^b = 0$), twist-free ($\omega_{ab} = 0$) and expansion-free

($\vartheta = 0$) but having non-vanishing shear ($\sigma_{ab} \neq 0$). It thus follows from Chap. 1 that $u_{a;b} = \sigma_{ab}$. With respect to this congruence we define the *electric* part of the Riemann tensor by

$$\mathcal{E}_{ab} = R_{apbq}\, u^p\, u^q = \mathcal{E}_{ba}\ , \tag{B.1}$$

and the *magnetic* part of the Riemann tensor by

$$\mathcal{H}_{ab} = {}^*R_{apbq}\, u^p\, u^q = \mathcal{H}_{ba}\ . \tag{B.2}$$

In the latter definition we have used the left dual of the Riemann tensor which is identical to the right dual because we are assuming that the Ricci tensor vanishes. Hence we see that

$$\mathcal{E}^a{}_a = 0 = \mathcal{E}_{ab}\, u^b = 0 \quad \text{and} \quad \mathcal{H}^a{}_a = 0 = \mathcal{H}_{ab}\, u^b = 0\ . \tag{B.3}$$

We shall require the projections in the direction of u^a and orthogonal to u^a (using the projection tensor $h^a_b = \delta^a_b - u^a\, u_b$) of the Ricci identities

$$u_{a;bc} - u_{a;cb} = u_d\, R^d{}_{abc}\ . \tag{B.4}$$

These can be obtained from [4] after specialising to a vacuum space-time and to the particular time-like congruence described above. The results are

$$\mathcal{E}_{ab} = -\sigma_{ab;c}\, u^c + \sigma_{ac}\, \sigma^c{}_b - \frac{1}{3}\, h_{ab}\, \sigma_{cd}\, \sigma^{cd}\ , \tag{B.5}$$

$$\mathcal{H}_{ab} = -\eta_{qpl(a}\, \sigma_{b)}{}^{l;p}\, u^q - \eta_{qpl(a}\, u_{b)}\, \sigma^{mp}\, \sigma_m{}^l\ . \tag{B.6}$$

and

$$\sigma^{ab}{}_{;b} = 0\ . \tag{B.7}$$

As always round brackets around indices denote symmetrisation. Similarly the Bianchi identities yield

$$\mathcal{E}^{ab}{}_{;b} = -\eta^{abpq}\, u_b\, \sigma^d{}_p\, \mathcal{H}_{qd} - u^a\, \sigma_{qd}\, \mathcal{E}^{qd}\ , \tag{B.8}$$

$$\mathcal{H}^{ab}{}_{;b} = \eta^{abpq}\, u_b\, \sigma^d{}_p\, \mathcal{E}_{qd} - u^a\, \sigma_{qd}\, \mathcal{H}^{qd}\ , \tag{B.9}$$

$$\mathcal{E}^{ab}{}_{;c}\, u^c = -\eta^{lpq(a}\, u_q\, \mathcal{H}^{b)}{}_{p;l} + u^{(a}\, \eta^{b)lpq}\, u_q\, \sigma_{nl}\, \mathcal{H}^n_p + 3\, \mathcal{E}^{(a}{}_l\, \sigma^{b)l}$$

$$-h^{ab}\, \mathcal{E}^{nl}\, \sigma_{nl}\ , \tag{B.10}$$

$$\mathcal{H}^{ab}{}_{;c}\, u^c = \eta^{lpq(a}\, u_q\, \mathcal{E}^{b)}{}_{p;l} - u^{(a}\, \eta^{b)lpq}\, u_q\, \sigma_{nl}\, \mathcal{E}^n_p + 3\, \mathcal{H}^{(a}{}_l\, \sigma^{b)l}$$

$$-h^{ab}\, \mathcal{H}^{nl}\, \sigma_{nl}\ . \tag{B.11}$$

If the space-time is a small perturbation of Minkowskian space-time with $g_{ab} = \eta_{ab}$, vanishing Riemannian connection, $\eta_{abcd} = \epsilon_{abcd}$ and $u^a = \delta_0^a$ tangent to the t-lines, then $\mathcal{E}^{ab}$, $\mathcal{H}^{ab}$ and σ_{ab} are small of first oder. Neglecting second order small quantities (B.5)–(B.11) reduce to

$$\mathcal{E}_{ab} = -\frac{\partial \sigma_{ab}}{\partial t} \; , \; \mathcal{H}_{ab} = -\epsilon_{0pl(a} \sigma_{b)}{}^{l,p} \; , \; \sigma^{ab}{}_{,b} = 0 \; , \tag{B.12}$$

$$\mathcal{E}^{ab}{}_{,b} = 0 \; , \; \mathcal{H}^{ab}{}_{,b} = 0 \; , \; \frac{\partial \mathcal{E}_{ab}}{\partial t} = -\epsilon_{0lp(a} \mathcal{H}_{b)}{}^{p,l} \; , \; \frac{\partial \mathcal{H}_{ab}}{\partial t} = \epsilon_{0lp(a} \mathcal{E}_{b)}{}^{p,l} \; . \tag{B.13}$$

Since $\eta_{ab} = \mathrm{diag}(1, -1, -1, -1)$ and $\sigma_{ab} u^b = 0 = \sigma^a{}_a$ we have

$$\sigma_{0a} = \sigma_{a0} = 0 \text{ and } \sigma_{\alpha\alpha} = 0 \; , \tag{B.14}$$

with Greek indices taking values 1, 2, 3 (and repeated indices summed over their range as usual). Similarly, from (B.3),

$$\mathcal{E}_{0a} = \mathcal{E}_{a0} = 0 = \mathcal{E}_{\alpha\alpha} \text{ and } \mathcal{H}_{0a} = \mathcal{H}_{a0} = 0 = \mathcal{H}_{\alpha\alpha} \; . \tag{B.15}$$

Using (B.14) and (B.15) along with $\epsilon_{0123} = -1$ the equations (B.12) simplify to

$$\mathcal{E}_{\alpha\beta} = -\frac{\partial \sigma_{\alpha\beta}}{\partial t} \; , \; \mathcal{H}_{\alpha\beta} = -\epsilon_{\sigma\rho(\alpha} \sigma_{\beta)\sigma,\rho} \text{ and } \sigma_{\alpha\beta,\beta} = 0 \; . \tag{B.16}$$

Since the first two equations in (B.13) reduce to $\mathcal{E}_{\alpha\beta,\beta} = 0 = \mathcal{H}_{\alpha\beta,\beta}$ we see from (B.16) that they are automatically satisfied. The remaining two equations in (B.13) reduce to

$$\frac{\partial \mathcal{E}_{\alpha\beta}}{\partial t} = \epsilon_{\rho\sigma(\alpha} \mathcal{H}_{\beta)\sigma,\rho} \text{ and } \frac{\partial \mathcal{H}_{\alpha\beta}}{\partial t} = -\epsilon_{\rho\sigma(\alpha} \mathcal{E}_{\beta)\sigma,\rho} \; , \tag{B.17}$$

and these in turn are consistent with $\mathcal{E}_{\alpha\beta,\beta} = 0 = \mathcal{H}_{\alpha\beta,\beta}$. Equations (B.16) and (B.17) are essential for constructing explicit linear perturbations of Minkowskian space-time describing gravitational waves responsible for introducing shear or distortion into the t-lines.

References

1. L. Mariot, C. R. Acad. Sci. **238**, 2055 (1954)
2. I. Robinson, A. Trautman, J. Math. Phys. **24**, 1425 (1983)
3. I. Robinson, J. Math. Phys. **2**, 290 (1961)
4. G.F.R. Ellis, *Relativistic Cosmology* (Gordon and Breach, London, 1971)

Gravitational (Clock) Compass

C

C.1 Coordinates for Minkowskian Space-Time

We provide some supplementary material on the construction of a coordinate system for Minkowskian space-time based on a family of space-like hypersurfaces. This construction leads to a particularly useful form of the Minkowskian line element which is important for the derivation of the frequency ratio in flat space-time in Sect. 5.3. In addition basic formulas (Eqs. (C.57)–(C.61)) in the Minkowskian case play an important role in the neighbourhood of a time-like world line in the general curved space-time.

We begin with the Minkowskian line element in rectangular Cartesian coordinates and time $X^i = (X, Y, Z, T)$:

$$ds^2 = -(dX)^2 - (dY)^2 - (dZ)^2 + (dT)^2 = \eta_{ij}\, dX^i\, dX^j \; . \tag{C.1}$$

Writing

$$X^i = w^i(u) + r\, p^i(u) \; , \tag{C.2}$$

with

$$p_i\, p^i = -1 \; , \tag{C.3}$$

we see that p^i is a unit space-like vector field defined along the world line, which we take to be time-like with u taken to be proper-time or arc length along it. Hence

$$v^i(u) = \frac{dw^i}{du} \quad \text{with} \quad v_i\, v^i = +1 \; . \tag{C.4}$$

© The Author(s), under exclusive license to Springer Nature Switzerland AG 2021
P. A. Hogan, D. Puetzfeld, *Frontiers in General Relativity*, Lecture Notes
in Physics 984, https://doi.org/10.1007/978-3-030-69370-1

Fig. C.1 Construction of the
coordinates of a point X in
the vicinity of the time-like
world line $w(u)$ parametrized
by the proper time u. The
parameter r is centered on the
world line, and the space-like
vector p^i is chosen to be
orthogonal to the velocity v^i
along the world line

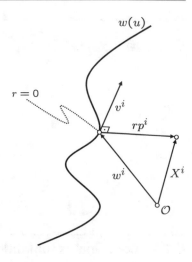

Thus $v^i(u)$ is the unit time-like tangent vector field to $r = 0$ and is therefore the
4-velocity of an observer with world line $r = 0$, see Fig. C.1 for a sketch of the
setup. The 4-acceleration of an observer with world line $r = 0$ is

$$a^i(u) = \frac{dv^i}{du} , \tag{C.5}$$

and this satisfies $a_i v^i = 0$ on account of the second of (C.4). The unit space-like
vector field p^i defined along $r = 0$ is assumed to be orthogonal to $r = 0$ at each of
its points and thus

$$v_i p^i = 0 . \tag{C.6}$$

We are free to choose the transport law for p^i along $r = 0$ subject to ensuring
that (C.3) and (C.6) are preserved at all points of $r = 0$. For our present purposes
we construct the transport law for p^i as follows: Begin by defining an orthonormal
tetrad $\{\lambda^i_{(a)}\}$ with $a = 1, 2, 3, 4$, at a point of $r = 0$ with

$$\eta_{ij} \lambda^i_{(a)} \lambda^j_{(b)} = \eta_{(a)(b)} = \mathrm{diag}(-1, -1, -1, +1) . \tag{C.7}$$

Tetrad indices (or labels) will be those indices with round brackets around them.
They will be raised and lowered with $\eta^{(a)(b)}$ and $\eta_{(a)(b)}$ respectively with the former
defined by $\eta_{(a)(b)} \eta^{(b)(c)} = \delta^c_a$. We shall choose

$$\lambda^i_{(4)} = v^i , \tag{C.8}$$

and we can invert (C.7) to read

$$\eta^{ij} = \eta^{(a)(b)} \lambda^i_{(a)} \lambda^j_{(b)} = -\lambda^i_{(\alpha)} \lambda^j_{(\alpha)} + v^i v^j , \tag{C.9}$$

with henceforth Greek indices taking values 1, 2, 3. On account of (C.8) $\lambda^i_{(4)}$ is extended to a field on $r = 0$ since we shall take (C.8) to hold at all points of $r = 0$ and thus

$$\lambda^i_{(4)}(u) = v^i(u) , \tag{C.10}$$

for all u. We define $\lambda^i_{(\alpha)}(u)$ along $r = 0$ by requiring this orthonormal triad to be transported according to the law:

$$\frac{d\lambda^i_{(\alpha)}}{du} = -v^i a^j \lambda_{(\alpha)j} + \omega^{ij} \lambda_{(\alpha)j} , \tag{C.11}$$

for $\alpha = 1, 2, 3$. Here $\omega^{ij}(u) = -\omega^{ji}(u)$ and $\omega^{ij} v_j = 0$. The first term on the right hand side of (C.11) represents Fermi–Walker transport while the second term represents transport with rigid rotation. The transport law (C.11) preserves the scalar products

$$\eta_{ij} \lambda^i_{(\alpha)} \lambda^j_{(\beta)} = -\delta_{\alpha\beta} \quad \text{and} \quad v_i \lambda^i_{(\alpha)} = 0 , \tag{C.12}$$

along $r = 0$. Multiplying (C.9) by p_j we have, on account of (C.6),

$$p^i = -p_{(\alpha)} \lambda^i_{(\alpha)} = p^{(\alpha)} \lambda^i_{(\alpha)} , \tag{C.13}$$

and multiply (C.9) by a_j results in

$$a^i = -a_{(\alpha)} \lambda^i_{(\alpha)} = a^{(\alpha)} \lambda^i_{(\alpha)} , \tag{C.14}$$

with $p_{(\alpha)} = p_i \lambda^i_{(\alpha)}$ and $a_{(\alpha)} = a_i \lambda^i_{(\alpha)}$. For future reference we note that

$$\delta_{\alpha\beta} p^{(\alpha)} p^{(\beta)} = +1 \quad \text{and} \quad a^i p_i = -a_{(\alpha)} p_{(\alpha)} = a_{(\alpha)} p^{(\alpha)}. \tag{C.15}$$

We will now take $p^{(\alpha)}$ to be independent of the proper time u and parametrize $p^{(\alpha)}$ with the stereographic variables x, y (with $-\infty < x, y < +\infty$) as

$$p^{(1)} = P_0^{-1}x , \quad p^{(2)} = P_0^{-1}y , \quad p^{(3)} = P_0^{-1}\left(\frac{1}{4}(x^2 + y^2) - 1\right) , \tag{C.16}$$

with

$$P_0 = 1 + \frac{1}{4}(x^2 + y^2) \, . \tag{C.17}$$

It now follows from (C.11), (C.13) and (C.16) that p^i obeys, along $r = 0$, the transport law

$$\frac{\partial p^i}{\partial u} = -v^i \, (a^j \, p_j) + \omega^{ij} \, p_j \, . \tag{C.18}$$

Here again the first term on the right hand side is Fermi–Walker transport while the second term represents a rigid rotation.

We now have

$$\frac{\partial p^i}{\partial x} \frac{\partial p_i}{\partial x} = -\delta_{\alpha\beta} \frac{\partial p^{(\alpha)}}{\partial x} \frac{\partial p^{(\beta)}}{\partial x} = -P_0^{-2} \, , \tag{C.19}$$

$$\frac{\partial p^i}{\partial y} \frac{\partial p_i}{\partial y} = -\delta_{\alpha\beta} \frac{\partial p^{(\alpha)}}{\partial y} \frac{\partial p^{(\beta)}}{\partial y} = -P_0^{-2} \, , \tag{C.20}$$

$$\frac{\partial p^i}{\partial x} \frac{\partial p_i}{\partial y} = -\delta_{\alpha\beta} \frac{\partial p^{(\alpha)}}{\partial x} \frac{\partial p^{(\beta)}}{\partial y} = 0 \, . \tag{C.21}$$

Using (C.18) we find that

$$\frac{\partial p_i}{\partial x} \frac{\partial p^i}{\partial u} = \omega_{(\alpha)(\beta)} \frac{\partial p^{(\alpha)}}{\partial x} p^{(\beta)} \, ,$$

$$\frac{\partial p_i}{\partial y} \frac{\partial p^i}{\partial u} = \omega_{(\alpha)(\beta)} \frac{\partial p^{(\alpha)}}{\partial y} p^{(\beta)} \, . \tag{C.22}$$

It will also be useful to have the formulae:

$$P_0^2 \left(\frac{\partial p^{(\alpha)}}{\partial x} \frac{\partial p^{(\beta)}}{\partial x} + \frac{\partial p^{(\alpha)}}{\partial y} \frac{\partial p^{(\beta)}}{\partial y} \right) = \delta^{\alpha\beta} - p^{(\alpha)} \, p^{(\beta)} \, , \tag{C.23}$$

and

$$\frac{\partial^2 p^{(\alpha)}}{\partial x^2} = -P_0^{-2} p^{(\alpha)} - P_0^{-1} \frac{\partial P_0}{\partial x} \frac{\partial p^{(\alpha)}}{\partial x} + P_0^{-1} \frac{\partial P_0}{\partial y} \frac{\partial p^{(\alpha)}}{\partial y} \, ,$$

$$\tag{C.24}$$

$$\frac{\partial^2 p^{(\alpha)}}{\partial y^2} = -P_0^{-2} p^{(\alpha)} + P_0^{-1} \frac{\partial P_0}{\partial x} \frac{\partial p^{(\alpha)}}{\partial x} - P_0^{-1} \frac{\partial P_0}{\partial y} \frac{\partial p^{(\alpha)}}{\partial y} \, ,$$

$$\tag{C.25}$$

$$\frac{\partial^2 p^{(\alpha)}}{\partial x \partial y} = -P_0^{-1} \frac{\partial P_0}{\partial y} \frac{\partial p^{(\alpha)}}{\partial x} - P_0^{-1} \frac{\partial P_0}{\partial x} \frac{\partial p^{(\alpha)}}{\partial y} , \tag{C.26}$$

which can be verified by direct substitution from (C.16) or otherwise.

Now (C.2) gives X^i in terms of x, y, r, u. To obtain the Minkowskian line element in coordinates x, y, r, u we first have from (C.2)

$$dX^i = \left(v^i + r \frac{\partial p^i}{\partial u} \right) du + p^i \, dr + r \frac{\partial p^i}{\partial x} dx + r \frac{\partial p^i}{\partial y} dy . \tag{C.27}$$

Now using the scalar products (C.19)–(C.22) the Minkowskian line element (C.1) becomes

$$ds^2 = \eta_{ij} \, dX^i \, dX^j$$

$$= -r^2 P_0^{-2} (dx^2 + dy^2) + 2 r^2 \omega_{(\alpha)(\beta)} \frac{\partial p^{(\alpha)}}{\partial x} p^{(\beta)} \, du \, dx$$

$$+ 2 r^2 \omega_{(\alpha)(\beta)} \frac{\partial p^{(\alpha)}}{\partial y} p^{(\beta)} \, du \, dy - dr^2$$

$$+ \{ (1 - h_0 r)^2 - r^2 \omega_{(\sigma)(\alpha)} \omega_{(\sigma)(\beta)} p^{(\alpha)} p^{(\beta)} \} du^2 , \tag{C.28}$$

with

$$h_0 = a_i \, p^i = a_{(\alpha)} \, p^{(\alpha)} , \tag{C.29}$$

by (C.15). We can rewrite (C.28) in the neat form

$$ds^2 = -r^2 P_0^{-2} \{ (dx + a_0 \, du)^2 + (dy + b_0 \, du)^2 \}$$
$$- dr^2 + (1 - h_0 r)^2 \, du^2 , \tag{C.30}$$

with

$$a_0 = -P_0^2 \omega_{(\alpha)(\beta)} \frac{\partial p^{(\alpha)}}{\partial x} p^{(\beta)} ,$$

$$b_0 = -P_0^2 \omega_{(\alpha)(\beta)} \frac{\partial p^{(\alpha)}}{\partial y} p^{(\beta)} , \tag{C.31}$$

since

$$P_0^2 \left(\omega_{(\alpha)(\beta)} \frac{\partial p^{(\alpha)}}{\partial x} p^{(\beta)} \right)^2 + P_0^2 \left(\omega_{(\alpha)(\beta)} \frac{\partial p^{(\alpha)}}{\partial x} p^{(\beta)} \right)^2$$

$$= \delta^{\sigma \rho} \omega_{(\sigma)(\alpha)} \omega_{(\rho)(\beta)} p^{(\alpha)} p^{(\beta)} . \tag{C.32}$$

The last equation here is a consequence of (C.23). Using (C.24) and (C.25) we see that

$$\frac{\partial a_0}{\partial x} = \frac{\partial b_0}{\partial y} \, , \tag{C.33}$$

and so it follows that we can write

$$a_0 = \frac{\partial q}{\partial y} \quad \text{and} \quad b_0 = \frac{\partial q}{\partial x} \, , \tag{C.34}$$

with the function q given by

$$q(x, y, u) = \frac{1}{2}(x^2 - y^2)\,\omega_{(1)(2)} + y\left(1 + \frac{1}{4}x^2 - \frac{1}{12}y^2\right)$$

$$\omega_{(1)(3)} + x\left(1 - \frac{1}{12}x^2 + \frac{1}{4}y^2\right)\omega_{(2)(3)}. \tag{C.35}$$

For future convenience we define the 3-vectors

$$\mathbf{p} = (p^{(1)}, p^{(2)}, p^{(3)}) \, , \quad \mathbf{a} = (a^{(1)}, a^{(2)}, a^{(3)}) \, ,$$

$$\boldsymbol{\omega} = (\omega^{(2)(3)}, \omega^{(3)(1)}, \omega^{(1)(2)}) \, . \tag{C.36}$$

Using the standard notation of the scalar product (or "dot product") of 3-vectors and for the vector product (or "cross product") of 3-vectors we have

$$h_0 = -\mathbf{a} \cdot \mathbf{p} \, , \quad a_0 = -P_0^2 \frac{\partial p^{(\alpha)}}{\partial x}\,(\boldsymbol{\omega} \times \mathbf{p})^{(\alpha)} \, ,$$

$$b_0 = -P_0^2 \frac{\partial p^{(\alpha)}}{\partial y}\,(\boldsymbol{\omega} \times \mathbf{p})^{(\alpha)} \, , \tag{C.37}$$

from which we find, using (C.23), that

$$P_0^{-2}(a_0^2 + b_0^2) = (\boldsymbol{\omega} \times \mathbf{p}) \cdot (\boldsymbol{\omega} \times \mathbf{p}) = |\boldsymbol{\omega} \times \mathbf{p}|^2 \, . \tag{C.38}$$

In the light of the foregoing we can now say that (C.2), written more explicitly, reads

$$X^i = w^i(u) + r\,p^i(x, y, u) \, , \tag{C.39}$$

which implicitly determines x, y, r, u as scalar functions of X^i on Minkowskian space-time. We will need the gradients of these functions with respect to X^i, denoted by a comma in each case. To obtain these we start by differentiating (C.39) with respect to X^j giving

$$\delta^i_j = \left(v^i + r \frac{\partial p^i}{\partial u} \right) u_{,j} + p^i r_{,j} + r \frac{\partial p^i}{\partial x} x_{,j} + r \frac{\partial p^i}{\partial y} y_{,j} . \tag{C.40}$$

Multiplying this by $v_i, p_i, \partial p_i/\partial x$ and $\partial p_i/\partial y$ yields successively

$$v_j = (1 - h_0 r) u_{,j} \quad \Rightarrow \quad u_{,j} = (1 - h_0 r)^{-1} v_j , \tag{C.41}$$

$$p_j = -r_{,j} \quad \Rightarrow \quad r_{,j} = -p_j , \tag{C.42}$$

$$\frac{\partial p_j}{\partial x} = r \frac{\partial p_i}{\partial x} \frac{\partial p^i}{\partial u} u_{,j} - r P_0^{-2} x_{,j}$$

$$\Rightarrow \quad x_{,j} + a_0 u_{,j} = -\frac{P_0^2}{r} \frac{\partial p_j}{\partial x} , \tag{C.43}$$

$$\frac{\partial p_j}{\partial y} = r \frac{\partial p_i}{\partial y} \frac{\partial p^i}{\partial u} u_{,j} - r P_0^{-2} y_{,j}$$

$$\Rightarrow \quad y_{,j} + b_0 u_{,j} = -\frac{P_0^2}{r} \frac{\partial p_j}{\partial y} , \tag{C.44}$$

with the final two cases relying on (C.22) and (C.31). We first note from (C.41)–(C.44) that

$$p^i \frac{\partial}{\partial X^i} = p^i \left(x_{,i} \frac{\partial}{\partial x} + y_{,i} \frac{\partial}{\partial y} + r_{,i} \frac{\partial}{\partial r} + u_{,i} \frac{\partial}{\partial u} \right) = \frac{\partial}{\partial r} . \tag{C.45}$$

Substituting (C.41)–(C.44) back into (C.40), using (C.18) and raising the covariant index using η^{ij}, we have

$$\eta^{ij} = -P_0^2 \left(\frac{\partial p^i}{\partial x} \frac{\partial p^j}{\partial x} + \frac{\partial p^i}{\partial y} \frac{\partial p^j}{\partial y} \right) - p^i p^j + v^i v^j$$

$$+ r(1 - h_0 r)^{-1} \left\{ \omega^{ik} p_k - a_0 \frac{\partial p^i}{\partial x} - b_0 \frac{\partial p^i}{\partial y} \right\} v^j . \tag{C.46}$$

However the final term here vanishes since

$$\omega^{ik} p_k - a_0 \frac{\partial p^i}{\partial x} - b_0 \frac{\partial p^i}{\partial y}$$

$$= \left(\delta^{\beta\gamma} \omega_{(\alpha)(\beta)} p_{(\gamma)} + a_0 \frac{\partial p_{(\alpha)}}{\partial x} + b_0 \frac{\partial p_{(\alpha)}}{\partial y} \right) \lambda^{i(\alpha)} , \tag{C.47}$$

and

$$a_0 \frac{\partial p^{(\alpha)}}{\partial x} + b_0 \frac{\partial p^{(\alpha)}}{\partial y}$$

$$= P_0^2 \omega_{(\gamma)(\beta)} p^{(\beta)} \left(\frac{\partial p^{(\gamma)}}{\partial x} \frac{\partial p^{(\alpha)}}{\partial x} + \frac{\partial p^{(\gamma)}}{\partial y} \frac{\partial p^{(\alpha)}}{\partial y} \right)$$

$$= \omega_{(\gamma)(\beta)} p^{(\beta)} (\delta^{\gamma\alpha} - p^{(\gamma)} p^{(\alpha)})$$

$$\overset{(C.23)}{=} -\omega^{(\alpha)(\beta)} p_{(\beta)}. \tag{C.48}$$

Hence (C.46) simplifies to

$$\eta^{ij} = -P_0^2 \left(\frac{\partial p^i}{\partial x} \frac{\partial p^j}{\partial x} + \frac{\partial p^i}{\partial y} \frac{\partial p^j}{\partial y} \right) - p^i p^j + v^i v^j . \tag{C.49}$$

The line element (C.30) can be written

$$ds^2 = -(\vartheta^{(1)})^2 - (\vartheta^{(2)})^2 - (\vartheta^{(3)})^2 + (\vartheta^{(4)})^2 , \tag{C.50}$$

with the basis 1-forms given by

$$\vartheta^{(1)} = r P_0^{-1}(dx + a_0 du) = r P_0^{-1}(x_{,i} + a_0 u_{,i})dX^i$$

$$= -P_0 \frac{\partial p_i}{\partial x} dX^i = -P_0 \frac{\partial p^{(\alpha)}}{\partial x} \lambda_{(\alpha)i} dX^i , \tag{C.51}$$

$$\vartheta^{(2)} = r P_0^{-1}(dy + b_0 du) = r P_0^{-1}(y_{,i} + b_0 u_{,i})dX^i$$

$$= -P_0 \frac{\partial p_i}{\partial y} dX^i = -P_0 \frac{\partial p^{(\alpha)}}{\partial y} \lambda_{(\alpha)i} dX^i , \tag{C.52}$$

$$\vartheta^{(3)} = dr = r_{,i} dX^i = -p_i dX^i$$

$$= -p^{(\alpha)} \lambda_{(\alpha)i} dX^i , \tag{C.53}$$

$$\vartheta^{(4)} = (1 - r h_0) du = (1 - r h_0) u_{,i} dX^i$$

$$= v_i dX^i . \tag{C.54}$$

Finally we shall also require $p_{i,j}$ and we can obtain this by rewriting (C.40), using (C.41) and (C.42), as

$$\delta^i_j = v^i u_{,j} + p^i r_{,j} + r p^i{}_{,j} = (1 - r h_0)^{-1} v^i v_j - p^i p_j + r p^i{}_{,j} . \tag{C.55}$$

Lowering the contravariant index with η_{ij} gives

$$p_{i,j} = \frac{1}{r}\{\eta_{ij} + p_i\, p_j - (1 - r\, h_0)^{-1} v_i\, v_j\}\,. \tag{C.56}$$

Using the basis 1-forms listed in (C.51)–(C.54) we derive the following formulae which will prove useful later:

$$p_{i,j}\, \vartheta^i_{(1)}\, \vartheta^j_{(1)} = -\frac{1}{r} = p_{i,j}\, \vartheta^i_{(2)}\, \vartheta^j_{(2)}\,, \tag{C.57}$$

$$p_{i,j}\, \vartheta^i_{(1)}\, \vartheta^j_{(2)} = 0 = p_{i,j}\, \vartheta^i_{(3)}\, \vartheta^j_{(3)}\,, \tag{C.58}$$

$$p_{i,j}\, \vartheta^i_{(4)}\, \vartheta^j_{(4)} = -\frac{h_0}{1 - r\, h_0}\,, \tag{C.59}$$

$$p_{i,j}\, \vartheta^i_{(1)}\, \vartheta^j_{(4)} = 0 = p_{i,j}\, \vartheta^i_{(2)}\, \vartheta^j_{(4)}\,, \tag{C.60}$$

from which, in particular, we find

$$p^i{}_{,i} = \frac{2 - 3\, r\, h_0}{r\,(1 - r\, h_0)} = \frac{2}{r} - h_0 + O(r)\,, \tag{C.61}$$

with the latter holding for small values of r.

C.2   Plane Gravitational Waves I

A particularly simple exact solution of Einstein's vacuum field equations provides a space-time model of the gravitational field of plane gravitational waves. This well known solution is given by the line element

$$ds^2 = -dX^2 - dY^2 - dZ^2 + dT^2 + 2\,H\,(dT - dZ)^2\,, \tag{C.62}$$

with

$$H = a(T - Z)\,(X^2 - Y^2) + 2\,b(T - Z)\,X\,Y\,. \tag{C.63}$$

A more general form for H, preserving the key properties for plane waves, namely, that H is a harmonic function in X, Y and the corresponding curvature tensor components are functions of $T - Z$ only, is required in Sect. C.4 below.

   The histories of the plane wave fronts in the space-time with line element (C.62) are the null hyperplanes

$$T - Z = \text{constant}\,. \tag{C.64}$$

The waves have two degrees of freedom of polarization reflected in the presence of the two arbitrary functions $a(T-Z)$ and $b(T-Z)$ and, in addition, their arbitrariness represents the freedom to choose the profile of the waves. In the coordinates $X^i = (X, Y, Z, T)$ the non-vanishing components of the Riemann curvature tensor are

$$R_{1414} = -R_{2424} = R_{1313} = -R_{2323} = -R_{1413}$$
$$= R_{2423} = -2\,a(T - Z)\,, \tag{C.65}$$

and

$$R_{1424} = -R_{1423} = R_{1323} = -R_{1324} = -2\,b(T - Z)\,. \tag{C.66}$$

From these it is clear that

$$R_{ijkm}\,k^m = 0 \ \text{ with } \ k^m = (0, 0, 1, 1)\,, \tag{C.67}$$

and so the curvature tensor is type N (purely radiative) in the Petrov classification with degenerate principal null direction k^i. The null vector field k^i is covariantly constant and its expansion-free, twist-free and shear-free geodesic integral curves generate the null hyperplanes (C.64).

From (C.62) and (C.63) we see immediately that the coordinate T is the arc length along the time-like world line $X = Y = Z = 0$. The parametric equations of an arbitrary time-like world line in the space-time with line element (C.62), with arc length s along it, are $X^i = X^i(s)$ with

$$-\left(\frac{dX}{ds}\right)^2 - \left(\frac{dY}{ds}\right)^2 - \left(\frac{dZ}{ds}\right)^2 + \left(\frac{dT}{ds}\right)^2$$
$$+2\,H\left(\frac{dT}{ds} - \frac{dZ}{ds}\right)^2 = +1\,. \tag{C.68}$$

Using

$$\mathbf{u} = (u^1, u^2, u^3) = \left(\frac{dX}{dT}, \frac{dY}{dT}, \frac{dZ}{dT}\right), \tag{C.69}$$

which is the 3-velocity of the observer with world line $X^i = X^i(s)$ measured by the observer with world line $X = Y = Z = 0$, we can rewrite (C.68) in the form

$$\left(\frac{ds}{dT}\right)^2 = 1 - |\mathbf{u}|^2 + 2\,H\left(1 - u^3\right)^2. \tag{C.70}$$

Using (C.63), (C.65) and (C.66) we see that

$$R_{A4B4}\, X^A\, X^B = -2\,H \qquad (A, B = 1, 2), \tag{C.71}$$

with $X^A = (X, Y)$. On account of the simplicity of the Riemann tensor (in particular that it has only two independent components) all of the information contained in it can be extracted using the observer with world line $X = Y = Z = 0$ and observers with world lines $Z = $ constant. The ratio of arc lengths or proper-times along such world lines is, by (C.70) and (C.71),

$$\left(\frac{ds}{dT}\right)^2 = 1 - u^A\, u^A - R_{A4B4}\, X^A\, X^B\,. \tag{C.72}$$

We note that in general the final term in (C.70) can be written

$$2\,H\left(1 - u^3\right)^2 = -R_{A4B4}\, X^A\, X^B - 2\,R_{A4B3}\, X^A\, X^B\, u^3$$
$$-R_{A3B3}\, X^A\, X^B\, (u^3)^2\,. \tag{C.73}$$

However this contains no more information on the Riemann tensor than the final term in (C.72) since

$$R_{A4B3}\, X^A\, X^B = -R_{A3B3}\, X^A\, X^B = 2\,H$$
$$= -R_{A4B4}\, X^A\, X^B\,. \tag{C.74}$$

C.3   Plane Gravitational Waves II

The function H in (C.62) and (C.63) has the property that it vanishes on the world line $X = Y = Z = 0$. Its essential analytical properties are that the vacuum field equations require it to be a harmonic function,

$$H_{XX} + H_{YY} = 0\,, \tag{C.75}$$

with the subscripts denoting partial derivatives, and the curvature tensor components must be functions of $T - Z$ so that

$$H_{XX} - H_{YY} = 4\,a(T - Z)\quad\text{and}\quad H_{XY} = 2\,b(T - Z)\,. \tag{C.76}$$

Hence we can have it vanish on the arbitrary time-like world line $X^i = w^i(u)$ by taking it to be

$$H = a(T - Z)\{(X - w^1(T - Z))^2 - (Y - w^2(T - Z))^2\}$$
$$+ 2b(T - Z)(X - w^1(T - Z))(Y - w^2(T - Z)).$$

With R_{ijkl} given by (C.65) and (C.66) we can write this as (again with capital indices taking values 1, 2)

$$2H = -R_{A4B4}(T - Z)[X^A - w^A(T - Z)][X^B - w^B(T - Z)]. \tag{C.77}$$

We now make the coordinate transformation

$$X^i = w^i + r\,p^i + \frac{1}{3}r^3(p^4 - p^3)(v^4 - v^3)$$
$$\times R_{A4B4}\,p^A\,p^B\,v^i + O(r^4), \tag{C.78}$$

which generalises (C.39) for small values of r and therefore applies in the neighborhood of the time-like world line $r = 0$. The effect of this on the line element (C.62) with H given by (C.77) is to transform it into

$$ds^2 = -r^2 P_0^{-2}\{(dx + a_0\,du)^2 + (dy + b_0\,du)^2\} - dr^2$$
$$+\left\{1 - 2h_0 r + h_0^2 r^2 - r^2(v^4 - v^3)^2\right.$$
$$\left.\times R_{A4B4}\,p^A\,p^B\right\}du^2, \tag{C.79}$$

neglecting $O(r^3)$-terms. Here P_0, a_0, b_0, h_0 are given by (C.17) and (C.38). To effect a closer comparison we note that

$$R_{(\alpha)(4)(\beta)(4)}\,p^{(\alpha)}\,p^{(\beta)} = R_{ijkl}\,p^i\,v^j\,p^k\,v^l$$
$$= R_{A4B4}\,U^A\,U^B, \tag{C.80}$$

with

$$U^A = (v^4 - v^3)\,p^A - (p^4 - p^3)\,v^A. \tag{C.81}$$

Hence if $v^A = 0$, so that the time-like world line $r = 0$ is the history of an observer accelerating in the direction of propagation of the gravitational waves (the Z-direction), then in this case (5.87) simplifies to

$$\left(\frac{ds}{du}\right)^2 = 1 - |\mathbf{u}|^2 + 2\{(\mathbf{a} \cdot \mathbf{p}) - \mathbf{u} \cdot (\boldsymbol{\omega} \times \mathbf{p})\} r$$

$$+ \left\{(\mathbf{a} \cdot \mathbf{p})^2 - |\boldsymbol{\omega} \times \mathbf{p}|^2 - R_{(\alpha)(4)(\beta)(4)} p^{(\alpha)} p^{(\beta)}\right\} r^2 + O(r^3).$$

$$\text{(C.82)}$$

The origin of the coordinate transformation (C.78) is to start with the line element

$$ds^2 = \eta_{ij} dX^i dX^j + 2H (dT - dZ)^2$$

$$= \eta_{ij} dX^i dX^j - R_{A4B4}(T - Z) \left\{X^A - w^A(T - Z)\right\}$$

$$\times \left\{X^B - w^B(T - Z)\right\} (dT - dZ)^2. \qquad \text{(C.83)}$$

Now in the final term here make the transformation (C.39). This involves

$$T - Z = w^4 - w^3 + r (p^4 - p^3) \Rightarrow$$

$$dT - dZ = (v^4 - v^3)du + (p^4 - p^3) dr + O(r), \qquad \text{(C.84)}$$

and

$$R_{A4B4}(T - Z) (X^A - w^A(T - Z)) (X^B - w^B(T - Z))$$

$$= r^2 R_{A4B4}(u) p^A p^B + O(r^3). \qquad \text{(C.85)}$$

Hence the final term in the line element (C.83) reads

$$-r^2 R_{A4B4} p^A p^B \left\{(v^4 - v^3)^2 du^2\right.$$

$$\left. +2 (v^4 - v^3)(p^4 - p^3) du\, dr\right\} + O(r^3). \qquad \text{(C.86)}$$

Now to calculate $\eta_{ij}\,dX^i\,dX^j$ we modify the transformation (C.39) to (C.78) in order to cancel the $du\,dr$-term in (C.86) when everything is substituted into the line element (C.83). From (C.78) it follows that

$$dX^i = (v^i + r\,\frac{\partial p^i}{\partial u})du + r\,\frac{\partial p^i}{\partial x}\,dx + r\,\frac{\partial p^i}{\partial y}\,dy$$

$$+ \left\{ p^i + r^2 v^i\,(v^4 - v^3)(p^4 - p^3)R_{A4B4}\,p^A\,p^B \right\}dr$$

$$+ O(r^3)\,. \tag{C.87}$$

Since v^i and p^i are orthogonal the only surviving Riemann tensor term in $\eta_{ij}\,dX^i\,dX^j$ is $2\,r^2(v^4 - v^3)(p^4 - p^3)R_{A4B4}\,p^A\,p^B\,du\,dr$ (neglecting $O(r^3)$-terms) and so when $\eta_{ij}\,dX^i\,dX^j$ is added to (C.86) now the result is the line element (C.79).

C.4   Waves Moving Radially Relative to $r = 0$

The plane gravitational waves have the property that their propagation direction in space-time is covariantly constant. Hence their propagation direction in space-time is, in particular, *non-expanding*. Arguably the simplest example of gravitational waves for which the propagation direction in space-time is not covariantly constant *and is expanding* are waves moving radially with respect to the observer with world line $r = 0$ in the present context. Such waves may, for example, be spherical fronted but the wave fronts cannot be centered on the observer with world line $r = 0$ since that would result in the Riemann curvature tensor being singular on $r = 0$ which emphatically is *not* the case here. It follows from (C.53) and (5.51) that the 3-direction is the radial direction relative to the world line $r = 0$. We thus consider gravitational waves whose propagation direction calculated on $r = 0$ is given by the 1-form

$$k_{(a)}\,\vartheta^{(a)} = -\vartheta^{(3)} + \vartheta^{(4)} \quad \Leftrightarrow \quad k^{(a)} = (0, 0, 1, 1)\,. \tag{C.88}$$

Thus for small values of r,

$$k_{(a)}\,\vartheta^{(a)} = \{-r_{,i} + (1 - r\,h_0)\,u_{,i} + O(r^2)\}\,dX^i = k_i\,dX^i\,, \tag{C.89}$$

and, using (C.41) and (C.42), we can write

$$k_i = -r_{,i} + (1 - r\,h_0)\,u_{,i} + O(r^2)$$

$$= p_i + v_i + O(r^2) \quad (\Rightarrow\ k^i\,k_i = O(r^2))\,, \tag{C.90}$$

and so the light-like propagation direction calculated *on* $r = 0$ is $k^i = p^i + v^i$. The vacuum field equations

$$R_{(a)(b)} = -R_{(\alpha)(a)(b)(\alpha)} + R_{(4)(a)(b)(4)} = 0 , \qquad \text{(C.91)}$$

and the radiative conditions on the Riemann tensor (that the Riemann tensor be type N in the Petrov classification with $k^{(a)}$ as degenerate principal null direction)

$$R_{(a)(b)(c)(d)} \, k^{(d)} = R_{(a)(b)(c)(3)} + R_{(a)(b)(c)(4)} = 0 , \qquad \text{(C.92)}$$

must be satisfied on $r = 0$ for substitution into (5.87). As a consequence of (C.91) and (C.92) there are only two independent non-vanishing components of the vacuum Riemann tensor calculated on $r = 0$, namely, $R_{(1)(4)(1)(4)} = -R_{(2)(4)(2)(4)}$ and $R_{(1)(4)(2)(4)}$. All remaining non-vanishing curvature components are given in terms of these by

$$R_{(1)(3)(1)(3)} = -R_{(2)(3)(2)(3)} = -R_{(1)(3)(1)(4)}$$
$$= R_{(2)(3)(2)(4)} = R_{(1)(4)(1)(4)} , \qquad \text{(C.93)}$$

and

$$R_{(1)(3)(2)(3)} = -R_{(1)(4)(2)(3)} = -R_{(2)(4)(1)(3)}$$
$$= R_{(1)(4)(2)(4)} . \qquad \text{(C.94)}$$

When these are substituted into the Riemann tensor terms in (5.87) we find that

$$R_{(\alpha)(4)(\beta)(4)} \, p^{(\alpha)} \, p^{(\beta)} = R_{(A)(4)(B)(4)} \, p^{(A)} \, p^{(B)} , \qquad \text{(C.95)}$$

$$R_{(\alpha)(4)(\beta)(\gamma)} \, p^{(\alpha)} \, u^{(\beta)} \, p^{(\gamma)}$$
$$= R_{(A)(4)(B)(4)} \, \{ \, u^{(3)} \, p^{(A)} - u^{(A)} \, p^{(3)} \} \, p^{(B)} , \qquad \text{(C.96)}$$

and

$$R_{(\alpha)(\beta)(\gamma)(\sigma)} \, u^{(\alpha)} \, p^{(\beta)} \, u^{(\gamma)} \, p^{(\sigma)} = R_{(A)(4)(B)(4)}$$
$$\times \{ u^{(A)} \, p^{(3)} - u^{(3)} \, p^{(A)} \} \{ u^{(B)} \, p^{(3)} - u^{(3)} \, p^{(B)} \}, \qquad \text{(C.97)}$$

where capital letters take values 1, 2.

Substituting (C.95)–(C.97) into (5.87) we find

$$\left(\frac{ds}{du}\right)^2 = 1 - |\mathbf{u}|^2 + 2\{(\mathbf{a} \cdot \mathbf{p}) - \mathbf{u} \cdot (\boldsymbol{\omega} \times \mathbf{p})\} r$$

$$+ \left\{ (\mathbf{a} \cdot \mathbf{p})^2 - |\boldsymbol{\omega} \times \mathbf{p}|^2 - R_{(A)(4)(B)(4)} \, p^{(A)} \, p^{(B)} \right.$$

$$+ \frac{4}{3} R_{(A)(4)(B)(4)} \{u^{(3)} \, p^{(A)} - u^{(A)} \, p^{(3)}\} \, p^{(B)}$$

$$- \frac{1}{3} R_{(A)(4)(B)(4)} \{u^{(A)} \, p^{(3)} - u^{(3)} \, p^{(A)}\}$$

$$\left. \times \{u^{(B)} \, p^{(3)} - u^{(3)} \, p^{(B)}\} \right\} r^2 + O(r^3) \,. \tag{C.98}$$

It is interesting to note that while k^i given by (C.90) when $r = 0$ is the propagation direction of the radial gravitational waves relative to the observer with world line $r = 0$ it cannot be the propagation direction of gravitational waves in the neighbourhood of $r = 0$ (i.e. for small, non-zero, values of r). The reason for this is because the Goldberg–Sachs [1] theorem requires the propagation direction in space-time of gravitational waves propagating in a vacuum to be geodesic and shear-free. Using (C.41) and (C.56) we have

$$k_{i,j} = \frac{1}{r}(\eta_{ij} + p_i \, p_j - v_i \, v_j) - h_0 \, v_i \, v_j + a_i \, v_j + O(r) \,, \tag{C.99}$$

from which we conclude that

$$k_{i,j} \, k^j = -h_0 \, v_i + O(r) \,, \tag{C.100}$$

and so k^i is not even approximately geodesic for small r if $a^i \neq 0$ (i.e. if $r = 0$ is not a time-like geodesic). However we can construct an approximately null vector field K^i in the neighbourhood of $r = 0$, which coincides with k^i on $r = 0$, and which is approximately geodesic and shear-free. Such a vector field is given by

$$K^i = p^i + v^i - \frac{1}{2}\{a^i + h_0 \, p^i\} r + O(r^2) \quad \Rightarrow \quad K^i \, K_i = O(r^2) \,. \tag{C.101}$$

When differentiating this with respect to X^i using (C.41), (C.42) and (C.56) it is useful to note that the partial derivative of $h_0 = a_i \, p^i$ reads

$$\frac{\partial h_0}{\partial X^i} = \frac{1}{r}(a_i + h_0 \, p_i) + O(r^0) \,. \tag{C.102}$$

In particular we calculate that

$$K_{i,j} + K_{j,i} = \lambda \, \eta_{ij} + \xi_i \, K_j + \xi_j \, K_i + O(r) \,, \tag{C.103}$$

with

$$\lambda = \frac{2}{r} - h_0 + O(r) \,, \tag{C.104}$$

and

$$\xi_i = \frac{1}{r}(p_i - v_i) + \frac{1}{2}(a_i - h_0 \, v_i) + O(r) \,. \tag{C.105}$$

The appearance of the algebraic form of the right hand side of (C.103) ensures that K^i is geodesic and shear-free in the neighbourhood of $r = 0$ (i.e. K^i is geodesic and shear-free if $O(r)$-terms are neglected). This characterization of "geodesic and shear-free" is due to Robinson and Trautman [2]. It is useful for discussing these geometrical properties when, (a) not using a null tetrad and (b) not assuming an affine parameter along the integral curves of the null vector field. We note in particular that it follows from (C.103) that

$$K_{i,j} \, K^j = -h_0 \, K_i + O(r) \,, \tag{C.106}$$

demonstrating that K^i is approximately geodesic (without an affine parameter if $h_0 \neq 0$).

References

1. J.N. Goldberg, R.K. Sachs, Acta Phys. Polon. Suppl. **22**, 13 (1962)
2. I. Robinson, A. Trautman, J. Math. Phys. **24**, 1425 (1983)

de Sitter Cosmology

D

D.1   Properties of Generalised Kerr–Schild Metrics

The generalised Kerr–Schild metric tensor has the algebraic form [1]

$$g_{ij} = \underset{(0)}{g}_{ij} + 2\,H\,k_i\,k_j\,, \tag{D.1}$$

with

$$\underset{(0)}{g}^{ij}\,k_i\,k_j = k^j\,k_j = 0 \quad \text{and} \quad k^i = \underset{(0)}{g}^{ij}\,k_j\,. \tag{D.2}$$

1. If the determinants of the matrices (g_{ij}) and $(\underset{(0)}{g}_{ij})$ are denoted $\det(g_{ij})$ and $\det(\underset{(0)}{g}_{ij})$ respectively then

$$\det(g_{ij}) = \det(\underset{(0)}{g}_{ij})\,. \tag{D.3}$$

The determinant of the matrix (g_{ij}) may be written, using a standard definition of the determinant of a matrix, in the form

$$\det(g_{ij}) = (\underset{(0)}{g}_{p0} + 2\,H\,k_p\,k_0)\,(\underset{(0)}{g}_{q1} + 2\,H\,k_q\,k_1)\,(\underset{(0)}{g}_{r2} + 2\,H\,k_r\,k_2)\,\times$$

$$(\underset{(0)}{g}_{s3} + 2\,H\,k_s\,k_3)\,\epsilon_{pqrs}$$

$$= \det(\underset{(0)}{g}_{ij}) + 2\,H\left[\underset{(0)}{g}_{p0}\,\underset{(0)}{g}_{q1}\,(\underset{(0)}{g}_{r2}\,k_s\,k_3 + \underset{(0)}{g}_{s3}\,k_2\,k_r)\right.$$

© The Author(s), under exclusive license to Springer Nature Switzerland AG 2021
P. A. Hogan, D. Puetzfeld, *Frontiers in General Relativity*, Lecture Notes
in Physics 984, https://doi.org/10.1007/978-3-030-69370-1

$$+ \underset{(0)}{g}_{r2} \underset{(0)}{g}_{s3} \left(\underset{(0)}{g}_{p0} k_q k_1 + \underset{(0)}{g}_{q1} k_p k_0 \right) \Big] \epsilon_{pqrs}$$

$$= \det(\underset{(0)}{g}_{ij}) + 2 H \Big[k_0 k_p \underset{(0)}{g}_{r2} \underset{(0)}{g}_{s3} \underset{(0)}{g}_{q1} + k_1 k_q \underset{(0)}{g}_{r2} \underset{(0)}{g}_{s3} \underset{(0)}{g}_{p0}$$

$$+ k_2 k_r \underset{(0)}{g}_{p0} \underset{(0)}{g}_{q1} \underset{(0)}{g}_{s3} + k_3 k_s \underset{(0)}{g}_{p0} \underset{(0)}{g}_{q1} \underset{(0)}{g}_{r2} \Big] \epsilon_{pqrs} . \tag{D.4}$$

However, for example,

$$k_p \underset{(0)}{g}_{r2} \underset{(0)}{g}_{s3} \underset{(0)}{g}_{q1} \epsilon_{pqrs} = \underset{(0)}{g}_{pl} k^l \underset{(0)}{g}_{r2} \underset{(0)}{g}_{s3} \underset{(0)}{g}_{q1} \epsilon_{pqrs}$$

$$= \underset{(0)}{g}_{p0} k^0 \underset{(0)}{g}_{r2} \underset{(0)}{g}_{s3} \underset{(0)}{g}_{q1} \epsilon_{pqrs}$$

$$= k^0 \det(\underset{(0)}{g}_{ij}) , \tag{D.5}$$

and thus we can rewrite (D.4) neatly as

$$\det(g_{ij}) = \det(\underset{(0)}{g}_{ij}) + 2 H \det(\underset{(0)}{g}_{ij}) k_p k^p . \tag{D.6}$$

Hence we have from (D.2) that

$$\det(g_{ij}) = \det(\underset{(0)}{g}_{ij}) . \tag{D.7}$$

We note that the inverse of (g_{ij}), namely (g^{ij}) such that $g^{ij} g_{jk} = \delta^i_k$, is given by

$$g^{ij} = \underset{(0)}{g}^{ij} - 2 H k^i k^j , \tag{D.8}$$

with $(\underset{(0)}{g}^{ij})$ the inverse of $(\underset{(0)}{g}_{ij})$. Also writing $\det(g_{ij}) = g$ and $\det(\underset{(0)}{g}_{ij}) = g_0$ we have, in particular, the useful expression

$$k^i{}_{|i} = k^i{}_{;i} \quad \Leftrightarrow \quad \frac{1}{\sqrt{-g}} \frac{\partial}{\partial x^i} (\sqrt{-g} \, k^i) = \frac{1}{\sqrt{-g_0}} \frac{\partial}{\partial x^i} (\sqrt{-g_0} \, k^i) , \tag{D.9}$$

where a stroke denotes covariant differentiation with respect to the Riemannian connection calculated with the metric tensor g_{ij} and a semicolon denotes covariant differentiation with respect to the Riemannian connection calculated with the metric tensor $\underset{(0)}{g}_{ij}$. This last equation relies on a well known property of the components of the Riemannian connection:

2. The components Γ^i_{ij} of the Riemannian connection calculated with the metric tensor g_{ij} can be written in the form

$$\Gamma^i_{ij} = \Gamma^i_{ji} = \frac{1}{2} g^{ik} g_{ik,j} = \frac{1}{2g} g_{,j} , \qquad (D.10)$$

with $g = \det(g_{ij})$ and consequently if $g < 0$ then we can write

$$\Gamma^i_{ij} = \frac{(\sqrt{-g})_{,j}}{\sqrt{-g}} . \qquad (D.11)$$

To see this we first note that the determinant of the metric tensor g_{ij} can be written

$$g = \det(g_{ij}) = g_{p0} \, g_{q1} \, g_{r2} \, g_{s3} \, \epsilon_{pqrs} . \qquad (D.12)$$

Then given the matrix $(g_{ij}) = (g_{ji})$ its cofactor matrix is $(\mathcal{G}_{ij})$ with

$$\mathcal{G}_{0p} = g_{q1} \, g_{r2} \, g_{s3} \, \epsilon_{pqrs} = \mathcal{G}_{p0} , \qquad (D.13)$$

$$\mathcal{G}_{1q} = g_{p0} \, g_{r2} \, g_{s3} \, \epsilon_{pqrs} = \mathcal{G}_{q1} , \qquad (D.14)$$

$$\mathcal{G}_{2r} = g_{p0} \, g_{q1} \, g_{s3} \, \epsilon_{pqrs} = \mathcal{G}_{r2} , \qquad (D.15)$$

$$\mathcal{G}_{3s} = g_{p0} \, g_{q1} \, g_{r2} \, \epsilon_{pqrs} = \mathcal{G}_{s3} . \qquad (D.16)$$

But

$$g^{ik} g_{ik,j} = g^{0p} g_{0p,j} + g^{1q} g_{1q,j} + g^{2r} g_{2r,j} + g^{3s} g_{3s,j} . \qquad (D.17)$$

Substituting here for the inverse $(g^{ij}) = (g_{ij})^{-1}$ of the metric tensor using the cofactor matrix with components (D.13)–(D.16) results in

$$g \, g^{ik} g_{ik,j} = (g_{p0,j} \, g_{q1} \, g_{r2} \, g_{s3} + g_{q1,j} \, g_{p0} \, g_{r2} \, g_{s3}$$

$$+ g_{r2,j} \, g_{p0} \, g_{q1} \, g_{s3} + g_{s3,j} \, g_{p0} \, g_{q1} \, g_{r2}) \, \epsilon_{pqrs}$$

$$= (g_{p0} \, g_{q1} \, g_{r2} \, g_{s3})_{,j} \, \epsilon_{pqrs}$$

$$= g_{,j} , \qquad (D.18)$$

and thus (D.10) is established.

Reference

1. A. Trautman, *Recent Developments in General Relativity* (Pergamon Press Inc., New York, 1962), p. 459

Small Magnetic Black Hole

E.1 The Electromagnetic Field

For an exact tetrad given via the 1-forms (7.27)–(7.30) and an exact potential 1-form given by (7.40) the tetrad components of the perturbed Maxwell tensor are given by

$$F_{(1)(2)} = \frac{\hat{p}^2}{r^2}\left(\frac{\partial \hat{M}}{\partial x} - \frac{\partial \hat{L}}{\partial y}\right), \tag{E.1}$$

$$F_{(1)(3)} = \frac{\hat{p}}{r}\left(-e^{-\hat{\alpha}}\cosh\hat{\beta}\,\frac{\partial \hat{L}}{\partial r} + e^{\hat{\alpha}}\sinh\hat{\beta}\,\frac{\partial \hat{M}}{\partial r}\right), \tag{E.2}$$

$$F_{(2)(3)} = \frac{\hat{p}}{r}\left(e^{-\hat{\alpha}}\sinh\hat{\beta}\,\frac{\partial \hat{L}}{\partial r} - e^{\hat{\alpha}}\cosh\hat{\beta}\,\frac{\partial \hat{M}}{\partial r}\right), \tag{E.3}$$

$$\begin{aligned}
F_{(1)(4)} = &-\frac{\hat{p}\,\hat{b}}{r}\left(\frac{\partial \hat{M}}{\partial x} - \frac{\partial \hat{L}}{\partial y}\right) + \frac{\hat{p}}{r}\,e^{-\hat{\alpha}}\cosh\hat{\beta}\left(\frac{\partial \hat{K}}{\partial x} - \frac{\partial \hat{L}}{\partial u}\right)\\
&-\frac{\hat{p}}{r}\,e^{\hat{\alpha}}\sinh\hat{\beta}\left(\frac{\partial \hat{K}}{\partial y} - \frac{\partial \hat{M}}{\partial u}\right) + \frac{\hat{p}\,\hat{c}}{2\,r}\,e^{-\hat{\alpha}}\cosh\hat{\beta}\,\frac{\partial \hat{L}}{\partial r}\\
&-\frac{\hat{p}\,\hat{c}}{2\,r}\,e^{\hat{\alpha}}\sinh\hat{\beta}\,\frac{\partial \hat{M}}{\partial r},
\end{aligned} \tag{E.4}$$

$$\begin{aligned}
F_{(2)(4)} = &\frac{\hat{p}\,\hat{a}}{r}\left(\frac{\partial \hat{M}}{\partial x} - \frac{\partial \hat{L}}{\partial y}\right) - \frac{\hat{p}}{r}\,e^{-\hat{\alpha}}\sinh\hat{\beta}\left(\frac{\partial \hat{K}}{\partial x} - \frac{\partial \hat{L}}{\partial u}\right)\\
&+\frac{\hat{p}}{r}\,e^{\hat{\alpha}}\cosh\hat{\beta}\left(\frac{\partial \hat{K}}{\partial y} - \frac{\partial \hat{M}}{\partial u}\right) - \frac{\hat{p}\,\hat{c}}{2\,r}\,e^{-\hat{\alpha}}\sinh\hat{\beta}\,\frac{\partial \hat{L}}{\partial r}
\end{aligned}$$

© The Author(s), under exclusive license to Springer Nature Switzerland AG 2021
P. A. Hogan, D. Puetzfeld, *Frontiers in General Relativity*, Lecture Notes
in Physics 984, https://doi.org/10.1007/978-3-030-69370-1

$$+\frac{\hat{p}\,\hat{c}}{2r}\,e^{\hat{\alpha}}\cosh\hat{\beta}\,\frac{\partial\hat{M}}{\partial r}\,,\tag{E.5}$$

$$F_{(3)(4)}=\frac{\partial\hat{K}}{\partial r}-e^{-\hat{\alpha}}(\hat{a}\,\cosh\hat{\beta}-\hat{b}\,\sinh\hat{\beta})\frac{\partial\hat{L}}{\partial r}$$

$$+e^{\hat{\alpha}}(\hat{a}\,\sinh\hat{\beta}-\hat{b}\,\cosh\hat{\beta})\frac{\partial\hat{M}}{\partial r}\,.\tag{E.6}$$

The exact Maxwell vacuum field equations read:

$$\frac{\partial f_1}{\partial r}-\frac{\partial f_4}{\partial y}+\frac{\partial f_5}{\partial x}=0\,,\tag{E.7}$$

$$\frac{\partial f_1}{\partial u}-\frac{\partial f_2}{\partial y}+\frac{\partial f_3}{\partial x}=0\,,\tag{E.8}$$

$$\frac{\partial f_2}{\partial r}-\frac{\partial f_4}{\partial u}-\frac{\partial f_6}{\partial x}=0\,,\tag{E.9}$$

$$\frac{\partial f_3}{\partial r}-\frac{\partial f_5}{\partial u}-\frac{\partial f_6}{\partial y}=0\,,\tag{E.10}$$

with

$$f_1=-r^2\hat{p}^{-2}F_{(3)(4)}\,,\tag{E.11}$$

$$f_2=\frac{1}{2}\hat{c}\,r\,\hat{p}^{-1}e^{\hat{\alpha}}\left(-\sinh\hat{\beta}\,F_{(1)(3)}+\cosh\hat{\beta}\,F_{(2)(3)}\right)+r\,\hat{p}^{-1}e^{\hat{\alpha}}(\sinh\hat{\beta}\,F_{(1)(4)}$$
$$-\cosh\hat{\beta}\,F_{(2)(4)})-r^2\,\hat{p}^{-2}e^{\hat{\alpha}}(-\hat{a}\,\sinh\hat{\beta}+\hat{b}\,\cosh\hat{\beta})\,F_{(3)(4)}\,,\tag{E.12}$$

$$f_3=\frac{1}{2}\hat{c}\,r\,\hat{p}^{-1}e^{-\hat{\alpha}}\left(-\cosh\hat{\beta}\,F_{(1)(3)}+\sinh\hat{\beta}\,F_{(2)(3)}\right)+r\,\hat{p}^{-1}e^{-\hat{\alpha}}(\cosh\hat{\beta}\,F_{(1)(4)}$$
$$-\sinh\hat{\beta}\,F_{(2)(4)})-r^2\,\hat{p}^{-2}e^{-\hat{\alpha}}(-\hat{a}\,\cosh\hat{\beta}+\hat{b}\,\sinh\hat{\beta})\,F_{(3)(4)}\,,\tag{E.13}$$

$$f_4=r\,\hat{p}^{-1}e^{\hat{\alpha}}(-\sinh\hat{\beta}\,F_{(1)(3)}+\cosh\hat{\beta}\,F_{(2)(3)})\,,\tag{E.14}$$

$$f_5=r\,\hat{p}^{-1}e^{-\hat{\alpha}}(-\cosh\hat{\beta}\,F_{(1)(3)}+\sinh\hat{\beta}\,F_{(2)(3)})\,,\tag{E.15}$$

$$f_6=F_{(1)(2)}+r\,\hat{p}^{-1}(\hat{b}\,F_{(1)(3)}-\hat{a}\,F_{(2)(3)})\,.\tag{E.16}$$

With the Maxwell tensor $F_{(a)(b)}$ given in (7.144)–(7.149) these functions acquire the approximate forms:

$$f_1=-r^2\,(P_0^{-2}K_1+O_1)+O(r^3)\,,\tag{E.17}$$

$$f_2 = \frac{1}{r}\left(-g\,P_0^{-2}\hat{a}_{-1} - 2\,g^2\,M_2 + O_3\right) + g\,\frac{\partial h_0}{\partial x}$$

$$+4\,m\,M_2 + O_2 - r\left(\frac{\partial K_1}{\partial y} + 2\,M_2 + O_1\right) + O(r^2)\,,$$

$$(E.18)$$

$$f_3 = \frac{1}{r}\left(-g\,P_0^{-2}\hat{b}_{-1} + 2\,g^2\,L_2 + O_3\right) + g\,\frac{\partial h_0}{\partial y}$$

$$-4\,m\,L_2 + O_2 + r\left(\frac{\partial K_1}{\partial x} + 2\,L_2 + O_1\right) + O(r^2)\,,$$

$$(E.19)$$

$$f_4 = r(-2\,M_2 + O_1) + O(r^3)\,,\qquad\qquad (E.20)$$

$$f_5 = r(2\,L_2 + O_1) + O(r^3)\,,\qquad\qquad (E.21)$$

$$f_6 = \frac{1}{r^2}\{g + O_3\} + P_0^2\left(\frac{\partial M_2}{\partial x} - \frac{\partial L_2}{\partial y}\right) + O_1 + O(r)\,.$$

$$(E.22)$$

Substituting these into the left hand sides of Maxwell's equations (E.7)–(E.10) we arrive at

$$\frac{\partial f_1}{\partial r} - \frac{\partial f_4}{\partial y} + \frac{\partial f_5}{\partial x} = -2\,r\,P_0^{-2}\left\{K_1 - P_0^2\left(\frac{\partial L_2}{\partial x} + \frac{\partial M_2}{\partial y}\right) + O_1\right\} + O(r^2)$$

$$= r \times O_1 + O(r^2)\,,\qquad\qquad (E.23)$$

using (7.53),

$$\frac{\partial f_1}{\partial u} - \frac{\partial f_2}{\partial y} + \frac{\partial f_3}{\partial x} = \frac{1}{r}\Bigg\{g\left(\frac{\partial}{\partial y}(P_0^{-2}\hat{a}_{-1}) - \frac{\partial}{\partial x}(P_0^{-2}\hat{b}_{-1})\right)$$

$$+2\,g^2\left(\frac{\partial L_2}{\partial x} + \frac{\partial M_2}{\partial y}\right) + O_3\Bigg\} - 4\,m\left(\frac{\partial L_2}{\partial x} + \frac{\partial M_2}{\partial y}\right)$$

$$+O_2 + r\left\{\frac{\partial^2 K_1}{\partial x^2} + \frac{\partial^2 K_2}{\partial y^2} + 2\left(\frac{\partial L_2}{\partial x} + \frac{\partial M_2}{\partial y}\right) + O_1\right\}$$

$$+O(r^2)\,,$$

$$= \frac{1}{r}\left\{6\,g^2\,P_0^{-2}F_{ij}k^i\,v^j + O_3\right\} - 4\,m\,P_0^{-2}F_{ij}\,k^i\,v^j$$

$$+O_2 + O_1 \times r + O(r^2)\,,$$

$$= O_2 \times \frac{1}{r} + O_1 + O_1 \times r + O(r^2)\,,\qquad\qquad (E.24)$$

using (7.51), (7.53), (7.52), (7.101) and (7.119),

$$\frac{\partial f_2}{\partial r} - \frac{\partial f_4}{\partial u} - \frac{\partial f_6}{\partial x} = \frac{1}{r^2} \left\{ g \, P_0^{-2} \hat{a}_{-1} + 2 \, g^2 \, M_2 + O_3 \right\}$$

$$- \frac{\partial K_1}{\partial y} - 2 \, M_2 - \frac{\partial}{\partial x} \left\{ P_0^2 \left(\frac{\partial M_2}{\partial x} - \frac{\partial L_2}{\partial y} \right) \right\}$$

$$+ O_1 + O(r) \, ,$$

$$= \frac{1}{r^2} \left\{ 3 \, g^2 \, F_{ij} \, k^i \, \frac{\partial k^j}{\partial y} + O_3 \right\} + O_1 + O(r) \, ,$$

$$= O_2 \times \frac{1}{r^2} + O_1 + O(r) \, , \tag{E.25}$$

using (7.56), (7.102) and (7.119), and

$$\frac{\partial f_3}{\partial r} - \frac{\partial f_5}{\partial u} - \frac{\partial f_6}{\partial y} = \frac{1}{r^2} \left\{ g \, P_0^{-2} \hat{b}_{-1} - 2 \, g^2 \, L_2 + O_3 \right\}$$

$$+ \frac{\partial K_1}{\partial x} + 2 \, L_2 - \frac{\partial}{\partial y} \left\{ P_0^2 \left(\frac{\partial M_2}{\partial x} - \frac{\partial L_2}{\partial y} \right) \right\}$$

$$+ O_1 + O(r) \, ,$$

$$= \frac{1}{r^2} \left\{ - 3 \, g^2 \, F_{ij} \, k^i \, \frac{\partial k^j}{\partial x} + O_3 \right\} + O_1 + O(r) \, ,$$

$$= O_2 \times \frac{1}{r^2} + O_1 + O(r) \, , \tag{E.26}$$

using (7.55), (7.99) and (7.119). Satisfying Maxwell's vacuum field equations approximately, with the degree of smallness of the coefficients of the various powers of r indicated here, is sufficient for us to determine the equations of motion of the small magnetic black hole in second approximation.

Index

© The Author(s), under exclusive license to Springer Nature Switzerland AG 2021
P. A. Hogan, D. Puetzfeld, *Frontiers in General Relativity*, Lecture Notes
in Physics 984, https://doi.org/10.1007/978-3-030-69370-1